Student Solutions Manual to Accompany

Introductory Statistics

Prepared by
John McGhee
Professor of Mathematics
California State University

D1472916

WEST PUBLISHING COMPANY
St. Paul • New York • Los Angeles • San Francisco

Table of Contents

Preface

This student guide is a supplement to Introductory Statistics by John W. McGhee. It has three important features which the reader should find useful.

First the guide contains a complete outline of the concepts and techniques of the main text. Additional notes, remarks, and well over 200 examples make up this outline. It is hoped that this will provide readers with a second resource for studying statistics.

The second feature is the calculator usage illustrations. A great deal of statistics is learned by doing and almost any problem of consequence requires a fair amount of computation. This burden is lessened considerably with a hand held calculator. Calculator usage illustrations with brief explanations and complete keystroke sequences are provided for important techniques. Additionally there is an appendix containing an introduction to the use of the statistical packages found on some calculators.

Finally this student guide contains solutions to all odd numbered exercises and to the chapter tests. These are detailed so that the readers may check their steps as the problems are worked.

In planning this guide, I asked myself what are the needs most often expressed by students in my statistic courses. I hope that I have identified these needs and have succeeded in providing you with a resource that will truly assist you in your studies.

Chapter I
Introduction

This chapter introduces the major areas of statistics and presents some
basic terms such as variable, population, and sample. Additionally, techniques
for obtaining a random sample using a table of random numbers are discussed.

Concepts/Techniques

1) Variables

A quantity whose value varies from subject to subject is called a variable.

Examples: 1) The height of a person.

2) The blood sugar level of a person.

3) The number of workers in a household.

Notation: A capital letter such as X is used for a variable. Its
lower case counterpart x is used for some particular
value. If there are several particular values subscripts
are used. Thus

x_1 = The first value of x (Read: "x sub 1")

x_2 = The second value of x (Read: "x sub 2")

etc.

Example: Five students were questioned as to the number of units they were taking. The results were reported as follows:

$$x_1 = 15, \ x_2 = 12, \ x_3 = 15, \ x_4 = 11, \ \text{and} \ x_5 = 12$$

Example: Using the data of the last example, find

$$x_1 + x_2 + x_3 + x_4 + x_5.$$

This is just the sum of the 5 observations. Thus

$$x_1 + x_2 + x_3 + x_4 + x_5 = 15 + 12 + 15 + 11 + 12 = 65$$

2) Discrete Variables

A variable whose values may be enumerated is called a discrete variable.
Often the values are whole numbers and represent the count of some item.

Example: The number X of automobiles owned by a family is a
discrete variable. If, for example, no family owns more than 100
autos, then the possible values of X are:

$$0,1,2,3,\ldots,99,100.$$

Note that these are possible "counts".

3) Continuous Variables

The values of continuous variables fill out continuous intervals on the
number line. A continuous variable is often associated with the measurement of some quantity.

Example: 1) The amount X of gasoline in an auto. If no auto
can hold more than, say, 70 gallons then $0 \leq x \leq 70$ for any auto.

Example: 2) The temperature X of a welding torch. If such
temperatures do not exceed $3500°$, then $0 \leq x \leq 3500$.

Note that in both of these examples the values of the variables must be
obtained by measuring and not by counting.

4) The Range of a Variable

The set of possible values of the variable is called the range of the
variable.

Notation: The range of X is denoted by S_X.

Example: Let X be the number of units being taken by a student.
If no student takes more than 21 units, then
$S_X = \{0,1,2,3,\ldots,21\}$.

Example: Let X be the length of tape on a newly purchased roll
of masking tape. If the manufacturer regulates these so that they
contain between 200 and 210 feet, then the range is the interval $200 \leq x \leq 210$.

5) Populations

A population is the objects or subjects of interest or the collection of
observation (values of the variable) corresponding to these subjects.
There are many reasons why it may not be practical to examine an entire
population

1) It may be undergoing continuous change.

2) It may be so large as to be impractical.

3) It may be so widely scattered as to be impractical.

4) The testing (or examination) may be destructuve.

5) Subjects may be uncooperative or unwilling to participate.

6) Samples

A sample is a subcollection of a population. Again we use this term both
for subjects or the observations made upon them.

Example: The weights of the book bags of five
students are given at the right. We can refer
to the sample as consisting of the bags of the
students or the weights 12, 8, 15, 6, and 9.

Student	Weight
Bob	12
Mary	8
Tom	15
Alice	6
Anne	9

7) Random Samples

A random sample is a sample selected by a process in which each population
member has an equally likely chance of being chosen to be in the sample.
There should be no bias or favortism involved in the selection. Often a
sample is not random because some segment of the population was systema-
tically excluded from the sample.

Example: The following do not produce random samples of students at a college.

1) Selecting students for the sample from those courses in which you are enrolled.

   (Any student not in one of your classes is excluded.)

2) Selecting students for the sample from those who are studying in the library.

   (Students who do not use the library are excluded from the sample.)

3) Selecting students for the sample as they enter the parking lot.

   (Students who do not use the parking lot or who do not drive are excluded from the sample.)

8) Stratified Samples

A sample having several stratas or segments in the same proportion as they exist in the population is called a stratified sample.

   Example: Suppose the student population at a university consists of 35% freshmen, 30% sophomores, 20% juniors, and 15% seniors. A stratified sample of 300 of these students would contain 105 freshmen (35% of 300), 90 sophomores (30% of 300), 60 juniors (20% of 300) and 45 seniors (15% of 300).

9) Table of Random Digits or Numbers

A table of the digits 0,1,2,3,4,5,6,7,8,9 prepared by a process which afforded each digit the same chance $(\frac{1}{10})$ of appearing in any position of the table is a table of random digits.

If the digits are grouped by pairs, we obtain 2 place numbers: 00,01, 02,03,...,98,99. Each of these has the same chance $(\frac{1}{100})$ of appearing in any position of the (2 digit) table.

If the digits are grouped by 3's, we obtain three digit numbers: 000,001,...,998,999. Each of these has the same chance $(\frac{1}{1000})$ of appearing in any position of the (3 digit) random number table.

10) Controlled and Observational Studies

Many studies are of a comparative nature wherein some new treatment (medication, technique, approach, etc.) is compared with some standard treatment. A study is said to be controlled if the experimenter selects the subjects which are to receive the new treatment and those which do not. The latters are called controls. In an observational experiment, the experimenter simply observes the results of the use of the two treatments. Those subjects which do or do not receive the new treatment are not determined by the design of the experiment. Basically the experimenter is just an interested spectator in some ongoing process.

11) <u>The Placebo Effect</u>

A worthless copy of some medication or treatment is called a placebo. In medical tests where subjects with some specific ailment have been given a placebo, some 30 to 35% of the subjects often show a positive reaction or an improvement in their condition. Such behavior is characterized as the placebo effect.

12) <u>Single and Double Blind Experiments</u>

An experiment where the subjects do not know whether or not they are receiving a placebo is an example of a single blind experiment. More generally, if the subjects are "blind" as to the type of treatment they are receiving, the experiment is a single blind experiment.

If both the subjects and the experimenters who make the initial evaluations are "blind" as to the type of treatment the subject received, the experiment is called a "double blind experiment."

A calculator will prove to be extremely useful in this course. Be certain
that the calculator you obtain has a square root key (or a y^x key)
and at least one memory location. Generally for only a few dollars more
you can get a calculator which has a minor statistical package permitting
automatic computation of the mean and standard deviation (and possibly
correlation and regression coefficients). Your instructor and your college
book store can suggest some good calculators. Texas Instruments, Casio,
and Sharps are but a few of the companies which market calculators suitable
for this course.

Notes on notation used with calculator illustrations.

1) Results are underlined (e.g. 12)

2) Operations are boxed (e.g. $\boxed{+}$)

3) No intermediate results are shown.

Example: The computation 36 ÷ 3 is shown as

$$36 \ \boxed{÷} \ 3 \ \boxed{=} \ \underline{\underline{12}}$$

Example: The computation $\dfrac{20 - 4}{8} = (20 - 4) ÷ 8$ is shown as

$$\boxed{(} \ 20 \ \boxed{-} \ 4 \ \boxed{)} \ \boxed{÷} \ 8 \ \boxed{=} \ \underline{\underline{2}}$$

or as

$$20 \ \boxed{-} \ 4 \ \boxed{=} \ \boxed{÷} \ 8 \ \boxed{=} \ \underline{\underline{2}}$$

Exercise Set I.1

1) a) Discrete. The number of correct answers on a test will be a whole
      number. The values of the variable can be enumerated as, say,
      0,1,2,3,...,10 if the test has 10 questions.

   b) Continuous. The time duration must be measured. The possible values
      of the variable fill out an interval, say, $0 \leq x \leq 3$ if the time
      duration never exceeds 3 hours.

   c) Discrete. The number of commercials is a whole number. The possible
      values of the variable can be enumerated.

   d) Continuous. Weights are obtained by measuring and not by counting.

3) a) Graduating Seniors since the sample is specifically from this group
      or the responses of this group.

   b) The 200 graduating seniors or their responses.

   c) Estimation. What per-cent of the graduating seniors are aware of the
      program?

5) Observational study. The population is being monitored. No variables
   (vitamin levels, age, etc.) are being controlled.

Part B

7) a) x_3 = 3rd grade = 90

   b) $x_1 + x_2 + x_3 + x_4$ = 50 + 60 + 90 + 80 = 280

   c) $\dfrac{x_1 + x_2 + x_3 + x_4}{4} = \dfrac{50 + 60 + 90 + 80}{4} = \dfrac{280}{4} = 70$

   d) The average of the 4 grades.

   e) $\dfrac{x_2 + x_3 + x_4}{3}$. The instructor throws out x_1.

9) a) Coronary prone adults or their aspiration levels

   b) The 100 adults or their aspiration levels

   c) Coronary prone adults have higher self aspiration levels than do adults
      in general.

Exercise Set I.2

1) No. Not all families have an equally likely chance of being selected. Those toward the other end of the street will be excluded.

3) Group the digits by threes and ignore the occurrence of 000,401,402,...,999. From the table we read off: 392 ~~843~~ 373 ~~742~~ ~~512~~ ~~864~~ 112 375 329 ~~690~~ 260 ~~968~~ 136 193 099 339

The page numbers selected are:

99, 112, 136, 193, 260, 329, 339, 373, 375, 392

5) Neither the patients nor the evaluators would know who received the new drug and who received the regular medication.

7) Both doctors and patients are often more enthusiastic about the potential of a "new" drug. As a result the reported effectiveness of a medication may be higher when it is relatively new on the market.

9) Twenty four families are to be selected from 8 floors. Thus $\frac{24}{8} = 3$ will be selected from each floor. Use the digits 0,1,2,3,...,9 as the unit numbers on, say, the 8th floor. Select a random sample of 3 of these using the table of random digits. For example, if we begin in row 41, column 1 we obtain 8, 2, ~~2~~, 0. Thus on the 8th floor we use units 0, 2, and 8. Now repeat this process for each of the 7 remaining floors.

11) We read off:

32 ~~94~~ 29 54 16 42

The random number 32 corresponds to student 11 (see table at right). Similarly 29 corresponds to student 10. Etc. The students selected are numbered 6, 10, 11, 15, and 19.

Student #	Random Numbers
1	0 – 2
2	3 – 5
3	6 – 8
4	9 – 11
5	12 – 14
6	15 – 17
7	18 – 20
8	21 – 23
9	24 – 26
10	27 – 29
11	30 – 32
12	33 – 35
13	36 – 38
14	39 – 41
15	42 – 44
16	45 – 47
17	48 – 50
18	51 – 53
19	54 – 56
20	57 – 60
Etc.	

13) i) People not listening are excluded from participating.

   ii) Only seriously motivated listeners will take the trouble to call in.

  iii) Special interest groups will marshall their forces to call in and bias the results.

Chapter Test

1) a) Continuous (a measurement). The possible temperatures fill out an inter-
      val.

   b) Discrete. The possible number of units completed can be enumerated: If
      say, 124 units are required, then the values of the variable are
      124,125,126,127,128,... . Note that this is also true if fractional
      units are allowed.

2) Discrete: 1) Number of windows   2) Number of seats

   Continuous: 1) Temperature   2) Blackboard area

3) a) Estimate the proportion of students who drink coffee.

   b) Is it true that coffee is an aphrodesiac?

   c) Is there a relationship between elevated pulse rates and coffee drinking?

4) a) Soldiers or their successes with the new weapon.

   b) The 56 marines or their results with the new weapon.

   c) No. The sample consists only of marines. And only of marines from this
      one particular base.

   d) Hypothesis testing. The hypothesis to be tested is the manufacturer's
      assertion that "any soldier can use the weapon after a brief introduction."

5) a) $\dfrac{x_1 + x_2 + x_3 + x_4 + x_5 + x_6}{6} = \dfrac{14 + 16 + 15 + 12 + 15 + 18}{6} = \dfrac{90}{6} = 15$

   b) $(x_1 - 15) + (x_2 - 15) + (x_3 - 15) + (x_4 - 15) + (x_5 - 15) + (x_6 - 15) =$
      $(14 - 15) + (16 - 15) + (15 - 15) + (12 - 15) + (15 - 15) + (18 - 15) =$
      $-1 + 1 + 0 - 3 + 0 + 3 = 0$

6) Group the digits by pairs and ignore 00,50,51,52,...,99. We read off
   12 65 16 16 46 11 76 91 51 09 86 99 69 76 69 25 75 73 25 35
   The students to be dropped are those numbered
   9, 11, 12, 16, 25, 35, 46

7) A positive response to a worthless medication or therapy.

8) The experimental subjects would be randomly assigned either the placebo or
   the actual medication and kept unaware of which they received. Similarly
   the evaluators who must make the determination as to whether or not the new
   medication is effective, would also be kept "in the dark" as to what the
   subjects received.

9) There are many ways to make this selection. For example, renumber the houses
   on the north side as 0,1,2,...,9. Now starting in, say row 6, column 1,
   we read off 9, 0, 1, 8. Thus we want the 2nd, 9th and 10th houses on the
   north side, i.e. those numbered 3, 17, and 19.

On the south side renumber the houses to 0,1,2,3,...,8. If we start in column 1, row 23, we read off 1, 4. Thus we want the 2 houses whose original numbers were 2 and 8.

An alternate approach is to group the digits by pairs to obtain 2 digit numbers. The first 3 odd numbers less than 20 and the first 2 even numbers less than 17 are the house numbers selected.

Chapter II
Descriptive Statistics: Tables and Graphs

The unorganized data of a sample often conveys little obvious information. This chapter is concerned with techniques for displaying data so that its important features, trends, and patterns are readily apparent. Important topics in this regard are the frequency distribution table and two graphs, the histogram and the frequency polygon.

Concepts/Techniques

1) The Frequency of an Observation

   The frequency of an observation in a sample is the number of times the observation occurred.

   Notation: f

2) The Relative Frequency of an Observation

   The relative frequency of an observation in a sample with n observations is the frequency of the observation divided by the sample size.

   Notation: rf

   Formula: $rf = \dfrac{f}{n}$

   Example: Consider the sample: 1, 2, 2, 3, 5, 5, 5, 6, 8, 8. The frequency of $x = 5$ is $f = 3$ and since there are $n = 10$ observations in all, its relative frequency is

$$rf = \frac{3}{10} = .3 = 30\%$$

A complete summary is given in the accompanying
table.

x	f	rf
1	1	.1 = 10%
2	2	.2 = 20%
3	1	.1 = 10%
5	3	.3 = 30%
6	1	.1 = 10%
8	2	.2 = 20%

Example: The frequency table for a sample is
given at the right. What was the sample size?
The sample size n is the sum of the frequencies.
Thus

$$n = 3 + 4 + 8 + 4 + 6 = 25$$

x	f
40	3
55	4
70	8
90	4
100	6

3) Frequency Distributions

The range of a variable may be divided up using a
sequence of adjacent intervals on the number line.
The observations of a sample may then be grouped or classified according
to the interval on which they fall. A table giving the class intervals
and the frequency or relative frequency of each is called a frequency
distribution or simply a distribution.

Example: Find a frequency distribution for the following sample:
94, 92, 83, 75, 64, 58, 67, 72, 84, 87, 82, 72, 79, 60, 74, 77, 73,
99, 85, 64, 51, 72, 77, 75, 44.

1) First select the number of intervals K to be used.
    Generally K is between 5 and 15 and for relatively
    small samples such as this, 6, 7, or 8 often proves a good
    choice. We choose K = 7.

2) Next find the interval length.

$$\text{Approximate length} = \frac{\text{Largest Observation - Smallest Observation}}{\text{Number of Intervals}}$$

$$= \frac{99 - 44}{7} = \frac{55}{7} \cong 7.8$$

This we could round to 8 or even 10. We choose to use a
length of 10. (Note - this will probably result in fewer than
7 intervals.

3) Next we choose the left end point of the first interval. This
    must be 44 or less but otherwise the choice is arbitrary.
    Since the observations are to the nearest unit (1), it is
    recommended that a end point ending in $\frac{1}{2}$ (1) = .5 be chosen.

In this way all the observations fall <u>within</u> the intervals. Choosing 40.5 as the left end point of the first interval we obtain the intervals:

$$40.5 - 50.5$$
$$50.5 - 60.5$$
$$60.5 - 70.5$$
$$70.5 - 80.5$$
$$80.5 - 90.5$$
$$90.5 - 100.5$$

4) Finally we tally the observations according to the interval on which they fall. We obtain

Interval	Tally	Frequency	Relative Frequency
$40.5 < x \le 50.5$	|	1	$\frac{1}{25} = .04 = 4\%$
$50.5 < x \le 60.5$	|||	3	$\frac{3}{25} = .12 = 12\%$
$60.5 < x \le 70.5$	|||	3	$\frac{3}{25} = .12 = 12\%$
$70.5 < x \le 80.5$	ﬤﬤ ﬤﬤ	10	$\frac{10}{25} = .40 = 40\%$
$80.5 < x \le 90.5$	ﬤﬤ	5	$\frac{5}{25} = .20 = 20\%$
$90.5 < x \le 100.5$	|||	3	$\frac{3}{25} = .12 = 12\%$

The next example shows an alternate approach to choosing the endpoints of the intervals.

<u>Example</u>: The times (in days) needed by 30 women to grow a new fingernail are as follows:

164, 212, 215, 173, 185, 186, 200, 195, 193, 210, 201, 182, 184, 170, 174, 177, 206, 184, 184, 180, 175, 192, 191, 187, 182, 196, 194, 181, 176, 188

Construct a frequency distribution using 6 intervals of length 10 with the left end point of the first being 160.

The intervals are now

160 - 170, 170 - 180, 180 - 190, 190 - 200, 200 - 210, and 210 - 220.

Some of the observations (170, 180, and 200) fall at the endpoint of two intervals. We agree that any such observation will be tallied in the interval immediately to the left of the observation. Thus 200 is tallied in 190 - 200, 180 in 170 - 180, and 170 in 160 - 170. The results of the tallying follows:

Interval	Tally	Frequency	Relative Frequency
$160 < x \leq 170$	||	2	$\frac{2}{30} = .0667 = 6.67\%$
$170 < x \leq 180$	⊬|	6	$\frac{6}{30} = .200 = 20.00\%$
$180 < x \leq 190$	⊬ ⊬	10	$\frac{10}{30} = .333 = 33.3\%$
$190 < x \leq 200$	⊬ ||	7	$\frac{7}{30} = .233 = 23.3\%$
$200 < x \leq 210$	|||	3	$\frac{3}{30} = .100 = 10.0\%$
$210 < x \leq 220$	||	2	$\frac{2}{30} = .0667 = 6.67\%$

The frequency distribution is the following table:

Interval	Relative Frequency
$160 < x \leq 170$	.0667
$170 < x \leq 180$	.2000
$180 < x \leq 190$	.3333
$190 < x \leq 200$	.2333
$200 < x \leq 210$	.1000
$210 < x \leq 220$	.0667

4)   Histograms

A histogram is a bar graph of a frequency distribution in which the height of the bar erected over an interval is proportional to the relative frequency.   In particular, either the relative frequency or the frequency may be used.   Also the density which we discuss later (item 6) may be used.

Example: Construct the histogram for the distribution below.

Interval	Relative Frequency
$32.5 < x \leq 36.5$	.12
$36.5 < x \leq 40.5$	.20
$40.5 < x \leq 44.5$	.40
$44.5 < x \leq 48.5$	.16
$48.5 < x \leq 52.5$	.10
$52.5 < x \leq 56.5$	.02

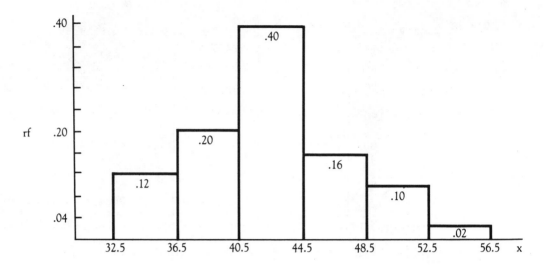

With discrete variables we sometimes choose not to group or classify
the observations if valuable information would be lost. Generally
this is the case if small differences in the observed values are
important. In the next example we construct the histogram of such
a distribution.

Example: The distribution of the number of courses taught during
the fall of 1984 by the 36 members of a university department is
given below followed by the histogram. Note that we have chosen
to use frequencies rather than relative frequencies. Also the
observations are used as the midpoints (class marks) of the intervals.

Courses Taught	Frequency
0	4
1	9
2	14
3	6
4	3

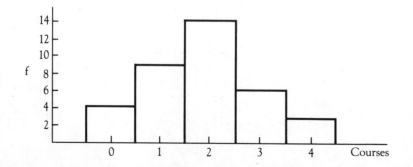

5) Stem and Leaf Diagrams

A stem and leaf diagram is a display of data in which the leaf or right most digit of each number is shown attached to the stem or left most digits of the number. In this diagram many numbers make use of the same stem. Each stem thus corresponds to a grouping or classification of the observations.

Example: For the number x = 153, the leaf is 3 and the stem is 15.

Example: Produce a stem and leaf diagram of the following weights:
153, 148, 141, 156, 162, 157, 155, 164, 172, 170
The stems are first written along a vertical line as shown (a).
The first leaf (3) of 153 is shown attached in (b).
The second leaf (8) of 148 is shown attached in (c).
The third leaf (1) of 141 is shown attached in (d).
The complete diagram is given in figure (e).

14		14		14	8	14	81	14	81
15		15	3	15	3	15	3	15	3675
16		16		16		16		16	24
17		17		17		17		17	20

   (a)   (b)   (c)   (d)   (e)

If a great many observations are present, a double stem and leaf diagram may be used. In this diagram the same stem is used twice, first for the digits 0 - 4, and then for the digits 5 - 9.

Example: Give the observations corresponding to the following stem and leaf diagram.

16.	04
16.	576
17.	3
17.	76859
18.	241
18.	59
19.	2
19.	7

The observations are:

16.0, 16.4, 16.5, 16.7, 16.6, 17.3, 17.7, 17.6, 17.8, 17.5, 17.9,
18.2, 18.4, 18.1, 18.5, 18.9, 19.2, 19.7

A little ingenuity will allow the use of intervals as stems.

Example: For the data below, prepare a stem and leaf diagram where
the stems are the intervals 10.0 - 10.4 (inclusive), 10.5 - 10.9
(inclusive), and so forth.

12.6, 10.4, 10.3, 11.5, 11.7, 12.2, 11.3, 11.0, 11.4, 12.1, 10.7,
10.8, 11.3, 11.1, 11.8, 12.0, 12.1, 11.0, 11.2, 11.5, 11.9, 11.6,
10.6, 10.5

Interval (Stem)	Leaf
10.0 - 10.4	43
10.5 - 10.9	7865
11.0 - 11.4	3043102
11.5 - 11.9	578596
12.0 - 12.4	2101
12.5 - 12.9	6

6) Frequency Polygons

A frequency polygon is a line graph of a distribution. It is produced
by plotting the points

(Interval Midpoint, Relative Frequency)

for each interval and joining these with line segments. The actual
frequency may be used in place of the relative frequency.

Example: Give the frequency polygon of the distribution below.

Interval	Relative Frequency
$5 < x \le 15$	.15
$15 < x \le 25$	.30
$25 < x \le 35$	.20
$35 < x \le 45$	.15
$45 < x \le 55$	.10
$55 < x \le 65$	.10

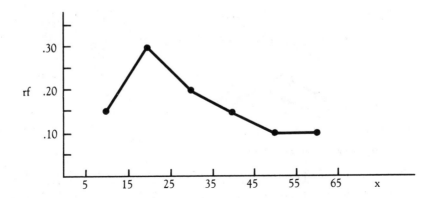

Some authors recommend starting and ending the polygon on the x   axis
by using two addit·ional intervals as shown below. Depending on the
range of x,   this may or may not make sense.

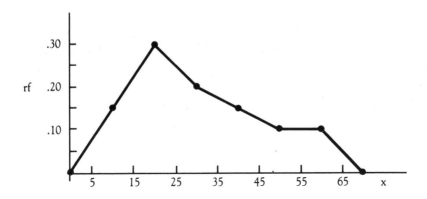

Frequency polygons are very useful for comparing two distributions.

Example: The frequency polygons of the distributions of the number
of "close misses" at two metropolitan airports for a sample of 30
week days are given. Which airport has the best record?

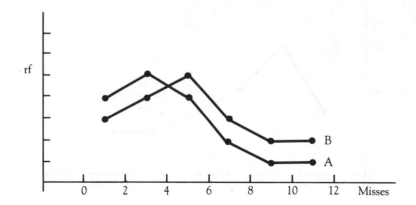

A low number of misses is desirable. Airport A has the highest
frequencies of these. On the other hand, when there is a large
number of near misses, say $4 \leq x \leq 6$, $6 \leq x < 8$, $8 \leq x < 10$,
$10 \leq x \leq 12$, airport B has the higher frequencies and again the
worst record. Thus airport A has the best record.

7) The Average Density Scale

In relating statistical distributions to probability distributions, the
notion of the average density of an interval proves useful.

Notation: d

Formula: $d = \dfrac{\text{relative frequency of the interval}}{\text{length of the interval}}$

Example: A class interval of a distribution has a relative frequency of .30 and a length of 20. Find its average density
and interpret the result.

$$d = \frac{.30}{20} = .015 = 1.5\%$$

If the observations on this interval were
uniformly "scattered" over this interval,
then each unit subunit (interval of length 1)
would contain .015 = 1.5% of the distribution.

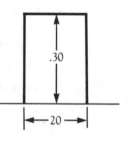

Example: A distribution is given. Compute the average densities
and use these to construct a histogram.

Interval	rf	d
$.5 < x \leq 20.5$	.10	.0050
$20.5 < x \leq 40.5$	.25	.0125
$40.5 < x \leq 60.5$	.40	.0200
$60.5 < x \leq 80.5$	.15	.0075
$80.5 < x \leq 100.5$	.10	.0050

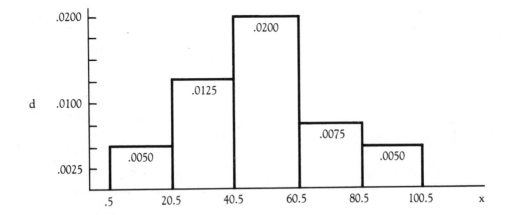

It is important to note that we may interpret areas under a density histogram as relative frequencies.

Example: Refer to the last example. For the rectangle over the interval $40.5 \leq x \leq 60.5$

$$Area = base \times height = (20)(.0200)$$
$$= .4000$$
$$= relative\ frequency$$

Also note that the total area bounded by a (density) histogram is 1 since the relative frequencies sum to 1.

8) Distribution Curves

A distribution curve for a population is the continuous analog of the sample histogram. While there are many distribution curves, there are four forms that are often encountered:

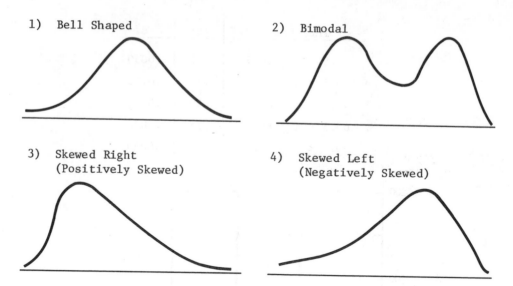

1) Bell Shaped   2) Bimodal

3) Skewed Right   4) Skewed Left
    (Positively Skewed)   (Negatively Skewed)

Bell shaped distribution curves arise from distributions which are symmetric and show a strong tendency to cluster about some central value. For example we would expect the birth weights of infants (at term) to cluster around some central value such as 7.0 pounds, i.e. there would be many weights at or near 7.0. Further, while weights above and below this central value occur, their frequency drops off quite rapidly as we get farther and farther away from this central value. The resulting distribution is pictured below.

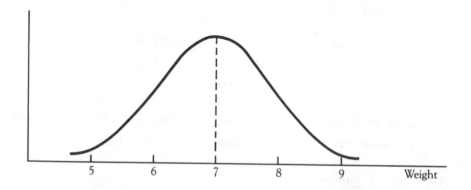

Bimodal distributions result from a tendency to cluster about two central values. Often these are the result of two competing stratas (with different central values) within the population. For example, the distribution of the weights of 35 year old adults should show a bimodal tendency since the womens weight and those of the mean have different central values.

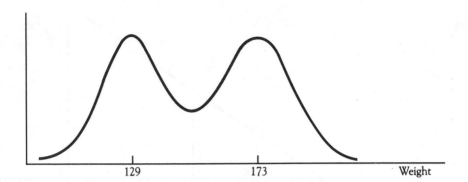

Skewed distributions result from abnormally high frequencies near the
beginning or end of the range of the variable.

For example, most students at a university live close to campus. The
distribution of distances traveled to school should show high frequencies
for small distances and then gradually reduce or taper off as the distance
increases. A distribution somewhat like that below would be expected.

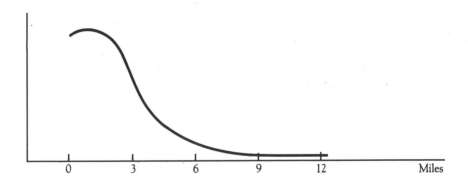

The vertical scale used with a distribution curve is the density scale.
As with (density) histograms the total area bounded by a distribution
curve is 1 and areas under this curve correspond to relative
frequencies.

    Example: For the distribution curve below, find the relative
    frequency for   a) $10 \le x \le 20$ b) $30 \le x \le 50$

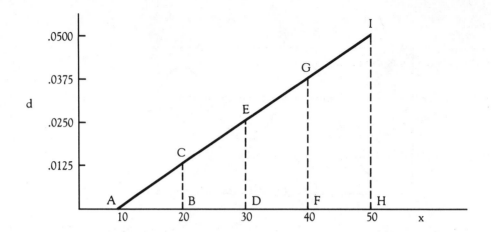

a) The area of the triangle $\triangle ABC$ is $\frac{1}{2}$ base × height
   $= \frac{1}{2}$ (10)(.0125) = .0625.
   Thus the relative frequency is

$$.0625 = 6.25\%.$$

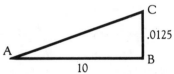

b) We need the area of the trapezoid DHIE.
   Rather than use the formula for this we
   will simply subtract the area $\triangle ADE$
   from the area of the total figure $\triangle AHI$
   which we know is 1. Thus the desired
   area is

$$A = 1 - \frac{1}{2} (20)(.025)$$

$$= 1 - .25 = .75$$

Thus the relative frequency of the
interval $30 \le x \le 50$ is .75 = 75%.

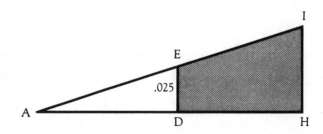

Calculator Usage Illustration

Compute the relative frequency of an observation as a percent.
Find $\frac{18}{32} \times 100$.

Method 1)   Arrange the computation as $18 \times 100 \div 32$.

$$18 \boxed{\times} \ 100 \ \boxed{\div} \ 32 \ \boxed{=} \ \underline{56.25}$$

Method 2) $(\frac{1}{32}) \times 18 \times 10$ (provided your calculator has the reciprocal

$\frac{1}{x}$ key)

$$32 \ \boxed{\frac{1}{x}} \ \boxed{\times} \ 18 \ \boxed{\times} \ 100 \ \boxed{=} \ \underline{56.25}$$

Calculator Usage Illustration

Compute the square root of a number.
Find $\sqrt{158.76}$.

Method 1)   Using the square root key (if available)

$$158.76 \ \boxed{\sqrt{x}} \ \underline{12.6}$$

Method 2)   Using the exponent y^x key. Note $\sqrt{y} = y^{1/2} = y^{.5}$.

$$158.76 \ \boxed{y^x} \ .5 \ \boxed{=} \ \underline{12.6}$$

Calculator Usage Illustration

Computing a quotient involving a square root.

Find $\dfrac{37.57}{\sqrt{6.76}}$.

Method 1) $37.57 \div \sqrt{6.76}$

$$37.57 \;\boxed{\div}\; 6.76 \;\boxed{\sqrt{x}}\; \boxed{=}\; \underline{14.45}$$

Method 2) $\left(\dfrac{1}{\sqrt{6.76}} \right) \times 37.57$

$$6.76 \;\boxed{\sqrt{x}}\; \boxed{\tfrac{1}{x}}\; \boxed{\times}\; 37.57 \;\boxed{=}\; \underline{14.45}$$

Method 3)   Make use of the memory to hold $\sqrt{x}$

$$6.76 \;\boxed{\sqrt{x}}\; \boxed{STO}\; 37.57 \;\boxed{\div}\; \boxed{RCL}\; \boxed{=}\; \underline{14.45}$$

Note: STO and RCL represent store to memory and recall from memory. Other notations such as M+ and MR are also used. See the guide that came with your calculator.

Exercise Set II.1

1) a) .18 = 18%   b) .28 + .18 + .06 + .06 + .02 = .60 = 60%

   c)

Interval	Relative Frequency
12500 – 13500	.04
13500 – 14500	.08
14500 – 15500	.18
15500 – 16500	.06
16500 – 17500	.04
17500 – 18500	.28
18500 – 19500	.18
19500 – 20500	.06
20500 – 21500	.06
21500 – 22500	.02

   d) 4% of 400 = .04 × 400 = 16

3) Smallest observation = 2.6, Largest observation = 8.3

Approximate width $w = \dfrac{8.3 - 2.6}{6} = \dfrac{5.7}{6} = .95$. Use a width of 1.0.

   a)

Interval	Tally	Frequency	Relative Frequency
2.55 – 3.55	⊬⊬ |	6	$\dfrac{6}{25} = .24$
3.55 – 4.55	⊬⊬ |||	8	$\dfrac{8}{25} = .32$
4.55 – 5.55	⊬⊬	5	$\dfrac{5}{25} = .20$
5.55 – 6.55	|||	3	$\dfrac{3}{25} = .12$
6.55 – 7.55	||	2	$\dfrac{2}{25} = .08$
7.55 – 8.55	|	1	$\dfrac{1}{25} = .04$

5) Smallest observation = .01, Largest observation = 1.92

Choosing K = 8 intervals, the approximate width is found as

$\dfrac{1.92 - .01}{8}$ = .23875. This we round to .25.

As the left endpoint of the first interval we choose .005

Interval	Tally	Frequency	Relative Frequency
.005 - .255	ⵜⵜⵜ ⵜⵜⵜ ⵜⵜⵜ ⵜⵜⵜ||	22	$\frac{20}{58}$ = .379 = 37.9%
.255 - .505	ⵜⵜⵜ ⵜⵜⵜ ||	12	$\frac{14}{58}$ = .207 = 20.7%
.505 - .755	ⵜⵜⵜ ||	7	$\frac{7}{58}$ = .121 = 12.1%
.755 - 1.005	||||	4	$\frac{4}{58}$ = .069 = 6.9%
1.005 - 1.255	|||	3	$\frac{3}{58}$ = .052 = 5.2%
1.255 - 1.505	||	2	$\frac{2}{58}$ = .034 = 3.4%
1.505 - 1.755	|||	3	$\frac{3}{58}$ = .052 = 5.2%
1.755 - 2.005	ⵜⵜⵜ	5	$\frac{5}{58}$ = .086 = 8.6%

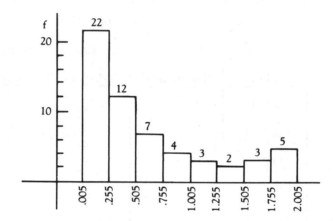

7) Choosing K = 6 as the approximate number of intervals, an approximate width is found by $\frac{3295 - 1955}{6}$ = 223.3. This we round to 225. Choosing the left endpoint of the first interval as 1950, we follow the convention that an observation falling on an endpoint is to be tallied in the interval to the left of the point.

Interval	Tally	Frequency
1950 - 2175	|||	3
2175 - 2400	||||	4
2400 - 2625	ⵜⵜⵜ ||	7
2625 - 2850	||||	4
2850 - 3075	ⵜⵜⵜ	5
3075 - 3300	|||	3

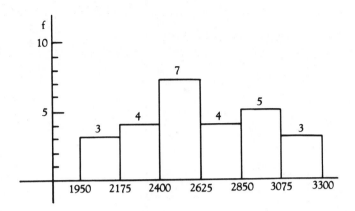

9) Use the observations as the interval midpoints

Observation	Tally	Frequency
0	||||	4
1	⫲⫲⫲ ⫲⫲⫲ |	11
2	⫲⫲⫲ |	6
3	⫲⫲⫲	5
4	|||	3
5	|	1

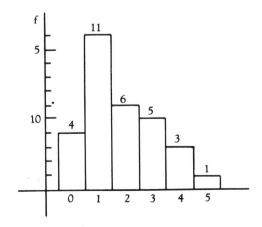

11)   8.0 – 8.1 |
       8.2 – 8.3 | 3
       8.4 – 8.5 | 55555554
       8.6 – 8.7 | 6677666
       8.8 – 8.9 | 88999
       9.0 – 9.1 | 00011
       9.2 – 9.3 | 2333
       9.4 – 9.5 | 445545454555
       9.6 – 9.7 |
       9.8 – 9.9 |

13)   Smallest observation = –1.46, Largest observation = 14.81

       Using K = 6, the approximate interval width is $\dfrac{14.81 - -1.46}{6}$ = 2.72.

       This we adjust to 2.75.

Interval	Tally	Frequency
$-1.50 \le x < 1.25$	|||	3
$1.25 \le x < 4.00$	||	2
$4.00 \le x < 6.75$	|||	3
$6.75 \le x < 9.50$	⊬⊬	5
$9.50 \le x < 12.25$	⊬⊬ |||	8
$12.25 \le x < 15.50$	||||	4

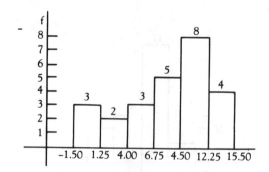

15)

Interval	f	rf
$379.5 < x \leq 399.5$	27	.108
$399.5 < x \leq 419.5$	65	.260
$419.5 < x \leq 439.5$	60	.240
$439.5 < x \leq 459.5$	45	.180
$459.5 < x \leq 479.5$	38	.152
$479.5 < x \leq 499.5$	13	.052
$499.5 < x \leq 519.5$	2	.008

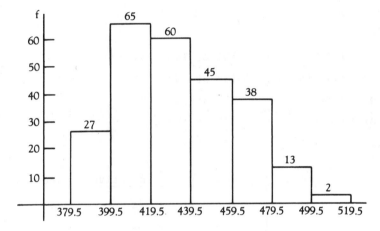

17) a) The <u>observed</u> number has certainly increased. However this would be
        expected as the number of laboratories has increased.

    b) Yes

    c) Most likely the cause of the increase in the number of cataloged
        earthquakes since 1974.

19) a) 30   b) 36   c) 40 − 36 = 4

Interval	Frequency
$1.05 < x \leq 2.05$	8
$2.05 < x \leq 3.05$	10
$3.05 < x \leq 4.05$	12
$4.05 < x \leq 5.05$	6
$5.05 < x \leq 6.05$	2
$6.05 < x \leq 7.05$	2

21) a) For A: .20 + .15 + .10 = .45

        For B: .20 + .30 + .20 = .70

    b) Make B

c)   Make A

Interval	rf
18 – 19	.05
19 – 20	.20
20 – 21	.30
21 – 22	.20
22 – 23	.15
23 – 24	.10

          Make B

Interval	rf
18 – 19	.05
19 – 20	.10
20 – 21	.15
21 – 22	.20
22 – 23	.30
23 – 24	.20

d)   For Make A

5% of 60 = 3

20% of 60 = 12

30% of 60 = 18

20% of 60 = 12

15% of 60 = 9

10% of 60 = 6

     For Make B

5% of 140 = 7

10% of 140 = 14

15% of 140 = 21

20% of 140 = 28

30% of 140 = 42

20% of 140 = 28

Interval	Frequency	Relative Frequency
18 – 19	3 + 7 = 10	$\frac{10}{200}$ = .05 = 5%
19 – 20	12 + 14 = 26	$\frac{26}{200}$ = .13 = 13%
20 – 21	18 + 21 = 39	$\frac{39}{200}$ = .195 = 19.5%
21 – 22	12 + 28 = 40	$\frac{40}{200}$ = .200 = 20.0%
22 – 23	9 + 42 = 51	$\frac{51}{200}$ = .255 = 25.5%
23 – 24	6 + 28 = 34	$\frac{34}{200}$ = .170 = 17.0%

e)

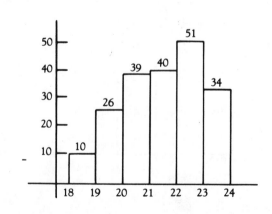

Exercise Set II.2

1) a)

Interval	Density	Relative Frequency = Area Area = length × Density
0 – 30	.004	30(.004) = .12
30 – 40	.030	10(.030) = .30
40 – 50	.040	10(.040) = .40
50 – 60	.010	10(.010) = .10
60 – 80	.004	20(.004) = .08

The interval 40 – 50 has the highest relative frequency.

b) rf = Area = (length of interval)(density)

$$= (43 - 42)(.040) = 1(.040) = .04 = 4\%$$

c) rf = Area = (length of interval)(density)

$$= (56 - 52)(.01) = 4(.01) = .04 = 4\%$$

d)

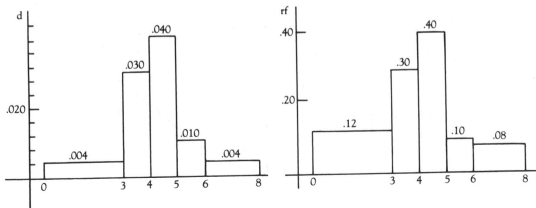

3)

Interval	Density
40 – 50	$\frac{.15}{10} = .0150$
50 – 60	$\frac{.20}{10} = .0200$
60 – 80	$\frac{.35}{20} = .0175$
80 – 100	$\frac{.30}{20} = .0150$

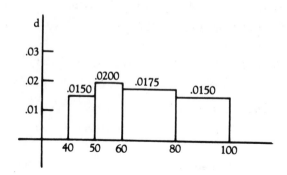

5) a) Bimodal (young and old have much higher rates)
   b) Symmetric
   c) Skewed right
   d) Skewed left
   e) Skewed right

7)

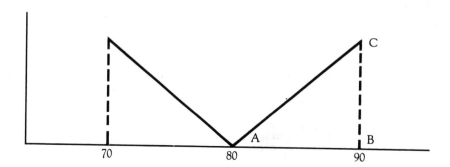

Area of the triangle ABC $= \frac{1}{2}$ (base)(height)

Area $= \frac{1}{2}$ (90 – 80)(.10) = .5

or by symmetry   rf $= \frac{1}{2}$ total area $= \frac{1}{2}$ (1) $= \frac{1}{2}$ = .5

9) a) The total area under the curve is 1. The area shaded is .95. Thus
   the total area outside the shaded region is 1 – .95 = .05. One half
   of this is under each tail. Thus under one tail,

$$A = \frac{1}{2} \ (.05) = .025.$$

   b) 2.5%

11) a) For the interval   30.5 - 50.5,

$$rf = \text{area} = \text{base} \times \text{height} = (20)(.012) = .24 = 24\%$$
$$24\% \text{ of } 300 = (.24)(300) = 72$$

b) $rf = (70.5 - 50.5)(.014) + (90.5 - 70.5)(.018)$

$= .28 + .36 = .64$

64% of $300 = (.64)(300) = 192$

Exercise Set II.3

1)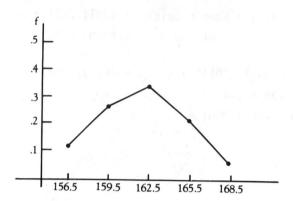

3)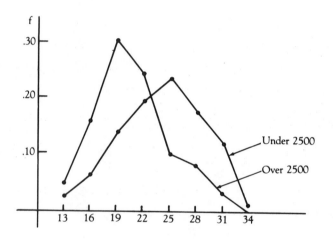

5)

Interval	(Section I)	f	rf	(Section II)	f	rf
50.5 – 60.5		5	.161		2	.056
60.5 – 70.5		6	.194		5	.139
70.5 – 80.5		13	.419		12	.333
80.5 – 90.5		4	.129		9	.250
90.5 – 100.5		3	.097		8	.222
		31			36	

Section II has higher frequencies with the higher grades and performed better.

7)

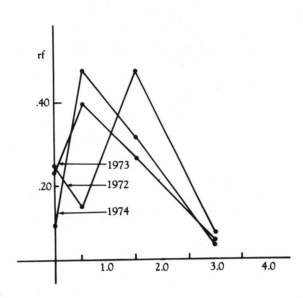

9) a) 15-70 roughly

    b) From figure (2), of the smokers the females have the greater incidence
        of respiratory symptoms. Of the non smokers, the men have the greater
        incidence of respiratory symptoms.

    c) $\dfrac{\text{\% Smoking Women with Symptoms}}{\text{\% Non Smoking Women with Symptoms}} = \dfrac{26.5}{8.1} = 3.27 \approx 3.3$

        $\dfrac{\text{\% Smoking Men with Symptoms}}{\text{\% Non Smoking Men with Symptoms}} = \dfrac{25.4}{16.0} = 1.59 \approx 1.6$

        Thus females run a higher risk than the males of obtaining respiratory
        disorders when they smoke.

11) a) 16

    b) 5000

    c) For x = 30000, cumulative frequency $\cong$ 3000.

        For x - 40000, cumulative frequency $\cong$ 4000.

        There are about 1000 primes between 30000 and 40000.

        There are about 1000 primes less than 10000 also.

        The cumulative frequency curve is almost a straight line.

Chapter Test

1) Smallest observation = 182, largest observation = 218

    Approximate interval width is

    $$w = \frac{218 - 182}{7} = \frac{36}{7} = 5.1.$$

    If we choose w = 5, more than 7 intervals are needed. Try w = 6.

Intervals	Tally	Frequency	Relative Frequency
181.5 – 187.5	|||	3	.12
187.5 – 193.5	|||	3	.12
193.5 – 199.5	卌	5	.20
199.5 – 205.5	卌 |||	8	.32
205.5 – 211.5	|||	3	.12
211.5 – 217.5	||	2	.08
217.5 – 223.5			1	.04

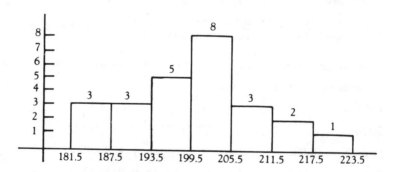

2) Environment B tends to produce lower times.

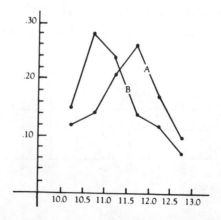

3) a) IQ scores of 10 year olds.

    b) Assembly times for 10 speed bicycles by untrained adults (see test
        example).

    c) The times until first failure of newly assembled engines.

    d) Salaries of professors at a well established university. (Most tend to
        be in the upper ranks.)

4) a) 20   b)

x	f	rf
0	2	2/20 = .10 = 10%
1	3	3/20 = .15 = 15%
2	5	5/20 = .25 = 25%
3	5	5/20 = .25 = 25%
4	4	4/20 = .20 = 20%
5	1	1/20 = .05 = 5%

c) 25% + 15% + 10% = 50%

    d) Use the observations as the midpoints (marks) of the intervals.

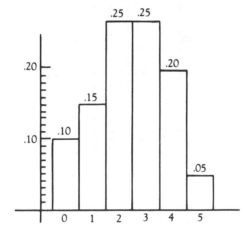

5) a)

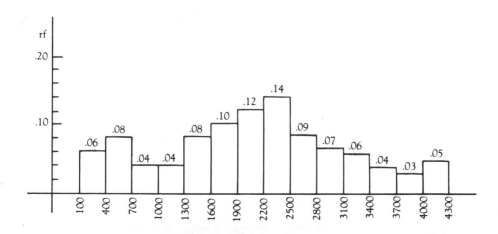

b)

Intervals	rf	d (for use with part c)
100 – 700	.06 + .08 = .14	$\frac{.14}{600}$ = .00023
700 – 1300	.04 + .04 = .08	$\frac{.08}{600}$ = .00013
1300 – 1900	.08 + .10 = .18	$\frac{.18}{600}$ = .00030
1900 – 2500	.12 + .14 = .26	$\frac{.26}{600}$ = .00043
2500 – 3100	.09 + .07 = .16	$\frac{.16}{600}$ = .00027
3100 – 3700	.06 + .04 = .10	$\frac{.10}{600}$ = .00017
3700 – 4300	.03 + .05 = .08	$\frac{.08}{600}$ = .00013

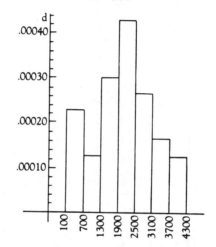

6) rf = (density)(length)

   a) rf = (.008)(20) = .16 = 16%

   b) rf = (.008)(15) = .12 = 12%

7) The total area is 1.

8) a)

```
0 | 75246
1 | 705984285634
2 | 40402233685657842
3 | 027435
```

   b)

```
0 | 24
0 | 756
1 | 04234
1 | 7598856
2 | 4040223342
2 | 6856578
3 | 0243
3 | 75
```

9) a) 2 | 50025862

3 | 55000674220

4 | 5302575

5 | 506895002423

6 | 5078230

7 | 620

b) 2 | 0022

2 | 5586

3 | 0004220

3 | 5567

4 | 302

4 | 5575

5 | 0002423

5 | 56895

6 | 0230

6 | 578

7 | 20

7 | 6

10)

Interval	f	rf
$5.00 \leq x < 5.20$	0	0.000
$5.20 \leq x < 5.40$	2	0.031
$5.40 \leq x < 5.60$	9	0.141
$5.60 \leq x < 5.80$	18	0.281
$5.80 \leq x < 6.00$	27	0.422
$6.00 \leq x < 6.20$	8	0.125

Chapter III
Descriptive Statistics: Numerical Measures

This chapter is concerned with numerical measures that describe important features of a distribution. The mode, median, and mean are introduced as measures of the center, the variance and standard deviation as measures of the variation of the data about the mean. Percentiles, percentile ranks, and z scores serve as measures of individual locations. Finally an empirical rule is formulated which describes the fraction of the observations that can be expected to be within one, two, and three standard deviations of the mean when the distribution is bell shaped.

<u>Concepts/Techniques</u>

1) <u>Measures of the Center</u>

There are three commonly used measures of the center of a set of data: the mode, the median, and the mean.

The <u>mode</u> is the most frequently occurring observation (if one exists). When the observations are arranged as to size and there is an odd number of observations, the <u>median</u> is the middle observation. If the number of observations is even, the <u>median</u> is the average of the two most central observations.

<u>Notation</u>: $\tilde{x}$

The <u>mean</u> is the arithmetic average of the observations.

Notation: $\bar{x}$.

If the observations are $x_1, x_2, x_3, \ldots, x_n$, then $\bar{x} = \dfrac{x_1 + x_2 + \cdots + x_n}{n}$.

Example: Find the mean, median, and mode of the sample 4, 5, 1, 4, 12, 10. The mean is

$$\bar{x} = \frac{4 + 5 + 1 + 4 + 12 + 10}{6} = \frac{36}{6} = 6.$$

To find the median, first arrange the observations as to size: 1, 4, 4, 5, 10, 12. The two most central observations of this list are 4 and 5. Thus

$$\tilde{x} = \frac{4 + 5}{2} = 4.5.$$

Since 4 is the most frequently occurring observation,

$$\text{mode} = 4.$$

Note that neither the mean nor the median must belong to the sample.

Example: Find the mean, median, and mode of the following data: 6, 6, 4, 2, 1, 1, 5

$$\bar{x} = \frac{6 + 6 + 4 + 2 + 1 + 1 + 5}{7} = \frac{25}{7} = 3.6 \text{ (rounded)}.$$

There are $n = 7$ observations (an odd number). The median is the middle observation of the ordering 1, 1, 2, 4, 5, 6, 6. Thus

$$\tilde{x} = 4.$$

There is no single mode. We speak of the distribution as being bimodal since $x = 6$ and $x = 1$ have equal frequencies which are greater than those of the other observations.

2) The Σ (Sigma) Notation for Sums

The notation $\displaystyle\sum_{i=1}^{n} x_i$ is used for the sum of the observations $x_1, x_2, \ldots, x_n$. The numbers 1 and n on $\displaystyle\sum_{i=1}^{n}$ specify the range of the subscript i.

Example: $\displaystyle\sum_{i=1}^{5} x_i = x_1 + x_2 + x_3 + x_4 + x_5$

$\displaystyle\sum_{i=1}^{3} x_i = x_1 + x_2 + x_3$

The sigma notation readily extends to more complicated terms.

Example: $\displaystyle\sum_{i=1}^{4} x_i^2 = x_1^2 + x_2^2 + x_3^2 + x_4^2$

$\displaystyle\sum_{i=1}^{5} x_i f_i = x_1 f_1 + x_2 f_2 + x_3 f_3 + x_4 f_4 + x_5 f_5$

$\displaystyle\frac{\sum_{i=1}^{3} x_i}{\sum_{i=1}^{3} f_i} = \frac{x_1 + x_2 + x_3}{f_1 + f_2 + f_3}$

Notes:

a) Subscripts other than i may be used. For example, there is no difference between $\displaystyle\sum_{i=1}^{n} x_i$ and $\displaystyle\sum_{j=1}^{n} x_j$. $\displaystyle\sum_{i=1}^{n} x_i = \sum_{j=1}^{n} x_j =$ $x_1 + x_2 + \cdots + x_n$.

b) Later we shall have need to consider the square of a sum $\displaystyle\sum_{i=1}^{n} x_i$. Note that

$$\left(\sum_{i=1}^{n} x_i\right)^2 \neq \sum x_i^2.$$

Example:

$$\left(\sum_{i=1}^{2} x_i\right)^2 = (x_1 + x_2)^2 = x_1^2 + 2x_1 x_2 + x_2^2$$

$$\neq x_1^2 + x_2^2 = \sum_{i=1}^{2} x_i^2$$

c) The mean $\bar{x}$ of $x_1, x_2, x_3, \ldots, x_n$ is

$$\bar{x} = \frac{\sum_{i=1}^{n} x_i}{n}$$

d) If it is clear that a summation is to extend over all the observations, it is permissible to leave off the subscript and its limits:

Example: We write $\sum x$ in place of $\displaystyle\sum_{i=1}^{n} x_i$.

We write $\Sigma\, x^2$ in place of $\displaystyle\sum_{i=1}^{n} x_i^2$.

We write $\Sigma\, xf$ in place of $\displaystyle\sum_{i=1}^{n} x_i f_i$.

3) <u>The Mean of a Frequency Distribution</u>

Suppose we have observations with assigned frequencies as in the table at the right. Then

x	f
x_1	f_1
x_2	f_2
$\vdots$	$\vdots$
x_k	f_k

$$\bar{x} = \frac{x_1 f_1 + x_2 f_2 + \cdots + x_k f_k}{n} = \frac{\displaystyle\sum_{i=1}^{k} x_i f_i}{n}$$

where $n = \displaystyle\sum_{i=1}^{k} f_i$.

Example: Find the mean of the sample:

1, 1, 2, 2, 2, 3, 4, 4, 4, 4, 4, 5 whose frequency table is at the right. The mean is

x	f
1	2
2	3
3	1
4	5
5	1

$$\bar{x} = \frac{(1+1) + (2+2+2) + 3 + (4+4+4+4+4) + 5}{12} = \frac{36}{12} = 3.0.$$

Note this sum may be written as

$$\bar{x} = \frac{1 \cdot 2 + 2 \cdot 3 + 3 \cdot 1 + 4 \cdot 5 + 5 \cdot 1}{2 + 3 + 1 + 5 + 1}$$

$$= \frac{x_1 f_1 + x_2 f_2 + x_3 f_3 + x_4 f_4 + x_5 f_5}{f_1 + f_2 + f_3 + f_4 + f_5} = \frac{\displaystyle\sum_{i=1}^{5} x_i f_i}{\displaystyle\sum_{i=1}^{5} f_i}$$

The mean of a frequency distribution is computed by using the class marks as the observations. Note that this will not be the mean of the original data. In most cases, however, these two means will be quite close.

Example: Find the mean of the distribution below.

Interval	Mark	Frequency
5 – 15	10	2
15 – 25	20	4
25 – 35	30	8
35 – 45	40	5
45 – 55	50	2
55 – 65	60	1

$$\bar{x} = \frac{10 \cdot 2 + 20 \cdot 4 + 30 \cdot 8 + 40 \cdot 5 + 50 \cdot 2 + 60 \cdot 1}{22}$$

$$= \frac{20 + 80 + 240 + 200 + 100 + 60}{22} = 31.8 \text{ (rounded)}$$

4) Deviations

The deviation of an observation x from the mean $\bar{x}$ is $d = x - \bar{x}$. If this deviation is negative, then x is less than $\bar{x}$ and located to the left of $\bar{x}$ on the number line. If the deviation is positive, then x is greater than $\bar{x}$ and located to the right of $\bar{x}$ on the number line.

5) The Variance

The variance of the observations $x_1, x_2, \ldots, x_n$ is denoted by s^2 and is defined by

$$s^2 = \frac{\sum_{i=1}^{n} (x_i - \bar{x})^2}{n - 1} = \frac{(x_1 - \bar{x})^2 + (x_2 - \bar{x})^2 + \cdots + (x_n - \bar{x})^2}{n - 1}.$$

Apart from using $n - 1$ in place of n, the variance is simply an average of the squares of the deviations.

6) The Standard Deviation

The standard deviation of $x_1, x_2, \ldots, x_n$ is the square root of their variance and is denoted by s. Thus

$$s = \sqrt{\text{variance}} = \sqrt{\frac{\sum_{i=1}^{n} (x_i - \bar{x})^2}{n - 1}}$$

The standard deviation s is a measure of the average deviation of the observations from the mean.

Example: Find the mean, variance, and standard deviation of the sample: 1, 7, 4, 5, 3.

First find the mean:

$$\bar{x} = \frac{1 + 7 + 4 + 5 + 3}{5} = 4.$$

The variance is

$$s^2 = \frac{(1 - 4)^2 + (7 - 4)^2 + (4 - 4)^2 + (5 - 4)^2 + (3 - 4)^2}{5 - 1}$$

$$= \frac{(-3)^2 + 3^2 + 0^2 + 1^2 + (-1)^2}{4}$$

$$= \frac{9 + 9 + 0 + 1 + 1}{4} = \frac{20}{4} = 5$$

The standard deviation is

$$s = \sqrt{5} = 2.2 \text{ (rounded)}.$$

Thus, on the average, the observations are approximately 2.2 units from the mean. These computations may be arranged in table form as follows:

i	x	$x - \bar{x}$	$(x - \bar{x})^2$
1	1	$1 - 4 = -3$	9
2	3	$3 - 4 = -1$	1
3	4	$4 - 4 = 0$	0
4	5	$5 - 4 = 1$	1
5	7	$7 - 4 = 3$	9
	$\Sigma x = 20$		$\Sigma(x-x)^2 = 20$

$$\bar{x} = \frac{\Sigma x}{5} = \frac{20}{5} = 4, \quad s^2 = \frac{\Sigma(x - \bar{x})^2}{5 - 1} = \frac{20}{4} = 5, \quad s = \sqrt{5} = 2.2$$

7) Rule for Rounding

The following rule for rounding is suggested. Round the mean $\bar{x}$ and standard deviation s to one more place than justified by the data. Round the variance to twice as many places as used in the standard deviation.

Example: For the sample 2.3, 2.5, 2.7, 2.8, 2.8, 3.0 computations yield

$$\bar{x} = \frac{\Sigma\ x}{6} = 2.68333\ldots$$

$$s^2 = 0.061666\ldots$$

$$s = 0.2483277\ldots$$

Since one decimal place data is involved, we use two places with the mean and standard deviation and four (2×2) with the variance. Thus

$$\bar{x} = 2.68$$

$$s^2 = 0.0617$$

$$s = 0.25$$

8) Alternate Formulas for the Variance and Standard Deviation

By means of some simple algebra we find

$$s^2 = \frac{\Sigma\ x^2 - \frac{(\Sigma\ x)^2}{n}}{n-1} = \frac{(x_1^2 + x_2^2 + \cdots + x_n^2) - \frac{(x_1 + x_2 + \cdots + x_n)^2}{n}}{n-1}$$

This formula and its square root are preferred formulas for computing the variance and standard deviation. In using these it is convenient to construct a table with the columns x and x^2 and use their sums.

Example: The number of cups of coffee drunk daily by a random sample of 8 office workers is given below. Find the mean, variance, and standard deviation.

Worker	x (Cups)	x^2
1	3	9
2	5	25
3	7	49
4	0	0
5	4	16
6	2	4
7	2	4
8	6	36

$$\Sigma\ x = 29 \quad \Sigma\ x^2 = 143$$

$$\bar{x} = \frac{\Sigma\ x}{8} = \frac{29}{8} = 3.625 = 3.6 \text{ (rounded)}$$

$$s^2 = \frac{\Sigma\ x^2 - \frac{(\Sigma\ x)^2}{8}}{8 - 1} = \frac{143 - \frac{(29)^2}{8}}{7} = \frac{143 - 105.125}{7} = 5.411$$

$$\cong 5.41 \text{ (rounded)}$$

$$s = \sqrt{5.411} = 2.3262 = 2.3 \text{ (rounded)}$$

9) <u>Data with Assigned Frequencies</u>

When the observations $x_1, x_2, \ldots, x_k$ have frequencies $f_1, f_2, \ldots, f_k$ the mean is computed by

$$\bar{x} = \frac{\Sigma\ xf}{n} \quad \text{where} \quad n = \Sigma\ f.$$

The variance is computed by either of the following:

$$s^2 = \frac{\sum\limits_{i=1}^{k} (x_i - \bar{x})^2 f_i}{n - 1} \quad \text{or} \quad s^2 = \frac{\sum\limits_{i=1}^{k} x_i^2 f_i - \frac{\left(\sum\limits_{i=1}^{k} x_i f_i\right)^2}{n}}{n - 1}$$

The second formula is preferred. When this is used, it is convenient to prepare a table with columns x, f, xf, and $x^2 f$.

Example: The distribution of the yearly number of visits x to a medical doctor or clinic by a sample of college students is given below. Compute the mean, variance and standard deviation.

Visit x	Frequency f	xf	$x^2 f$
0	2	0	0
1	6	6	6
2	8	16	32
3	6	18	54
4	5	20	80
6	3	18	108
$n = \Sigma\ f = 30$		78	280

$$\overline{x} = \frac{\Sigma \ xf}{n} = \frac{78}{30} = 2.6$$

$$s^2 = \frac{\Sigma \ x^2 f - \frac{(\Sigma \ xf)^2}{30}}{30 - 1} = \frac{280 - \frac{(78)^2}{30}}{29} = 2.66$$

$$s = \sqrt{2.66} = 1.6$$

10) The Empirical Rule

The standard deviation s of a sample is a measure of the variation
in the data or the spread of the data about the mean.   The empirical
rule states that if the distribution has a bell shaped appearance, then

a) Within one standard deviation of the mean will be found approximately
    68% of the observations.

b) Within two standard deviations of the mean will be found approximately
    95% of the observations.

c) Within three standard deviations of the mean will be found approxi-
    mately 99.7% (virtually all) of the observations.

Don't expect these predictions to be followed exactly.   Think of them
as "ball park" estimates.

Example: The distribution of grades of 40 students on an
examination is bell shaped with mean $\overline{x} = 60$ and standard
deviation s = 15.   What does the empirical rule say about the
grades?

One standard deviation is s = 15.   Those grades that are within 1
standard deviation of the mean lie on the interval 60 ± 15,   that
is between 45 and 75.   Since 68% of 40 = 27.2,   about 27
of the 40 grades should be between 45 and 75.

Two standard deviations is $2s = 2 \cdot 15 = 30$.   Thus about 95%
of the grades are predicted to be between 60 - 30 = 30 and
60 + 30 = 90.   Since 95% of 40 = 38, approximately 38 of the
40 grades are predicted to be between 30 and 90.

Example: The results of two auto mileage studies by two testing
laboratories are presented below.   Which study indicates a greater
reliability?

Laboratory A	Laboratory B
n = 80 Autos	n = 80 Autos
$\bar{x}$ = 30 m.p.g.	$\bar{x}$ = 30 m.p.g.
s = 2.3 m.p.g.	s = 7.6 m.p.g.

<u>Laboratory A</u>: 95% (76 of 80) cars performed between 30 - 2(2.3) and 30 + 2(2.3), that is between 23.4 and 34.6 m.p.g.

<u>Laboratory B</u>: 95% (76 of 80) cars performed between 30 - 2(7.6) and 30 + 2(7.6), that is between 14.8 and 45.2.

Obviously the data of laboratory B has considerably more variation in it than that of laboratory A. Hence the data from laboratory B is less reliable.

Remember

1) The larger the standard deviation, the more variation there is in the observations.

2) When there is no variation at all, (i.e. all observations are the same), the standard deviation is 0.

11) Standardized Z Scores

Given a set of observations with mean $\bar{x}$ and standard deviation s, the z score corresponding to x is

$$z = \frac{x - \bar{x}}{s} .$$

The z score is the distance of x from the mean $\bar{x}$ measured in standard deviation units.

<u>Example</u>: The scores on a mathematics placement test had a mean $\bar{x}$ = 30 and standard deviation s = 4. Compute the z scores corresponding to x_1 = 38, x_2 = 24, x_3 = 30 and interpret. For x_1 = 38

$$z_1 = \frac{x_1 - \bar{x}}{s} = \frac{38 - 30}{4} = \frac{8}{4} = 2.$$

Thus x_1 = 38 is 2 standard deviations above the mean.
For x_2 = 24

$$z_2 = \frac{x_2 - \bar{x}}{s} = \frac{24 - 30}{4} = \frac{-6}{4} = -1.5.$$

Thus $x_2 = 24$ is 1.5 standard deviations <u>below</u> the mean.

For $x_3 = 30$

$$z_3 = \frac{x_3 - \bar{x}}{s} = \frac{30 - 30}{4} = \frac{0}{4} = 0.$$

Thus $x_3 = 30$ is located at the mean.

Note that the empirical rule predicts that 68% of the observations of a bell shaped distribution have z scores between -1 and $+1$, 95% between -2 and $+2$, and virtually all between -3 and $+3$.

<u>Example</u>: Assuming $\bar{x} = 150$ and $s = 20$, find the observation x corresponding to

a) $z = 1.6$ b) $z = -2.4$

We know x and z are related by the formula

$$z = \frac{x - 150}{20}.$$

a) When $z = 1.6$, $1.6 = \frac{x - 150}{20}$.

Thus

$$x - 150 = (1.6)(20) = 32$$
$$x = 32 + 150 = 182$$

Alternatively, we can interpret $z = 1.6$ to mean that x is located 1.6 standard deviations above the mean and write

$$x = 150 + 1.6(20) = 182$$

b) When $z = -2.4$

$$-2.4 = \frac{x - 150}{20}$$

$$x - 150 = -2.4(20) = -48$$
$$x = -48 + 150 = 102$$

Alternatively, we can interpret $z = -2.4$ to mean that x is 2.4 standard deviations below the mean and write

$$x = 150 - 2.4(20) = 102.$$

Notes:

1) z scores are generally rounded to 2 decimal places.

2) If we have a set of data $x_1, x_2, \ldots, x_n$ with mean $\bar{x}$ and standard deviation s then the corresponding z scores $z_1, z_2, \ldots, z_n$ have a mean $\bar{z} = 0$ and standard deviation of 1.

12) <u>Notation</u>

The mean of a population is denoted by the Greek letter μ (mu).

The standard deviation of a population is denoted by the Greek letter σ (sigma) and the variance by its square σ^2.

> <u>Example</u>: Suppose family incomes x in the U.S. have mean $\mu = \$18500$ and standard deviation $\sigma = \$4200$. Find the z score corresponding to the income x = \$25000. The formula is
>
> $$z = \frac{x - \mu}{\sigma} = \frac{x - 18500}{4200} \, .$$

When x = 25000

$$z = \frac{25000 - 18500}{4200} = 1.55.$$

13) <u>Percentiles</u>

The Kth percentile of a distribution is a number P_K having K% of the observations less than it and (100 − K)% greater than it.

> <u>Example</u>: Suppose the 90th percentile for distances in the discus throw is 528 feet. Then this means that 90% of the distances recorded were below 528 feet and 10% were above 528 feet.
>
> The lower or first quartile is the 25th percentile.
>
> <u>Notation</u>: Q_1 or P_{25}
>
> The upper or third quartile is the 75th percentile.
>
> <u>Notation</u>: Q_3 or P_{75}
>
> The median is the 50th percentile.
>
> <u>Notation</u>: P_{50} or $\tilde{x}$.

14)   Interpolation to Find Percentiles

A process called interpolation is used to locate percentiles (approximately) of a frequency distribution. Once the interval containing the Kth percentile is located, the formula is (loosely) as follows:

$$P_k = \begin{pmatrix} \text{Lower Limit} \\ \text{of the Interval} \end{pmatrix} + \begin{pmatrix} \dfrac{\text{Additional \% Needed}}{\text{\% of Observations Present}} \end{pmatrix}\begin{pmatrix} \text{Length of the} \\ \text{Interval} \end{pmatrix}$$

Example: Find the upper quartile of the distribution below.

Interval	Relative Frequency
20 – 30	10%
30 – 40	20%
40 – 50	30%
50 – 60	25%
60 – 70	15%

We see $P_{60} = 50$ and $P_{85} = 60$. Thus the 75th percentile is between 50 and 60. This interval is of length 10, has lower limit 50, and contains 25% of the observations. Since an additional $75 - 60 = 15$ percent is needed, we have

$$P_{75} = 50 + \frac{15}{25}(10) = 56.$$

Example: Estimate the median of the distribution of the last example. The 50th percentile P_{50} or $\tilde{x}$ is located on the interval $40 - 50$ since $P_{30} = 40$ and $P_{60} = 50$. This interval is of length 10, has lower limit 40, and contains 30% of all observations. An additional $20 = (50 - 30)$ percent is needed. Thus

$$\tilde{x} = 40 + \frac{20}{30}(10) = 46.67.$$

15)   Percentile Ranks

The percentile rank of an observation x, denoted by $PR(x)$, locates x relative to the other observations. The formula for the rank is

$$PR(x) = \frac{\begin{pmatrix} \text{\# Observations} \\ \text{Below } x \end{pmatrix} + \dfrac{1}{2}\begin{pmatrix} \text{\# Observations} \\ \text{Equal to } x \end{pmatrix}}{\text{Total Number of Observations}} \times 100.$$

Example: You received a score of 14 on a quiz for which the grade
distribution is shown at the right. Find the
percentile rank of your score.

There are $5 + 7 + 11 + 6 = 29$ scores below
14 and 9 at 14 and there are 50 scores
in all. Thus

x	f
6	5
8	7
12	11
13	6
14	9
17	8
20	4

$$PR(14) = \frac{29 + \frac{1}{2}(9)}{50} \times 100 = \frac{33.5}{50} \times 100 = 67\%$$

Thus the percentile rank of 14 is 67.

Example: For the distribution of the last example, find PR(6).
There are 0 observations below 6 and 5 at 6. Thus

$$PR(6) = \frac{0 + \frac{1}{2}(5)}{50} \times 100 = 5.$$

Calculator Usage Illustration

Compute the mean of a sample.

Find $\dfrac{4 + 9 + 11}{3}$.

Method: Direct summation to find the numerator followed by a division.

4 $\boxed{+}$ 9 $\boxed{+}$ 11 $\boxed{=}$ $\boxed{\div}$ 3 $\boxed{=}$ <u>8</u>

Calculator Usage Illustration

Compute the mean when the observations have assigned frequencies.

Find $\dfrac{2 \cdot 5 + 4 \cdot 8 + 3 \cdot 7}{20}$.

Method 1) Direct summation to find the numerator followed by a division.

2 $\boxed{\times}$ 5 $\boxed{+}$ 4 $\boxed{\times}$ 8 $\boxed{+}$ 3 $\boxed{\times}$ 7 $\boxed{=}$ $\boxed{\div}$ 20 $\boxed{=}$ <u>3.15</u>

Method 2) Accumulate the products in the memory, recall the sum and do the division.

2 $\boxed{\times}$ 5 $\boxed{=}$ $\boxed{SUM}$ 4 $\boxed{\times}$ 8 $\boxed{=}$ $\boxed{SUM}$ 3 $\boxed{\times}$ 7 $\boxed{=}$ $\boxed{SUM}$ $\boxed{RCL}$ $\boxed{\div}$ 20 $\boxed{=}$ <u>3.15</u>

Calculator Usage Illustration

Compute the variance of a sample using the formula $s^2 = \dfrac{\Sigma(x - \bar{x})^2}{n - 1}$

Find $\dfrac{(4 - 8)^2 + (9 - 8)^2 + (11 - 8)^2}{2}$.

Method 1) Direct summation to find the numerator followed by a division.

$\boxed{(}$ 4 $\boxed{-}$ 8 $\boxed{)}$ $\boxed{x^2}$ $\boxed{+}$ $\boxed{(}$ 9 $\boxed{-}$ 8 $\boxed{)}$ $\boxed{x^2}$ $\boxed{+}$ $\boxed{(}$ 11 $\boxed{-}$ 8 $\boxed{)}$ $\boxed{x^2}$ $\boxed{=}$ $\boxed{\div}$ 2 $\boxed{=}$ <u>13</u>

Method 2) Accumulate the squares in the memory, recall the sum and do the division.

4 $\boxed{-}$ 8 $\boxed{=}$ $\boxed{x^2}$ $\boxed{SUM}$ 9 $\boxed{-}$ 8 $\boxed{=}$ $\boxed{x^2}$ $\boxed{SUM}$ 11 $\boxed{-}$ 8 $\boxed{=}$ $\boxed{x^2}$ $\boxed{SUM}$ $\boxed{RCL}$ $\boxed{\div}$ 2 $\boxed{=}$ <u>13</u>

Calculator Usage Illustration

Compute the standard deviation using the alternate formula.

Find $\sqrt{\dfrac{218 - \dfrac{(24)^2}{3}}{2}}$.

Method 1) Direct evaluation.

218 $\boxminus$ 24 $\boxed{x^2}$ $\boxed{\div}$ 3 $\boxed{=}$ $\boxed{\div}$ 2 $\boxminus$ $\boxed{\sqrt{}}$ <u>3.6055513</u>

Note: This is the sample of the first illustration.

Note: Some calculators have a different hierarchy of operations
and relations may yield an erroneous result when the above
is used. If so, try the following method.

Method 2) Reverse the order of computation in the numerator of the fraction,
then change its sign, do the division, and take the square root.

24 $\boxed{x^2}$ $\boxed{\div}$ 3 $\boxminus$ $\boxminus$ 218 $\boxminus$ $\boxed{\pm}$ $\boxed{\div}$ 2 $\boxminus$ $\boxed{\sqrt{}}$ <u>3.6055513</u>

Note: On most calculators the $\pm$ key changes the sign of the number
in the register.

Calculator Usage Illustration

Compute the sum of squares when the observations have assigned frequencies.
Find $2^2 \cdot 5 + 4^2 \cdot 8 + 3^2 \cdot 7$.

Method 1) Direct summation (no parentheses).

2 $\boxed{x^2}$ $\boxed{\times}$ 5 $\boxed{+}$ 4 $\boxed{x^2}$ $\boxed{\times}$ 8 $\boxed{+}$ 3 $\boxed{x^2}$ $\boxed{\times}$ 7 $\boxed{=}$ <u>211</u>

Method 2) Accumulate the sum in the memory.

2 $\boxed{x^2}$ $\boxed{\times}$ 5 $\boxed{\text{SUM}}$ 4 $\boxed{x^2}$ $\boxed{\times}$ 8 $\boxed{\text{SUM}}$ 3 $\boxed{x^2}$ $\boxed{\times}$ 7 $\boxed{\text{SUM}}$ $\boxed{\text{RCL}}$ <u>211</u>

Calculator Usage Illustration

Compute the Z score corresponding to an observation.

Find $\dfrac{10 - 8}{3.61}$.

Method: Direct evaluation.

10 $\boxminus$ 8 $\boxminus$ $\boxdiv$ 3.61 $\boxminus$ <u>.5540166</u>

Calculator Usage Illustration

Compute the percentile rank of an observation.

Find $\dfrac{5 + \dfrac{1}{2}\,(3)}{11} \times 100.$

Method: Direct evaluation of the fraction followed by the multiplication
         by 100.

5 $\boxplus$ 3 $\boxdiv$ 2 $\boxminus$ $\boxdiv$ 11 $\boxminus$ $\boxtimes$ 100 $\boxminus$ <u>59.090909</u>

Note: This is the percentile rank of x = 8 in the sample:
       1, 2, 2, 3, 5, 8, 8, 8, 11, 12, 15.

Note: Better results are generally obtained if rounding is done only
after all computations are complete. In the solution sets, <u>unrounded
quantities</u> are used in the intermediate computations.

Exercise Set III.1

1) First arrange by size.

   a) 1.0, 2.4, 2.4, 3.8, 4.6, 5.6, 7.2

      mode = 2.4

      $\tilde{x}$ = 3.8 (middle observation)

   b) 1, 2, 3, 3, 4, 7, 8, 10

      mode = 3

      $\tilde{x} = \dfrac{3+4}{2}$ = 3.5 (the average of the two most central observations)

3) Ordered as to size, the observations are

   25, 30, 30, 30, 32, 35, 40, 40, 42, 46

   mode = 30 (most frequently occurring observation)

   median $\tilde{x} = \dfrac{32+35}{2}$ = 33.5 (average of the two most central observations)

   mean $\bar{x} = \dfrac{25 + 30 + 30 + 30 + 32 + 35 + 40 + 40 + 42 + 46}{10}$ = 35

5)

x	f	xf	$(x-\bar{x})$	$(x-\bar{x})^2$	$(x-\bar{x})^2 f$
0	16	0	−2.2	4.84	77.44
1	10	10	−1.2	1.44	14.40
2	14	28	−0.2	.04	.56
3	20	60	0.8	.64	12.80
4	8	32	1.8	3.24	25.92
5	7	35	2.8	7.84	54.88

$\Sigma f = 75$ $\Sigma xf = 165$ $\Sigma(x-\bar{x})^2 f = 186.00$

$$\bar{x} = \frac{165}{75} = 2.2$$

$$s = \sqrt{\frac{186.00}{75-1}} = 1.59 \approx 1.6$$

7)

x (Interval Midpoint)	f	xf	$x^2 f$
16	1	16	256
17	4	68	1156
18	8	144	2592
19	12	228	4332
20	3	60	1200
22	1	22	484
23	1	23	529
	30	561	10549

a) mean $\bar{x} = \dfrac{\Sigma\ xf}{n} = \dfrac{561}{30} = 18.7$ (rounded)

b) $s^2 = \dfrac{\Sigma\ x^2 f - \dfrac{(\Sigma\ xf)^2}{n}}{n-1} = \dfrac{10549 - \dfrac{(561)^2}{30}}{29}$

$$= \dfrac{10549 - 10490.7}{29}$$

$$= 2.0101345$$

$$= 2.01 \text{ rounded}$$

The standard deviation is

$$s = \sqrt{\text{variance}} = \sqrt{2.01} = 1.4177921$$

$$= 1.4 \text{ rounded.}$$

9) a)

x	f	xf	$x^2 f$
5	5	25	125
6	4	24	144
7	7	49	343
8	10	80	640
9	13	117	1053
10	12	120	1200
11	6	66	726
12	3	36	432

$n = \Sigma\ f = 60 \quad \Sigma\ xf = 517 \quad \Sigma\ x^2 f = 4663$

$\bar{x} = \dfrac{\Sigma\ xf}{n} = \dfrac{517}{60} = 8.6$

$s = \sqrt{\dfrac{\Sigma\ x^2 f - \dfrac{(\Sigma\ xf)^2}{n}}{n-1}}$

$\quad = \sqrt{\dfrac{4663 - \dfrac{(517)^2}{60}}{60 - 1}} = 1.9$ rounded

b) The mean is $\bar{x} = 8.6$.

1(S.D.) = 1.8

1(S.D.) above the mean is $8.6 + 1.8 = 10.4$.

1(S.D.) below the mean is $8.6 - 1.8 = 6.8$.

Thus we are interested in those observations x which are between 6.8
and 10.4, i.e., $6.8 < x < 10.4$. These are

$$x = 7, \ f = 7$$
$$x = 8, \ f = 10$$
$$x = 9, \ f = 13$$
$$x = 10, \ f = 12$$

Thus $7 + 10 + 13 + 12 = 42$ of the 60 observations are within 1(S.D.) of $\bar{x}$.

Thus $\dfrac{42}{60} = .70 = 70\%$ of the sample is within 1(S.D.) of the mean.

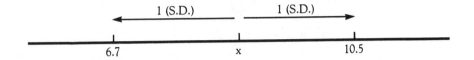

All (100%) of the observations are within 2(S.D.) of the mean.

11) This would be used if the sample were in fact in the whole population.

    1) Let the population be <u>all</u> scores on the 1974 S.A.T.

    2) Let the population be the yearly salaries of all active players on the NFL player roster.

13) a) If each observation is reduced by 16 the mean is reduced by 16. Thus for the data, $\bar{x} = 16 - 16 = 0$.

    Subtracting 16 from each observation does not change the variation between the observations. Thus the standard deviation is unchanged.

    b) If each observation is reduced by 16, there is no change in the variation. Hence the standard deviation is still 3.

15) Changing each observation by $2020 - 1975 = 45$ years does not change the variation between the observations. Thus the S.D. is unchanged. S.D. = 70. Since each age is increased by 45 years, the mean age is increased by 45 years.

Thus now $\bar{x} = 1260 + 45 = 1305$ years.

17) a) First (The observations of this first distribution tend to be farther from the mean.)

    b) First (The observations of this distribution tend to be farther from the mean.)

19) $\bar{x} = \dfrac{\Sigma x}{50} = 27.3$

Thus $\Sigma x = 50 \times 27.3 = 1365$ (million).

21) a) The median is $\dfrac{x_4 + x_5}{2}$ in each case. Thus $\tilde{x} = 5$

for all 3 sets.

b) For the first (symmetric) set $\bar{x} = x_4 = 5 = \tilde{x}$ by symmetry

For the skewed right set,

$$\bar{x} = \frac{2 + 3 + 4 + 5 + 6 + 7 + 11}{7} = 5.4 > \tilde{x}.$$

For the skewed left set

$$\bar{x} = \frac{0 + 3 + 4 + 5 + 6 + 7 + 8}{7} = 4.7 < \tilde{x}.$$

c) First set: 2, 2, 3, 4, 6, 7, 8, 8

$$x_1 \ x_2 \ x_3 \ x_4 \ x_5 \ x_6 \ x_7 \ x_8$$

Median: $\tilde{x} = \dfrac{x_4 + x_5}{2} = 5$,   Mean: $\bar{x} = 5$ by symmetry

Second set (Skewed Right)

2, 2, 3, 4, 6, 7, 8, 10

$$\tilde{x} = \frac{x_4 + x_5}{2} = 5$$

$$\bar{x} = \frac{2 + 2 + 3 + 4 + 7 + 7 + 8 + 10}{8} = \frac{43}{8} = 5.4 > \tilde{x}$$

Third set (Skewed Left)

1, 2, 3, 4, 6, 7, 8, 8

$$\tilde{x} = \frac{x_4 + x_5}{2} = \frac{4 + 6}{2} = 5$$

$$\bar{x} = \frac{1 + 2 + 3 + 4 + 6 + 7 + 8 + 8}{8} = \frac{39}{8} = 4.9 < \tilde{x}$$

d) These results suggest that

a) For a symmetric distribution $\bar{x} = \tilde{x}$.

b) For a skewed right distribution, $\bar{x} > \tilde{x}$.

c) For a skewed left distribution, $\bar{x} < \tilde{x}$.

23) 0 The sum of the derivations of the observations about their mean is 0.

Exercise Set III.2

1) a) $2(S.D.) = 2 \times 10 = 20$

   Since x is $2(S.D.)$ above the mean $x = 112 + 20 = 132$.

   b) $3(S.D.) = 3 \times 10 = 30$

   Since x is $3(S.D.)$ below the mean $x = 112 - 30 = 82$.

   c) $1.4(S.D.) = 1.4 \times 10 = 14$

   Since x is $1.4(S.D.)$ above the mean $x = 112 + 14 = 126$.

   d) $\frac{1}{2}(S.D.) = \frac{1}{2} \times 10 = 5$

   Since x is $\frac{1}{2}(S.D.)$ <u>below</u> the mean, $x = 112 - 5 = 107$.

3) a) $1.5 \ S.D. = 1.5 \times 6 = 9$ $18 - 9 < x < 18 + 9$ or $9 < x < 27$

   b) $2 \ S.D. = 2 \times 6 = 12$ $18 - 12 < x < 18 + 12$ or $6 < x < 30$

   c) $3 \ S.D. = 3 \times 6 = 18$ $x > 18 + 18 = 36$ or $x > 36$

   d) $1 \ \underline{S.D.} = 6, \ 2 \ S.D. = 12$ $18 - 12 < x < 18 - 6$ or $6 < x < 12$

5) $z = \dfrac{x - \bar{x}}{s} = \dfrac{x - 175}{16}$

   a) When $x = 200$, $z = \dfrac{200 - 175}{16} = \dfrac{25}{16} = 1.5625 = 1.56$

   b) When $x = 163$, $z = \dfrac{163 - 175}{16} = \dfrac{-12}{16} = -.75$

   c) When $x = 175$, $z = \dfrac{175 - 175}{16} = \dfrac{0}{16} = 0$

   d) When $z = 2.3$, $2.3 = \dfrac{x - 175}{16}$

$$(2.3) \cdot 16 = x - 175$$

$$36.8 = x = 175$$

$$175 + 36.8 = x$$

$$211.8 = x$$

   e) When $z = -3.6$

$$-3.6 = \dfrac{x - 175}{16}$$

$$-57.6 = x - 175$$

$$175 - 57.6 = x$$

$$117.4 = x$$

   f) When $z = 1.5$

$$1.5 = \dfrac{x - 175}{16}$$

$$(1.5)16 = x - 175$$

$$24 = x - 175$$

$$175 + 24 = x$$

$$199 = x$$

7) a)

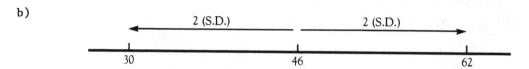

About 68% of the observations should be within 1 S.D. of $\bar{x} = 46$.
68% of 300 = 204.

b)

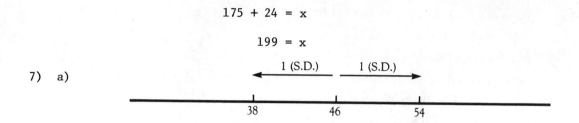

About 95% of the observations should be within 2 S.D. of $\bar{x}$.
95% of 300 = 285.

c)

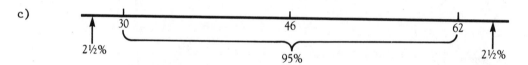

$2\frac{1}{2}\%$ of 300 = 7.5.   About   7   or   8   observations should exceed   62.

d)

Between   54   and   62   is located approximately   47.5% − 34% = 13.5% of
the sample.   13.5% of 300 = 40.5.   ∴   About   40   or   41.

e) $2\frac{1}{2}\%$ (see c)

$2\frac{1}{2}\%$ of 300 = 7.5.   Around   7   or   8   observations.

9) a)   45   and   85   are 2 S.D. below and above the mean respectively.   Within
2 S.D. is located approximately 95% of the observations.
95% of 60 = 57.

b)   Grades exceeding   85   are more than 2 S.D. above the mean.   On $\bar{x} \pm 2$ S.D.
is located about 95% of the observations.   Thus 5% of the grades are
more than 2 S.D. from the mean. $2\frac{1}{2}\%$ (half of these) should be 2 S.D.

<u>above</u> the mean.

$2\frac{1}{2}\% \times 60 = 1.5$

Thus somewhere around 1 or 2 grades is predicted.

c)   The distribution should be "more or less" bell shaped.

11)   The range is approximately 6 S.D. in length.

a)   6 S.D. $\tilde{=}$ 320 - 110 = 210

$S.D. \tilde{=} \frac{210}{6} = 35$ $\bar{x} \approx 230$

b)   6 S.D. $\tilde{=}$ 68 - 52 = 16

$S.D. \tilde{=} \frac{16}{6} = 2.7$ $\bar{x} \approx 60$

13)   a) $z_1 = \frac{200 - 190}{20} = .5$

$z_2 = \frac{210 - 190}{20} = 1$

$z_3 = \frac{184 - 190}{20} = -.30$

$z_4 = \frac{170 - 190}{20} = -1.$

b)   All observations are within 1 S.D. of the mean of the original medication.
This interval should contain about 68% of the observations.
There is no evidence of an improved effective time with this new
medication.

15)   a) $\bar{x} = \frac{119}{9} = 13.2$

$s = \sqrt{\dfrac{2357 - \dfrac{(119)^2}{9}}{8}} = 9.9$

b) L_1: $z_1 = \frac{65 - 13.2}{9.9} = 5.2$

L_2: $z_2 = \frac{24 - 13.2}{9.9} = 1.1$

L_3: $z_3 = \frac{52 - 13.2}{9.9} = 3.9$

L_4: $z_4 = \frac{86 - 13.2}{9.9} = 7.4$

L_5: $z_5 = \frac{120 - 13.2}{9.9} = 10.8$

L_6: $z_6 = \frac{82 - 13.2}{9.9} = 6.9$

L_7: $z_7 = \frac{399 - 13.2}{9.9} = 39.0$

L_8: $z_8 = \frac{87 - 13.2}{9.9} = 7.5$

L_9: $z_9 = \dfrac{139 - 13.2}{9.9} = 12.7$

$\bar{z} = \dfrac{94.5}{9} = 10.5$

Thus the nickel concentration on the average was 10.5 standard deviations above the control group.

17) a)

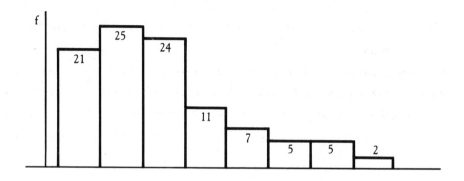

b) The distribution is highly skewed. The empirical predictions work best
·   with a symmetric bell shaped distribution.

c)

x	f	xf	x^2f
1	21	21	21
2	25	50	100
3	24	72	216
4	11	44	176
5	7	35	175
6	5	30	180
7	5	35	245
8	2	16	128
	100	303	1241

$\bar{x} = \dfrac{303}{100} = 3.0$

$s = \sqrt{\dfrac{1241 - \dfrac{(303)^2}{100}}{99}} = 1.8$

d)

The x = 2, x = 3, and x = 4 are within 1 S.D. of x.

Their frequencies total 25 + 24 + 11 = 60.

$\dfrac{60}{100} = 60\%$

19) a) Within h = 3 S.D. of the mean will be found at least $1 - \dfrac{1}{3^2} = 1 - \dfrac{1}{9} =$
        $\dfrac{8}{9}$ = 88.89% of the sample.

   b) Within h = $1\frac{1}{2}$ = 1.5 S.D. of the mean will be found at least
        $1 - \dfrac{1}{(1.5)^2}$ = 1 - .4444 = .5556 = 55.56% of the sample.

   c) For h = 1

$$1 - \frac{1}{h^2} = 1 - \frac{1}{1^2} = 1 - 1 = 0.$$

   So the theorem does not guarantee any observations within 1 S.D. of
   the mean.

It should be noted that this is a theorem – the results are "guaranteed"
so to speak as opposed to the "predictions" of the empirical result. As a
result they must apply to any distribution regardless of how bizzare it may
be. Of necessity, its results are "somewhat" pessimistic in most cases.

21)

On the interval $\mu \pm 1$ S.D. is located about 68% of the observations. Thus
by symmetry on $\mu - (\mu + 1$ S.D.) is located about $\dfrac{68}{2}$ % = 34%. Similarly
on $(\mu - 1$ S.D.) $- \mu$ is located 34%.
On $\mu \pm 2$ S.D. is located about 95% of the observations. Within this
interval and outside the interval $\mu \pm 1$ S.D. must be located about
95% – 68% = 27% of the observations. Thus on $(\mu - 2$ S.D.) $- (\mu - 1$ S.D.)
is located about $\dfrac{27}{2}$ % = 13.5%. Similarly $(\mu + 1$ S.D.) $- (\mu + 2$ S.D.)
contains about 13.5%. Outside $\mu \pm 2$ S.D. and inside $\mu \pm 3$ S.D.
is located about 99.7% – 95% = 4.7% of the data. Thus the
remaining 2 intervals $(\mu - 3$ S.D.) $- (\mu - 2$ S.D.) and $(\mu + 2$ S.D.) $-$
$(\mu + 3$ S.D.) have relative frequency about $\dfrac{.047}{2}$ = .0235.

Interval	rf
$(\mu - 3$ S.D.) $- (\mu - 2$ S.D.)	.0235
$(\mu - 2$ S.D.) $- (\mu - 1$ S.D.)	.135
$(\mu - 1$ S.D.) $- \mu$	.340
$\mu - (\mu + 1$ S.D.)	.340
$(\mu + 1$ S.D.) $- (\mu + 2$ S.D.)	.135
$(\mu + 2$ S.D.) $- (\mu + 3$ S.D.)	.0235

Exercise Set III.3

1) a) Q_1, the 25th percentile, is located at $x = 65$ since 25% of the signals are less than 65 KHZ.

   b) $\tilde{x} = 65 + \dfrac{25}{32} (80 - 65) = 76.7$

   c) $Q_3 = 100 + \dfrac{4}{22} (130 - 100) = 105.5$

3) a) $\tilde{x} = P_{50} = 16.4 + \dfrac{25}{40} (1.5) = 17.34$

   b) $P_{90} = 19.4 + \dfrac{5}{10} (1.5) = 20.15$

   c) $Q_1 = P_{25} = 16.4$

   d) $Q_3 = P_{75} = 17.9 + \dfrac{10}{20} (1.5) = 18.65$

5) Percentile Rank $= \dfrac{480 + \frac{1}{2} (110)}{1000} (100) = 53.5$

7) $PR(0) = \dfrac{0 + \frac{1}{2} (7)}{20} (100) = 17.5$

   $PR(1) = \dfrac{7 + \frac{1}{2} (3)}{20} (100) = 42.5$

   $PR(2) = \dfrac{10 + \frac{1}{2} (5)}{20} (100) = 62.5$

   $PR(3) = \dfrac{15 + \frac{1}{2} (2)}{20} (100) = 80.0$

   $PR(4) = \dfrac{17 + \frac{1}{2} (3)}{20} (100) = 92.5$

9) It means it was high at the time compared to other individuals. This by itself does not mean that it is "dangerously" high.

11) The income tax you pay is at a low percentile for people in your income range.

13) $Q_3 - Q_1 = 27.5 - 22.86 = 4.64$

   50% of all graduting seniors are between 27.5 and 22.86 years of age. This interval is of length 4.64.

Chapter Test

1) Ordered as to size, the observations are: 2, 2, 2, 2, 3, 4, 5, 5, 5

   $\tilde{x}$ = 3   (the middle observation)

   mode = 2   (the most frequently occurring observation)

2) First obtain the frequencies and construct a table as shown.

x	f	xf	x^2f
2	2	4	8
3	4	12	36
5	6	30	150
6	6	36	216
9	2	18	162
	20	100	572

   a) n = 20   and $\bar{x} = \dfrac{100}{200} = 5$

   b) $s = \sqrt{\dfrac{572 - \dfrac{(100)^2}{20}}{19}} = 1.946 \approx 1.95$

   c) 1.5(S.D.) = (1.5)(1.95) = 2.93

   $\bar{x}$ - 1.5s = .5 - 2.93 = 2.07

   $\bar{x}$ + 1.5s = .5 + 2.93 = 7.93

   The four 3's, the six 5's and the six 6's are within 1.5(S.D.) of the mean. Thus there are 16 observations within 1.5 standard deviations of the mean.

3) a) The observations between 1050 and 1450 are within one standard deviation of the mean. By the empirical rule, approximately 68% of the observations are here.

   b) The observations exceeding 1650 are more than 2 standard deviations above the mean. Approximately 2.5% of the observations are located here.

4) a) x = 140 + (2.5)(14) = 175

This can also be found by solving $2.5 = \dfrac{x - 140}{14}$ for x.

b) x = 140 − (1.4)(14) = 120.4

Or solve $-1.4 = \dfrac{x - 140}{14}$ for x.

c) x is 1.8 standard deviations above the mean.   Thus
x = 140 + (1.8)(14) = 165.2.

d) x is 2.6 standard deviations below the mean.   Thus
x = 140 − (2.6)(14) = 103.6.

5)

x	f	xf	$x^2 f$
0	2	0	0
1	7	7	7
2	8	16	32
3	14	42	126
4	10	40	160
5	3	15	75
6	1	6	36
	45	126	436

n = 45

$\bar{x} = \dfrac{126}{45} = 2.8$

$s = \sqrt{\dfrac{436 - \dfrac{(126)^2}{45}}{45 - 1}} = 1.3751 \approx 1.4$

6) .

Class Mark

x	f	xf	$x^2 f$
1	15	15	15
2	20	40	80
3	15	45	135
4	20	80	320
5	30	150	750
	100	330	1300

$\bar{x} = \dfrac{330}{100} = 3.3$

$s = \sqrt{\dfrac{1300 - \dfrac{(330)^2}{100}}{99}} = 1.4599 \approx 1.5$

7) a) $\Sigma \; x^2 = x_1^2 + x_2^2 + x_3^2 + x_4^2$

   b) $(\Sigma \; x)^2 = (x_1 + x_2 + x_3 + x_4)^2$

   c) $\Sigma \; xf = x_1 f_1 + x_2 f_2 + x_3 f_3$

   d) $\Sigma \; x^2 f - (\Sigma \; xf)^2 = x_1^2 f_1 + x_2^2 f_2 + x_3^2 f_3 - (x_1 f_1 + x_2 f_2 + x_3 f_3)^2$

8) Range $= 798 - 390 - 408 \approx 6s$

   $s \approx \dfrac{408}{6} = 68$

9) a) The mean is also reduced by 66. Thus the mean of the new scores is
       $66 - 66 = 0$. The variation is not changed by subtracting or adding a
       fixed constant to all the grades. Thus the standard deviation is un-
       changed and $s = 10$ for the new grades.

   b) The new scores are of the form $z = \dfrac{x - 66}{10}$.
       Since $\bar{x} = 66$ and $s = 10$, the new z scores have a mean of 0 and a
       standard deviation of 1.

10)

miles run x	relative frequency
0 - 2	.03 = 3%
2 - 4	.12 = 12%
4 - 6	.30 = 30%
6 - 8	.20 = 20%
8 - 10	.14 = 14%
10 - 12	.06 = 6%
12 - 14	.05 = 5%
14 - 16	.10 = 10%

   a) $Q_1 = 4 + \dfrac{10}{30} \; (2) = 4.67$

       Note that 15% of the observations are below 4. Thus an additional 10%
       are needed and the interval 4-6 contains 30% of the scores. Thus the
       25th percentile is located $\dfrac{10}{30}$ th of the way into this interval.

   b) $Q_3 = 8 + \dfrac{10}{14} \; (2) = 9.4$

   c) $\tilde{x} = 6 + \dfrac{5}{20} \; (2) = 6.5$

       Note that 45% of the observations are below 6. Thus an additional 5%
       are needed and the next interval contains 20% of the scores.

   d) $P_{90} = 14$

11) $PR(6) = \dfrac{\frac{1}{2}(10)}{38} \times 100 = 13$ (rounded)

$PR(9) = \dfrac{10 + \frac{1}{2}(14)}{38} \times 100 = 45$ (rounded)

$PR(10) = \dfrac{24 + \frac{1}{2}(8)}{38} \times 100 = 74$ (rounded)

$PR(12) = \dfrac{32 + \frac{1}{2}(6)}{38} \times 100 = 92$ (rounded)

Chapter IV
Probability

Probabilities are used in inferential processes to express the certainty or confidence in the decisions and judgements which are made. This chapter is concerned with the basic techniques and terminology of probability theory.

Concepts/Techniques

1) Probability Experiments

An experiment in which there is an element of doubt or uncertainty about which of several outcomes will occur is called a probability experiment.

2) Sample Space of a Probability Experiment

The collection of possible outcomes of a probability experiment is called the sample space.

Notation: S

Example: A person is selected and questionned about his political party preference. A sample space is:

S = {Republican,Democrat,Independent,Other}.

Example: A box contains 4 slips of paper numbered 1, 2, 3, and 4. Two slips of paper are drawn in succession. Give a sample space if there is no replacement between draws.

Indicate an outcome by the ordered pair

(First Draw Result, Second Draw Result)

4 Choices x 3 Choices

$$S = \left\{ \begin{array}{l} (1,2),(1,3),(1,4),(2,1),(2,3),(2,4) \\ (3,1),(3,2),(3,4),(4,1),(4,2),(4,4) \end{array} \right\}$$

An outcome such as (3,2) indicates the slip with a 3 occurred on the first draw and the slip with a 2 on the second.

3)   Events

An event is a collection of outcomes, i.e., a subset of S, generally corresponding to some occurrence of interest.

Notation: Capital letters such as E,F,G,... are used for events.

Example: Consider again the last example where 2 slips of paper are to be drawn from the 4 slips numbered 1, 2, 3, and 4. Here

$$S = \{(1,2),(1,3),(1,4),(2,1),(2,3),(2,4),(3,1),(3,2),(3,4),(4,1),(4,2),(4,4)\}.$$

The event E "Sum of the numbers is even" is

$$E = \{(1,3),(2,4),(3,1),(4,2),(4,4)\}.$$

These are the pairs that produce an even sum. The event "Sum is even" can thus occur in 5 different ways.

Example: Suppose a card is drawn from a standard bridge deck. As a sample space we take the set of 52 cards of the deck since any one of these can occur. As examples of events we mention:

E: "Card Drawn is a Heart"
F: "Card Drawn is a Face Card".

Here the number of cards in E is $N(E) = 13$ and the number of cards in F is $N(F) = 12$. (Note: the Aces were not counted as face cards.)

4)   Equally Likely Outcomes

If there is no bias towards the occurrence of any outcome over any other outcome, then we say the outcomes are equally likely.

Example: A ball is to be selected by a blindfolded student from an urn containing 3 balls that are colored red (R), white (W) and blue (B). Each of the outcomes R, W, and B has the same chance of occurring. Thus the outcomes are equally likely.

Example: A sample of 25 college students is to be selected and the average weight determined. Since sample means around, say 140, are more likely than those above 200 or below 90, the outcomes are not equally likely.

5) Probability

The probability of an outcome or an event is a measure of the expected relative frequency of occurrence of the outcome or event when the experiment is performed many times.

Example: Suppose that of each 10 persons administered a certain medication, two have a reaction. Then the probability of a reaction is

$$P(\text{Reaction}) = \frac{2}{10} = .2 = 20\%.$$

Note that probabilities may be expressed as fractions, decimals, or percents.

When the outcomes of an experiment are equally likely, the probability P(e) of an outcome e is

$$P(e) = \frac{1}{\text{Number of Outcomes in the Sample Space}} = \frac{1}{N(S)} \ .$$

And the probability of an event E is

$$P(E) = \frac{\text{Number of Outcomes in the Event}}{\text{Number of Outcomes in the Sample Space}} = \frac{N(E)}{N(S)} \ .$$

Example: Two drawings with replacement between draws are to be made from an urn containing 3 slips of paper numbered 1, 2, and 3. The sample space of equally likely outcomes is formed by pairs of the form (First Draw Result, Second Draw Result). Thus

$$S = \{(1,1),(1,2),(1,3),(2,1),(2,2),(2,3),(3,1),(3,2),(3,3)\}.$$

The probability of each of these is $\frac{1}{9}$. Thus we could write

$$P(1,1) = \frac{1}{9} \ , \ P(2,3) = \frac{1}{9} \ , \ \text{etc.}$$

The event E "Sum is even" is represented by $E = \{(1,1),(1,3),(2,2),$
$(3,1),(3,3)\}$. It has probability $\frac{5}{9}$

$$P(\text{Sum is Even}) = \frac{N(E)}{N(S)} = \frac{5}{9} \ .$$

It should be clear that the probability of an event E is the sum of the
probabilities of the outcomes which are in E. This proves useful when
the outcomes are not equally likely.

Example: In the last example, suppose we had chosen as a sample
space the set of possible sums from the numbers on the 2 slips of
paper. Then the sample space is $S = \{2,3,4,5,6\}$. The sum 2 can
occur in only 1 way as $(1,1)$. Thus $P(2) = \frac{1}{9}$. Similarly, 3
can occur as $(1,2)$, or $(2,1)$ and has probability $P(3) = \frac{2}{9}$.
The probability distribution is given by the
table at the right. Consider the event E:
"Sum is less than 5." Since $E = \{2,3,4\}$,

x	P(x)
2	1/9
3	2/9
4	3/9
5	2/9
6	1/9

$$P(E) = P(2) + P(3) + P(4)$$
$$= \frac{1}{9} + \frac{2}{9} + \frac{3}{9} = \frac{6}{9} = \frac{2}{3} \ .$$

6) A Counting Technique

Often the outcomes of an experiment can be regarded as the composite
of several separate occurrences. If so, the outcomes can be listed as
ordered pairs, ordered triples, etc. The total number of outcomes is
then the product of the number of ways available to fill each entry
of the pair, triple, or whatever.

Example: Two draws are to be made from an urn containing 3 slips
of paper numbered 1, 2, and 3. Suppose the outcomes are ordered
pairs of the form (First Draw Result, Second Draw Result).
   a) How many outcomes are in the sample space S if there is no
       replacement between draws?
   b) How many outcomes are in S if there is replacement between
       draws?
Answer (a). On the first draw any one of the 3 numbers 1, 2, and

3 is possible. Thus we have 3 choices for the first entry of the
pair. On the second draw, there are only 2 possibilities (since
there is no replacement after the first draw). Thus we have 2
choices for the second entry. Thus there are $3 \times 2 = 6$ pairs.
We abbreviate all of this by writing

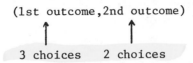

Answer (b). Since there is replacement after the first draw, there
are 3 possibilities on the second draw

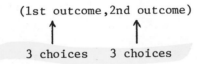

Thus there are $3 \times 3 = 9$ pairs in the sample space.

The same reasoning applies to counting the outcomes of an event.

Example: A room contains 5 women and 3 men which we indicate by
w_1, w_2, w_3, w_4, w_5, m_1, m_2, m_3. If 3 people are selected at random
for a committee, what is the probability that the 3 are women?
As a sample space, we use outcomes of the form

(1st person selected, 2nd person selected, 3rd person selected)

    8 choices   7 choices   6 choices

The sample space has $N(S) = 8 \cdot 7 \cdot 6 = 336$ outcomes.
The event consists of outcomes of the form

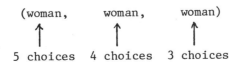

The event E "All 3 members are women" has $N(E) = 5 \cdot 4 \cdot 3 = 60$
outcomes. Thus

$$P(E) = \frac{N(E)}{N(S)} = \frac{60}{336} = \frac{5}{28} .$$

7) <u>The Events (not E), (E or F), and (E and F)</u>

The complement of the event   E   consists of those outcomes of the sample space that are not in   E.

<u>Notation</u>:   not E   or $\tilde{E}$.

The union of the events   E   and   F   consists of those outcomes which are in   E   or   F   or both.

<u>Notation</u>:   E   or   F   or   E ∪ F.

The intersection of the events   E   or   F   consists of those outcomes which are jointly in   E   and   F.

<u>Notation</u>:   E   and   F   or   E ∩ F.

<u>Example</u>:   Consider the experiment of randomly selecting a college student.

Let   E   be the event "student is a science major."

Let   F   be the event "student is a freshman."

The event   not E   occurs if the student selected is not a science major.

The event   E or F   occurs if the student selected is a science major, a frehman, or both.

The event   E and F   occurs if the student selected is both a science major and a freshman, i.e., a freshman science major.

8) <u>Mutually Exclusive Events</u>

The events   E and F   are said to be mutually exclusive if the occurrence of one excludes the occurrence of the others.   Viewed as sets of outcomes, this means that they have no outcomes in common, that is,   E ∩ F = ∅.

<u>Example</u>:   The following pairs of events are mutually exclusive.

   a)   E: "student selected is a freshman"
        F: "student selected is a graduate"
   b)   E: "sample mean exceeds   140   pounds"
        F: "sample mean is below   120   pounds"
   c)   E: "the card selected is a heart"
        F: "the card selected is a spade".

9) The Probability of the Event (not E)

If E is an event and the probability experiment is carried out, the outcome must be in either E or not E. As a result

$$P(E) + P(\text{not } E) = 1 \quad \text{or equivalently}$$
$$P(\text{not } E) = 1 - P(E).$$

Example: Suppose 40% of all adults smoke. If an adult is selected at random, then

$$P(\text{smoker selected}) = .40 \quad \text{and}$$
$$P(\text{non smoker selected}) = 1 - .40 = .60 = 60\%.$$

Example: Find the probability of 1 or more heads in 4 tosses of a coin.

$$P(1 \text{ or more heads}) + P(0 \text{ heads}) = 1$$

And $P(0 \text{ heads}) = \dfrac{1}{16}$.

Thus $P(1 \text{ or more heads}) = 1 - \dfrac{1}{16} = \dfrac{15}{16}$.

Note: If ordered 4 tuples are used for the outcomes, there are $2 \times 2 \times 2 \times 2 = 16$ outcomes. Only one of these has 0 heads, namely (T,T,T,T).

10) The Probability of the Event E or F

$$P(E \text{ or } F) = P(E) + P(F) - P(E \text{ and } F).$$

Example: The composition of a statistics course is given below.

	Freshman	Sophomore	Junior	Senior	Total
Psychology	5	9	4	2	20
Sociology	3	6	2	1	12
Business	2	3	3	0	8
Totals	10	18	9	3	40

Suppose a student is selected at random. Find the probability that the student is a freshman or a psychology major.

There are 40 students in all. Of these $20 + 10 - 5 = 25$ are freshmen or psychology majors or both.   Thus

$$P(\text{freshman or psychology major}) = \frac{25}{40} = \frac{5}{8}\ .$$

We also see

$$P(\text{psychology major}) = \frac{20}{40}$$

$$P(\text{freshman}) = \frac{10}{40}$$

$$P(\text{freshman and psychology major}) = \frac{5}{40}\ .$$

Applying the formula, we find again

$$P(\text{freshman or psychology major}) = \frac{10}{40} + \frac{20}{40} - \frac{5}{40} = \frac{25}{40} = \frac{5}{8}\ .$$

If E and F are mutually exclusive, then there are no outcomes in E and F and P(E and F) = 0.   We then have the so called addition formula:

$$P(E \text{ or } F) = P(E) + P(F).$$

Note that "or" suggests addition.

 Example: If one card is drawn from a standard bridge deck, then

$$P(\text{heart}) = \frac{13}{52} = \frac{1}{4} \quad \text{and} \quad P(\text{spade}) = \frac{13}{52} = \frac{1}{4}$$

and since "hearts" and "spade" are mutually exclusive events

$$P(\text{heart or spade}) = P(\text{heart}) + P(\text{spade})$$

$$= \frac{1}{13} + \frac{1}{13} + \frac{2}{13}\ .$$

11) Conditional Probability

If an event E is known to have occurred, then this may alter the chances of occurrence of some other event F.   The probability of the occurrence of F subject to the condition that E has occurred is the conditional probability of F given E.

Notation: $P(F|E)$.

Formula: $P(F|E) = \dfrac{P(E \text{ and } F)}{P(E)}$

In many instances the formula is not needed.

Example: The composition of a statistics class is given below. A student is selected at random. Let E and F be the events

E: "student selected is a freshman"

F: "student selected is a psychology major".

Find $P(F|E)$ in 2 ways.

	Freshman	Sophomore	Junior	Senior	Total
Psychology	5	9	4	2	20
Sociology	3	6	2	1	12
Business	2	3	3	0	8
Totals	10	18	9	3	40

a) We are given that the student selected is a freshman and there are ten of these. Of these, 5 are psychology majors. Thus

$$P(\text{psychology major}|\text{freshman}) = P(F|E) = \frac{5}{10} = \frac{1}{2}.$$

b) $P(\text{psychology major and freshman}) = P(F \text{ and } E) = \frac{5}{40}$

$$P(\text{freshman}) = P(F) = \frac{10}{40}$$

By the conditional probability formula

$$P(F|E) = \frac{P(E \text{ and } F)}{P(E)} = \frac{\frac{5}{40}}{\frac{10}{40}} = \frac{5}{10} = \frac{1}{2}$$

(as above).

12) The Probability of the Event E and F

Rearranging the conditional probability formula we obtain

$$P(E \text{ and } F) = P(E) \cdot P(F|E).$$

This is the "multiplication rule" and is used when two events are joined by "and". It is particularly easy to use when E and F are associated with outcomes or events at different stages of an experiment: The probability of F given E, $P(F|E)$ can generally be easily found.

Example: A cage contains 4 white and 3 black mice. Two are selected
in succession. Find the probability that both are white.

$$P(2 \text{ white}) = P(\text{1st is white and 2nd is white})$$
$$= P(\text{1st is white}) \cdot P(\text{2nd is white} | \text{1st is white})$$
$$= \frac{4}{7} \cdot \frac{3}{6} = \frac{2}{7} .$$

Example: A card is drawn from a standard bridge deck and a single
die is tossed. What is the probability that a six is obtained on
both the card and the die.

$$P(\text{six on die and six on card}) = \frac{1}{6} \cdot \frac{4}{52} = \frac{1}{78} .$$

Here $P(\text{six on card} | \text{six on die}) = P(\text{six on card})$ since the die and
the deck of cards are unrelated.

13) **Independent Events**

E and F are independent events if $P(F | E) = P(F),$ that is the
probability of F is not effected by the occurrence of E.

Example: Workers are being interviewed. Decide on an intuitive
basis if the following are independent:

F: "Salary is above the median income"
E: "Worker is a woman"

Granting the existence of an income inequity between men and women,
the probability of the worker's salary being above the median is
certainly changed if the worker is a woman, that is

$$P(F | E) \neq P(F).$$

The events are not independent.

Example: A ball is to be drawn from each of 2 boxes. The first box
contains 3 red and 2 white, the second 5 red and 5 white. Decide
on an intuitive basis if the following are independent.

E: "ball from box 1 is red"
F: "ball from box 2 is red"

What is drawn from box 1 does not effect what is drawn from box 2. Thus $P(F|E) = P(F)$ and the events are independent.

Do not make the mistake of equating the notion of independent events with that of mutually exclusive. These are best thought of as being unrelated concepts.

14) <u>The Multiplication Rule for Independent Events</u>

If E and F are independent, $P(F|E) = P(F)$ and the multiplication rule simplifies to

$$P(E \text{ and } F) = P(E) \cdot P(F).$$

<u>Example</u>: The probability that a test animal will show a reaction to a certain medication is .23. If 2 of these animals are randomly selected and given the medication, what is the probability that both will show a reaction?

P(both show a reaction) = P(first shows a reaction and second shows a reaction)

= P(first shows a reaction) $\cdot$ P(second shows a reaction)

= $(.23)(.23) = .0529 = 5.29\%$.

The equation $P(E \text{ and } F) = P(E) \cdot P(F)$ may be used as a check on the independence of E and F.

<u>Example</u>: An experiment has 5 outcomes e_1, e_2, e_3, e_4, and e_5, the probabilities for which are given at the right. Let $E = \{e_1, e_2\}$, $F = \{e_2, e_4, e_5\}$. Are E and F independent?

outcome	p
e_1	.1
e_2	.1
e_3	.4
e_4	.15
e_5	.25

$P(E) = P(e_1) + P(e_2) = .1 + .1 = .2$

$P(F) = P(e_2) + P(e_4) + P(e_5)$

$= .1 + .15 + .25 = .50$

(E and F) = $\{e_2\}$

$P(E \text{ and } F) = P(e_2) = .1$

Since $P(E) \cdot P(F) = (.2)(.5) = .1 = P(E \text{ and } F)$, the events E and F are independent.

15) Additional Examples

    Example: Three wheels of fortune are shown below. If each is spun, what is the probability that the sum of the individual outcomes is 4?

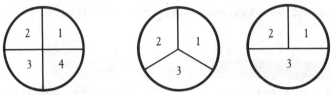

A four can occur either as (1,2,1), (1,1,2) or (2,1,1). Since these are mutually exclusive

$$P(\text{sum} = 4) = P((1,2,1) \text{ or } (1,1,2) \text{ or } (2,1,1))$$
$$= P(1,2,1) + P(1,1,2) + P(2,1,1)$$

Now use the multiplication rule to find each of these probabilities.

$$P(1,2,1) = \left(\tfrac{1}{4}\right)\left(\tfrac{1}{3}\right)\left(\tfrac{1}{4}\right) = \tfrac{1}{24}$$
$$P(1,1,2) - \left(\tfrac{1}{4}\right)\left(\tfrac{1}{3}\right)\left(\tfrac{1}{4}\right) = \tfrac{1}{24}$$
$$P(2,1,1) = \left(\tfrac{1}{4}\right)\left(\tfrac{1}{3}\right)\left(\tfrac{1}{4}\right) = \tfrac{1}{24}$$

$$P(\text{sum} = 4) = \tfrac{1}{24} + \tfrac{1}{24} + \tfrac{1}{24} = \tfrac{3}{24} = \tfrac{1}{8} = .125.$$

    Example: The first part of a test has 3 multiple choice questions each having 4 possible answers. If a student guesses at the answer on each question, what is the probability of obtaining 2 correct answers on this part of the test.

If we use I for incorrect and C for correct, the 2 correct answers can occur as either (C,C,I), (C,I,C), or (I,C,C).

Note the word "or"!

$$P(2 \text{ correct}) = P((C,C,I) \text{ or } (C,I,C) \text{ or } (I,C,C))$$
$$= P(C,C,I) + P(C,I,C) + P(I,C,C)$$
$$= \left(\tfrac{1}{4}\right)\left(\tfrac{1}{4}\right)\left(\tfrac{3}{4}\right) + \left(\tfrac{1}{4}\right)\left(\tfrac{3}{4}\right)\left(\tfrac{1}{4}\right) + \left(\tfrac{3}{4}\right)\left(\tfrac{1}{4}\right)\left(\tfrac{1}{4}\right) = \tfrac{9}{64} = .1406.$$

    Example: Refer to the last example. What is the probability of at least 2 correct answers?

At least 2 correct occurs as "exactly 2 or exactly 3". Note the or! Thus

P(at least 2 correct) = P(exactly 2 correct) + P(exactly 3 correct).

In the last example, we found

$$P(\text{exactly 2 correct}) = .1406.$$

Also

$$P(\text{exactly 3 correct}) = P(C,C,C)$$
$$= (\frac{1}{4})(\frac{1}{4})(\frac{1}{4})$$
$$= \frac{1}{64} = .0156$$

P(at least 2 correct) = .1406 + .0156 = .1562.

Exercise Set IV.1

1) As outcomes use ordered pairs of the form

    (affiliation of 1st person,affiliation of 2nd person)

$$S = \{(R,R),(R,D),(R,I),(D,D),(D,R),(D,I),(I,I),(I,D),(I,R)\}$$

3) a) The probability that a randomly selected tire produced by this method
        will fail to pass inspection is .12.

    b) The probability that a randomly selected person will find this new product
        superior to the present one is .58.

    c) The probability of a randomly selected freshman being on probation after
        one semester is .30.

    d) The probability that the mean of a randomly selected sample will be more
        than 2 standard deviations from the mean is .05.

5) a) Use ordered pairs of the form: (coin result,die result)

$$S = \{(1,1),(1,2),(1,3),(1,4),(1,5),(1,6),(2,1),(2,2),(2,3),(2,4),(2,5),(2,6)\}$$

b)

Sum	Pairs	Probability
2	$\{(1,1)\}$	$p = \frac{1}{12}$
3	$\{(1,2),(2,1)\}$	$p = \frac{2}{12} = \frac{1}{6}$
4	$\{(1,3),(2,2)\}$	$p = \frac{2}{12} = \frac{1}{6}$
5	$\{(1,4),(2,3)\}$	$p = \frac{2}{12} = \frac{1}{6}$
6	$\{(1,5),(2,4)\}$	$p = \frac{2}{12} = \frac{1}{6}$
7	$\{(1,6),(2,5)\}$	$p = \frac{2}{12} = \frac{1}{6}$
8	$\{(2,6)\}$	$p = \frac{1}{12}$

c) $p = \frac{1}{6}$

    $\frac{1}{6} \times 2400 = 400$

    About 400 of the 2400 tosses should produce a sum of 4.

7) a) As outcomes use ordered pairs of the form:

   (page number of 1st story,page number of 2nd story).

   There are 4 choices for each entry.   Thus there are $4 \times 4 = 16$ possible

   outcomes.

$$S = \begin{cases} (1,1), \ (1,2), \ (1,3), \ (1,4), \\ (2,1), \ (2,2), \ (2,3), \ (2,4), \\ (3,1), \ (3,2), \ (3,3), \ (3,4), \\ (4,1), \ (4,2), \ (4,3), \ (4,4) \end{cases}$$

   b) Of the 16 outcomes in part (a), there are 12 where stories would appear

   on different pages.   Thus

$$p = \frac{12}{16} = \frac{3}{4} = .75 .$$

9) a) $P(E) = P(e_1) + P(e_2)$ (by addition rule)

$$= \frac{1}{6} + \frac{1}{3}$$

$$= \frac{1}{2}$$

   b) $P(S) = 1$

   c) Have the urn contain 1 white (e_1), 2 blue (e_2), 2 red (e_3) and 1 green (e_4).
      If one ball is selected, $P(\text{white}) = P(e_1) = \frac{1}{6}$

$$P(\text{blue}) = P(e_2) = \frac{2}{6} = \frac{1}{3}$$

$$P(\text{red}) = P(e_3) = \frac{2}{6} = \frac{1}{3}$$

$$P(\text{green}) = P(e_4) = \frac{1}{6}$$

   d) A student population on a campus consists of 6000 freshmen (e_1), 12000

   sophomores (e_2), 12000 juniors (e_3) and 6000 seniors (e_4).

   If a student is selected at random:

$$P(\text{freshmen}) = P(e_1) = \frac{6000}{36000} = \frac{1}{6}$$

$$P(\text{sophomore}) = P(e_2) = \frac{12000}{36000} = \frac{1}{3}$$

$$P(\text{junior}) = P(e_3) = \frac{12000}{36000} = \frac{1}{3}$$

$$P(\text{senior}) = P(e_4) = \frac{6000}{36000} = \frac{1}{6}$$

11) Using outcomes of the form (die 1 result, die 2 result, die 3 result)
there are 6 possible entries for each position.

a)   There are $6 \times 6 \times 6 = 216$ outcomes in the sample space.

b)   The outcomes are of the form: (even result, even result, even result).
There are 3 possible entries for each position, namely 2, 4, and 6.
Thus there are $3 \times 3 \times 3 = 27$ outcomes where all entries are even.

c) $p = \dfrac{27}{216} = \dfrac{1}{8}$

13) As a sample space of equally likely outcomes, we have:

$$S = \{ (M_1,M_2),(M_1,W_1),(M_1,W_2),(M_2,M_1),(M_2,W_1),(M_2,W_2),$$
$$(W_1,M_1),(W_1,M_2),(W_1,W_2),(W_2,M_1),(W_2,M_2),(W_2,W_1) \}.$$

There are 12 pairs and 2 of these are both women.

Thus $p = \dfrac{2}{12} = \dfrac{1}{6} = .1667$

Note: There must be $(4)(3) = 12$ pairs in all and there are $(2)(1) = 2$
pairs where both are women. Thus the listing can be skipped.

15) There are, in all, $4 \times 3 \times 2 \times 1 = 24$ possible rankings for the 4 boxes
(4 choices for the first box, 3 for the second, two for the third, and one
for the fourth). Of these only one is correct. Thus

$$p = \frac{1}{24} = .0417.$$

17)

O',	T,	O,	A	(O' = onion)
O',	T,	O,	A	
O',	T,	A,	O	
O',	O,	T,	A	
O',	O,	A,	T	
O',	A,	O,	T	
O',	A,	T,	O	
T,	O',	O,	A	
T,	O',	A,	O	
T,	A,	O',	O	
T,	A,	O,	O'	
T,	O,	A,	O'	
T,	O,	O'	A	
O,	T,	O',	A	
O,	T,	A,	O'	
O,	O',	T,	A	
O,	O',	A,	T	
O,	A,	O',	T	
O,	A,	T,	O'	
A,	O',	O,	T	
A,	O',	T,	O	
A,	T,	O',	O	
A,	T,	O,	O'	
A,	O,	T,	O'	
A,	O,	O',	T	

The possible responses are recorded below the heading that gives the sequence in which the juices were presented. Of the $4 \times 3 \times 2 \times 1 = 24$ possible responses, 7 have 2 or more correct identifications. Thus

$$p = \frac{7}{24}.$$

19) We know $p(e_1) + p(e_2) + p(e_3) = 1$. Thus

$$\frac{1}{4} + p(e_2) + \frac{3}{8} = 1.$$

Solving for $P(e_2)$ yields

$$p(e_2) = 1 - \frac{1}{4} - \frac{3}{8} = \frac{3}{8}.$$

21) Let an outcome indicate the number of throws necessary to obtain the 1st head.

$$S = \{1,2,3,4,5,\ldots\}.$$

23) a) $2 \times 2 \times 2 \times 2 = 16$ (There are 2 choices for each position).

b) Think of the responses of a husband and wife as being a single response. We then have, in essence, only ordered triples:

(joint response, response of 1st child, response of 2nd child)
There are $2 \times 2 \times 2 = 8$ such triples.

Exercise Set IV.2

1) a) Not mutually exclusive. Being a history major does not exclude a student from being a senior.

  b) Not mutually exclusive. Being used does not exclude the possibility of a price exceeding $20.

  c) Mutually exclusive. If one card is red (King of Hearts), then both cannot be black.

  d) Mutually exclusive. If both dice show an even number, then their sum is even and thus cannot be 7.

  e) Mutually exclusive. If no components failed, then this excludes the possibility of at least one component failing.

3) a) Not Independent.
   Convertibles are generally more expensive. The probability of a price tag exceeding $15000 should be increased if it is known that the automobile is a convertible. Thus $P(E|F) > P(E)$.

  b) Not Independent.
   The probability of getting a second seven (given the occurrence of the first one) should be greater than the probability of getting two sevens in succession. Thus $P(F|E) > P(F)$.

  c) Independent.
   What happens on the die does not influence what coin is drawn. Thus $P(F|E) = P(F)$.

  d) Not Independent.
   Systolic pressure tends to increase with age and a higher pressure would be more likely if the individual's hair color was gray (one of the non-black colors). Thus $P(F|E) > P(F)$.

  e) Not Independent.
   One would expect a higher frequency of weights exceeding 150 pounds among males and possibly even more so among male P.E. majors. Thus $P(F|E) > P(F)$.

5) a) Twenty or fewer smoke or at most 20 smoke
     $E = \{21,22,23,24,25\}$, $\tilde{E} = \{0,1,2,3,\ldots,20\}$

  b) Three or more smoke or at least 3 smoke
     $E = \{0,1,2\}$, $\tilde{E} = \{3,4,5,\ldots,25\}$

  c) Fewer than 5 smoke or four or fewer smoke or at most 4 smoke
     $E = \{5,6,7,\ldots,25\}$, $\tilde{E} = \{0,1,2,3,4\}$

d) At least one smokes or 1 or more smoke

$E = \{0\}$, $\tilde{E} = \{1,2,3,\ldots,25\}$

e) None smoke

$E = \{1,2,3,\ldots,25\}$, $\tilde{E} = \{0\}$

7) a) The probability of a phrase being recognized when encountered a second time given that the phrase was previously encountered in reading.

b) The probability of a phrase being recognized when encountered a second time given that the phrase was previously heard.

c) The probability that the phrase is not recognized when encountered a second time given that it was previously read.

d) The probability that the phrase is recognized when encountered a second time given that it had not been previously read.

e) No (assuming there is a difference between oral and visual acquisition of data).

9) a) $(E \text{ and } F) = E \cap F = \{e_2, e_3\}$

$P(E \text{ and } F) = P(e_2) + P(e_3) = .20 + .50 = .70$

b) $(E \text{ or } F) = E \cup F = \{e_1, e_2, e_3, e_5\}$

$P(E \text{ or } F) = P(e_1) + P(e_2) + P(e_3) + P(e_5)$

$= .05 + .20 + .50 + .15 = .90$

c) $(\text{not } E) = \tilde{E} = \{e_4, e_5\}$

$P(\text{not } E) = P(e_4) + P(e_5)$

$= .10 + .15 = .25$

d) $P(E \mid F) = \dfrac{P(E \text{ and } F)}{P(F)} = \dfrac{.70}{.85} = .8235$

Note: $P(F) = P(e_2) + P(e_3) + P(e_5) = .20 + .50 + .15 = .85$

e) $P(F \mid E) = \dfrac{P(E \text{ and } F)}{P(E)} = \dfrac{.70}{.75} = .9333$

Note: $P(E) = P(e_1) + P(e_2) + P(e_3) = .05 + .20 + .50 = .75$

11) The reduced sample space consists of those outcomes having only one head. Thus

$$S' = \{(H,T,T),(T,H,T),(T,T,H)\}.$$

Of these only one has a H as the 1st result. Thus $p = \dfrac{1}{3}$.

13)

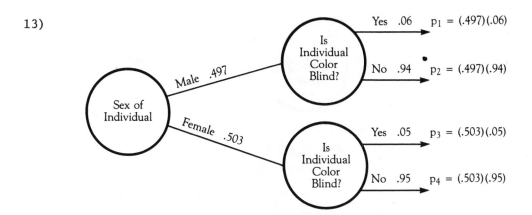

a) P(male and color blind) = p_1 = (.497)(.06) = .02982 = 2.929%
    Also by the product rule,

    P(male and color blind) = P(male) × P(color blind male)
                            = (.497) × (.06) = .02982

b) P(male and normal vision) = p_2 = (.497) × (.94) = .46718 = 46.718%
    Also by the product rule

    P(male and normal vision) = P(male) × P(normal vision male)
                              = (.497)(.94) = .46718

c) P(color blind) = P(color blind male or color blind female)
                    = $p_1 + p_3$ = (.497)(.06) + (.503) × (.05)
                    = .05497 = 5.497%

15) The event "The order is filled" occurs if "both A and B fill the order"
     or "A fills the order and B does not" or "A does not fill the order
     and B does". These are mutually exclusive events.

                    P = (.8)(.6) + (.8)(.4) + (.2)(.6)
                      = .48 + .32 + .12 = .92 = 92%

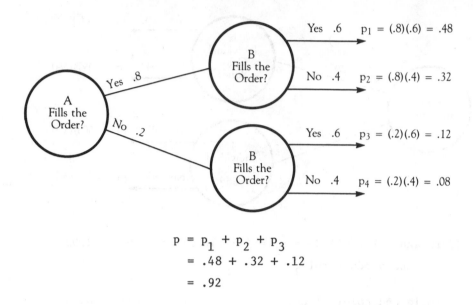

$$p = p_1 + p_2 + p_3$$
$$= .48 + .32 + .12$$
$$= .92$$

Alternatively one could work with the complement which is "neither order is filled". For this, q = (.2)(.4) = .08. Thus p = 1 - .08 = .92.

17)

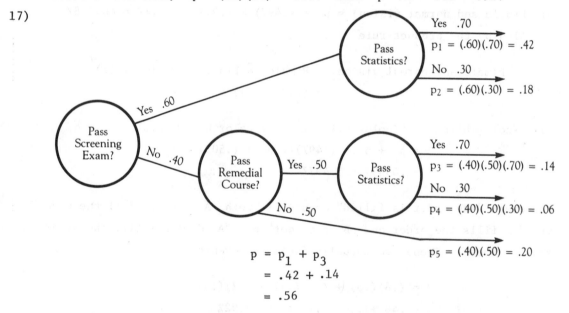

$$p = p_1 + p_3$$
$$= .42 + .14$$
$$= .56$$

A student may pass the statistics course by "passing the screening exam and the statistics course" or "failing the screening exam and passing the remedial course and passing the statistics course".

19)

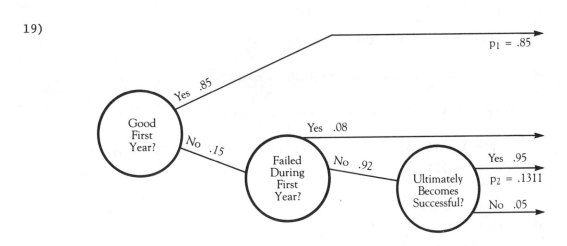

a) .15   b) .012   c) $p = p_1 + p_2$
$$= .85 + .1311$$
$$= .9811$$

21) a) A high probability by either the lonely or the not lonely group indicates
the problem was rated in the top 5 problems by many of the respondents.
A high probability for one of the groups and not by the other is
desirable for the investigator since they have them identified a
"tract" or "tendency" that is possessed or not possessed by the lonely
group. Both would be valuable.
A high probability for a problem by both groups might simply identify
a common social problem.

b) For the lonely, $p = .12$.
12% of 25 = 3
For the not lonely, $p = .07$.
7% of 45 = 3.15 $\tilde{=}$ 3

c) Left to chance, the probability for each category is $\frac{1}{2}$. There are 13
categories. Thus
$$p = \frac{1}{2} \times \frac{1}{2} \times \frac{1}{2} \times \cdots \times \frac{1}{2} = (\frac{1}{2})^{13}.$$

23) We use the notation (U,N,U) to indicate the 1st package is underweight,
the 2nd is not, and the third is underweight, and so forth.
a) $p(U,U,U) = (.25)(.25)(.25) = .015625$
(use the product rule with independent events)
b) $p(N,N,N) = (.75)(.75)(.75) = .421875$

c) Exactly one underwieght occurs by

$$(U,N,N) \quad \text{or} \quad (N,U,N) \quad \text{or} \quad (N,N,U).$$

These are mutually exclusive events. Thus

$$\begin{aligned}
p &= (.25)(.75)(.75) + (.75)(.75)(.25) + (.75)(.75)(.25) \\
&= 3(.25)(.75)^2 \\
&= .421875
\end{aligned}$$

25) a) $P(F) = .70$

P(not F) $= 1 - .70 = .30$

b) $P(R) = .60$

P(not R) $= 1 - .60 = .40$

c) P(R or F) $= P(R) + P(F) - P(R$ and $F)$

$$= .60 + .70 - .40 = .90$$

d) $P(R|F) = \dfrac{P(R \text{ and } F)}{P(F)} = \dfrac{.40}{.70} = \dfrac{4}{7}$

e) $P(F|R) = \dfrac{P(F \text{ and } R)}{P(R)} = \dfrac{.40}{.60} = \dfrac{2}{3}$

f) $P(\text{not } R|F) = 1 - P(R|F) = 1 - \dfrac{4}{7} = \dfrac{3}{7}$

27) If P(not E) is affected by the occurrence of F, then P(E) would be also.

Exercise Set IV.3

1)   Complement (not E) rule

   a) $P(\text{not E} \mid F) = 1 - P(E \mid F) = 1 - .85 = .15$

   b) $P(E \mid D) = 1 - P(\text{not E} \mid D) = 1 - .60 = .40$

3) $P(\text{not E}) = 1 - .76 = .24$

5) $P(D \mid \text{not E}) + P(F \mid \text{not E}) = 1$

   $P(D \mid \text{not E}) = 1 - P(F \mid \text{not E})$

   $$= 1 - \frac{1}{2} = \frac{1}{2} = .5 = 50\%$$

Chapter Test

1) a) i) A student is selected at random. Let E and F be defined as
follows:

    E: The student selected is a freshman

    F: The student selected is a senior

    ii) A pair of dice are tossed. Let E and F be defined by:

    E: The sum is even

    F: The sum is odd

    iii) The addition rule for mutually exclusive events states that

    $$P(E \text{ or } F) = P(E) + P(F)$$

    b) i) Suppose a coin and a die are tossed. Let E and F be defined by:

    E: The coin shows a head

    F: The die shows an even number

    ii) Suppose a man and a woman are selected at random. Let E and F be
    defined by:

    E: The age of the man exceeds 30

    F: The age of the woman exceeds 30

    iii) The multiplication rule for independent events states that

    $$P(E \text{ and } F) = P(E) \cdot P(F)$$

2) a) Observing the exposed subjects who received the vaccine.
   b) influenza contracted, influenza not contracted .
   c) $p = \dfrac{1050}{1440} = .7291$

3) a) $P(E) = P(e_1) + P(e_2) = .40 + .25 = .65$
   b) $P(\text{not } E) = 1 - P(E) = 1 - .65 = .35$
   c) $E \cup F = \{e_1, e_2, e_3, e_5\}$
      $$P(E \text{ or } F) = P(e_1) + P(e_2) + P(e_3) + P(e_5)$$
      $$= .40 + .25 + .20 + .05 = .90$$
      An alternate approach:
      $$P(E \text{ or } F) = P(E) + P(F) - P(E \text{ or } F)$$

d) $E \cap F = \{e_2\}$

$P(E \text{ and } F) = P(e_2) = .25$

4) As outcomes use ordered pairs of the form: (section of first student, section of second student)

$$S = \{(1,1),(1,2),(1,3),(1,4),(1,5),$$
$$(2,1),(2,2),(2,3),(2,4),(2,5),$$
$$(3,1),(3,2),(3,3),(3,4),(3,5),$$
$$(4,1),(4,2),(4,3),(4,4),(4,5),$$
$$(5,1),(5,2),(5,3),(5,4),(5,5)\}.$$

5) a) $P(E \text{ and } F) = P(E) \cdot P(F) = (.2)(.5) = .1$

b) $P(F|E) = P(F) = .5$

c) $P(E \text{ or } F) = P(E) + P(F) - P(E \text{ and } F)$

$\qquad = .2 + .5 - .1 = .6$

6) a) $P(\text{sum} < 6) = P(2) + P(3) + P(4) + P(5)$

$$= \frac{1}{36} + \frac{2}{36} + \frac{3}{36} + \frac{4}{36} = \frac{10}{36} = \frac{5}{18}$$

b) $P(\text{sum is even or a } 6) = P(2) + P(4) + P(6) + P(8) + P(10) + P(12)$

$$= \frac{1}{36} + \frac{3}{36} + \frac{5}{36} + \frac{5}{36} + \frac{3}{36} + \frac{1}{36} = \frac{18}{36} = \frac{1}{2}$$

Note that 6 is an even sum.

c) $P(\text{sum} < 5 | \text{sum is even}) = P(2) + P(4) = (\frac{1}{36} + \frac{3}{36}) \div \frac{1}{2} = \frac{2}{9}$

(Or: of the 18 possible pairs, only 4 have sums < 5.)

7) Use the multiplication rule.

$$P(\text{four aces}) = (\frac{4}{52})(\frac{3}{51})(\frac{2}{50})(\frac{1}{49}) = \frac{24}{6497400} = .0000037$$

8) a) $P(\text{man, woman}) = (\frac{5}{9})(\frac{4}{8}) = \frac{5}{18}$

b) $P(\text{man, man}) = (\frac{5}{9})(\frac{4}{8}) = \frac{5}{18}$

c) $p = \frac{3}{8}$. Note that if the first to be selected is a woman, then 3 of the remaining 8 people are women when the second selection is made.

d) $P((\text{man, woman}) \text{ or } (\text{woman, man})) = P(\text{man, woman}) + P(\text{woman, man})$

$$= (\frac{5}{9})(\frac{4}{8}) + (\frac{4}{9})(\frac{5}{8}) = \frac{5}{9}$$

9) a) The first row total is the number who made use of assistance. Thus

$$p = \frac{400}{1000} = .4.$$

b) Forty "young" men and sixty "young" women made use of assistance. Thus

$$p = \frac{40 + 60}{1000} = .1.$$

c) 120 of the "old" men and 180 "old" women made use of assistance. Thus

$$p = \frac{120 + 180}{1000} = .3.$$

d) Of the 260 "young" women, 200 did not use assistance. Thus

$$p = \frac{200}{260} = .7692.$$

10) a) $P(S,S) = (\frac{2}{5})(\frac{1}{4}) = \frac{1}{10}$

b) $P(L,L) = (\frac{3}{5})(\frac{2}{4}) = \frac{3}{10}$

c) $P((S,L) \text{ or } (L,S)) = P(S,L) + P(L,S)$

$$= (\frac{2}{5})(\frac{3}{4}) + (\frac{3}{5})(\frac{2}{4}) = \frac{3}{5}$$

Chapter V
Random Variables and Probability Distributions

This chapter is concerned with random variables and probability distributions. These concepts are closely related to the notion of the distribution of a variable in a population. The mean, variance, and standard deviation of a probability distribution are introduced and related to their statistical counterparts from Chapter III. Finally distributions of several random variables and the concept of independent random variables are explored.

Concepts/Techniques

1) Random Variables

A variable whose values are the outcomes of a probability experiment is called a random variable. Random variables may be discrete or continuous depending on the nature of their values.

Notation: Capital letters such as X, Y and Z are used for variables, their lower case equivalents, x, y, and z (possibly with subscripts) are used for values of the variable.

Example: A student is selected at random. Let X be the number of units completed. If the student has completed 112 units then x = 112 is the value of the variable which occurred.

Example: Let X be the sum when a pair of dice are tossed. The values of X are $x_1 = 2$, $x_2 = 3$, $x_3 = 4$, $x_4 = 5$, $x_5 = 6, \ldots, x_{11} = 12$.

2) <u>Probability Distributions</u>

A table or formula yielding probabilities for the values or sets of
values of a random variable X is called a probability distribution
for X.

   <u>Notation</u>: $P(X = x)$ is the probability that the outcome is x.
   $P(X < x)$ is the probability that the outcome is less than x.
   $P(x_1 < X < x_2)$ is the probability that the outcome is between x_1
   and x_2.

   <u>Example</u>: The probability distribution of the number of heads when
   4 coins are tossed is given at the right.

x	P
0	$\frac{1}{16}$
1	$\frac{4}{16} = \frac{1}{4}$
2	$\frac{6}{16} = \frac{3}{8}$
3	$\frac{4}{16} = \frac{1}{4}$
4	$\frac{1}{16}$

Find

   a) $P(X = 3)$ b) $P(X < 3)$

   c) $P(X \geq 2)$ d) $P(1 < X < 4)$

Answer:

   a) The probability of exactly 3 heads
   is $P(X = 3) = \frac{4}{16} = \frac{1}{4}$.

   b) The probability of fewer than 3
   heads is

$$P(X < 3) = P(X = 0) + P(X = 1) + P(X = 2)$$
$$= \frac{1}{16} + \frac{4}{16} + \frac{6}{16} = \frac{11}{16} .$$

   c) The probability of 2 or more heads is

$$P(X \geq 2) = P(X = 2) + P(X = 3) + P(X = 4)$$
$$= \frac{6}{16} + \frac{4}{16} + \frac{1}{16} = \frac{11}{16} .$$

   d) The probability of between 1 and 4 heads (less than 4
   and greater than 1) is

$$P(1 < X < 4) = P(X = 2) + P(X = 3)$$
$$= \frac{6}{16} + \frac{4}{16} = \frac{10}{16} .$$

3) <u>Probability Histograms</u>

The probability histogram for the distribution of a discrete random
variable is prepared exactly as with the histogram of a relative frequency
distribution.

Example: Draw the probability histogram of the probability distribution at the right. (See previous example.)

x	P
0	$\frac{1}{16}$
1	$\frac{4}{16}$
2	$\frac{6}{16}$
3	$\frac{4}{16}$
4	$\frac{1}{16}$

Note that the areas of the individual bars add to 1 and that the area of each bar is the probability of the corresponding outcome.

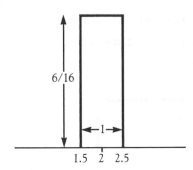

Area = base × height

$$= 1 \cdot (\frac{6}{16}) = \frac{6}{16}$$

$$= P(X = 2)$$

4) The Mean, Variance, and Standard Deviation

Let X be a discrete random variable with values $x_1, x_2, \ldots, x_n$.
The mean μ_X of X is defined by

$$\mu_X = x_1 P(x_1) + x_2 P(x_2) + \cdots + x_n P(x_n) = \sum_{i=1}^{n} x_i P(x_i).$$

The variance σ_X^2 of X is defined by

$$\sigma_X^2 = (x_1 - \mu_X)^2 P(x_1) + \cdots + (x_n - \mu_X)^2 P(x_n) = \sum_{i=1}^{n} (x_i - \mu_X)^2 P(x_i).$$

The standard deviation σ_X of X is defined by

$$\sigma_X = \sqrt{\text{Variance}} = \sqrt{\sigma_X^2} = \sqrt{\sum_{i=1}^{n} (x_i - \mu_X)^2 P(x_i)} \ .$$

The subscript X on μ and σ may be dropped when no other variables are involved in the discussion.

The mean μ of X is a measure of the center of the probability distribution of X. The variance σ^2 and standard deviation σ are measures of the variation or spread of the outcomes about the mean.

x	P(x)
0	$\frac{1}{16}$
1	$\frac{4}{16}$
2	$\frac{6}{16}$
3	$\frac{4}{16}$
4	$\frac{1}{16}$

Example: When 4 coins are tossed, the distribution of the number of heads X is given at the right. Find the mean, variance and standard deviation.

$$\mu = 0(\frac{1}{16}) + 1(\frac{4}{16}) + 2(\frac{6}{16}) + 3(\frac{4}{16}) + 4(\frac{1}{16})$$

$$= 0 + \frac{4}{16} + \frac{12}{16} + \frac{12}{16} + \frac{4}{16} = \frac{32}{16} = 2.$$

Examination of the histogram below shows $\mu = 2$ to be a good measure of the center.

$$\sigma^2 = (0 - 2)^2 \frac{1}{16} + (1 - 2)^2 \frac{4}{16} + (2 - 2)^2 \frac{6}{16} + (3 - 2)^2 \frac{4}{16} + (4 - 2)^2 \frac{1}{16}$$

$$= \frac{4}{16} + \frac{4}{16} + 0 + \frac{4}{16} + \frac{4}{16} = \frac{16}{16} = 1$$

$$\sigma = \sqrt{\sigma^2} = \sqrt{1} = 1.$$

An alternate formula for the variance is

$$\sigma^2 = \Sigma \; x^2 P(x) - (\Sigma \; xP(x))^2 = \Sigma \; x^2 P(x) - \mu^2$$

Example: Using the alternate formula, compute the variance of the distribution of the last example.

The simplest approach is to prepare a table with columns $x^2 P(x)$ and $xP(x)$ and find their sums.

x	P(x)	xP(x)	$x^2 P(x)$
0	$\dfrac{1}{16}$	0	0
1	$\dfrac{4}{16}$	$\dfrac{4}{16}$	$\dfrac{4}{16}$
2	$\dfrac{6}{16}$	$\dfrac{12}{16}$	$\dfrac{24}{16}$
3	$\dfrac{4}{16}$	$\dfrac{12}{16}$	$\dfrac{36}{16}$
4	$\dfrac{1}{16}$	$\dfrac{4}{16}$	$\dfrac{16}{16}$

$$\mu = \Sigma \; xP(x) = \frac{32}{16} = 2 \qquad \Sigma \; x^2 P(x) = \frac{80}{16}$$

$$\sigma^2 = \Sigma \; x^2 P(x) - \mu^2$$

$$= \frac{80}{16} - 2^2 = 1$$

Apart from rounding differences, the same results are obtained if the probabilities are expressed as decimals and a calculator is used.

5) The Expected Value of X

Suppose the values $x_1, x_2, \ldots, x_n$ of X are the outcomes of a game of chance and let $g(x_1), g(x_2), \ldots, g(x_n)$ be the corresponding payoffs for these outcomes. Then the expected value E of the game is given by

$$E = g(x_1)P(x_1) + g(x_2)P(x_2) + \cdots + g(x_n)P(x_n) = \Sigma \; g(x)P(x).$$

Literally each payoff is multiplied by the probability of the associated outcome and these products are summed.

The expected value E is the average "payoff" when the game is played many many times and the probability distribution is followed exactly.

Example: A card is drawn from a standard bridge deck. If it is an Ace (x = 1) the payoff is $10.00. If it is a face card (x = 2),

the payoff is $3.00. For any other card (x = 3) you pay $1.00.
What is the expected value of the game?
We summarize the data in the table at
the right. Note that money received
is treated as a positive payoff while
that paid out is negative. We see

x	g(x)	P(x)	xg(x)
1	10	$\frac{4}{52}$	$\frac{40}{52}$
2	3	$\frac{12}{52}$	$\frac{36}{52}$
3	-1	$\frac{36}{52}$	$\frac{-36}{52}$

$$E = \Sigma\ g(x)P(x) = \frac{40}{52} = .77.$$

$$\Sigma\ g(x)P(x) = \frac{40}{52}$$

Thus in many plays of this game the
average of all the payoffs should be 77¢. Thus, say, in 1000 plays
we should expect to net 1000(.77) = $770.

A game is said to be a fair game if the expected value is 0.
When there is a charge to play a game, each payoff should be adjusted
by this charge when computing the expected value.

Example: Consider the following game. A coin with faces numbered
1 and 2 and a die with faces numbered 1, 2, 3, 4, 5, and 6
are tossed. As a player you will receive in dollars the sum of the
numbers showing on the coin and the die. The charge to play is
$5.00. Is this a fair game?
A sample space of equally likely outcomes is

$$S = \left\{ \begin{array}{l} (1,1),(1,2),(1,3),(1,4),(1,5),(1,6) \\ (2,1),(2,2),(2,3),(2,4),(2,5),(2,6) \end{array} \right\}$$

From this we can determine the probabilities of the various sums.
The results are summarized on the following page:

Sum x	Net Payoff g(x)	P(x)	g(x)P(x)
2	$2 - 5 = -3$	$\frac{1}{12}$	$\frac{-3}{12}$
3	$3 - 5 = -2$	$\frac{2}{12}$	$\frac{-4}{12}$
4	$4 - 5 = -1$	$\frac{2}{12}$	$\frac{-2}{12}$
5	$5 - 5 = 0$	$\frac{2}{12}$	0
6	$6 - 5 = 1$	$\frac{2}{12}$	$\frac{2}{12}$
7	$7 - 5 = 2$	$\frac{2}{12}$	$\frac{4}{12}$
8	$8 - 5 = 3$	$\frac{1}{12}$	$\frac{3}{12}$

$$E = \frac{-3}{12} - \frac{4}{12} - \frac{2}{12} + 0 + \frac{2}{12} + \frac{4}{12} + \frac{3}{12} = 0$$

Since $E = 0$, this is a fair game.

6) **Properties of $E(g(X))$**

As we have seen, $E = E(g(X))$ is an average or expected value. In general, if we regard $g(X)$ as a random variable, then $E(g(X))$ is the mean of that variable.

Example: Let $g(X) = X$.

$$E = E(g(X)) = E(X) = \Sigma\, xP(x) = \mu.$$

Example: Let $g(X) = (X - \mu)^2$.

$$E = E(g(X)) = E((X - \mu)^2) = \Sigma(x - \mu)^2 P(x) = \sigma^2.$$

The following properties are important

a) Let $g(X) = c$ (a constant). Then

$$E = E(g(X)) = E(c) = c.$$

b) Let $g(X) = aX$ where a is a constant. Then

$$E = E(g(X)) = E(aX) = aE(X) = a\mu_X.$$

Example: $E(3x) = 3E(x) = 3\mu_X$

Example: $E(\frac{1}{4}x) = \frac{1}{4}E(x) = \frac{1}{4}\mu_X$

c) Let $g(X) = aX + bY$ where $Y = h(X)$ is a random variable. Then

$$E = E(g(X)) = E(aX + bY) = aE(X) + bE(Y).$$

Example: Let $Y = X^2$ and $g(X) = 3X + 4X^2$

$$E = E(3X + 4X^2) = 3E(X) + 4E(X^2).$$

Example: Let $Y = h(X) = b$ (a constant) and let $g(X) = aX + b$.
Then

$$E = E(ax + b) = aE(x) + E(b) = a\mu_X + b.$$

As we have seen z scores are produced from those of X by $z = \frac{x - \mu}{\sigma}$.
The above properties may be used to show that the variable Z has a
mean of 0 and standard deviation of 1.

Example: Find μ_Z

$$\mu_Z = E(Z) = E(\frac{X - \mu}{\sigma})$$

$$= \frac{1}{\sigma}E(X - \mu)$$

$$= \frac{1}{\sigma}[E(X) - E(\mu)]$$

$$= \frac{1}{\sigma}[\mu - \mu] = 0.$$

7) Probability Density Functions

If X is a continuous variable the probability $P(a \le X \le b)$ is found
as the area under a curve and above the interval $a \le x \le b$. This curve
is called a probability density curve and the function for which this
graph is called a probability density function.

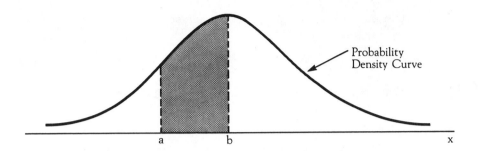

P(a $\leq$ x $\leq$ b) = area of the shaded region

Example: A probability density curve is given. Find P(2 $\leq$ x $\leq$ 4).

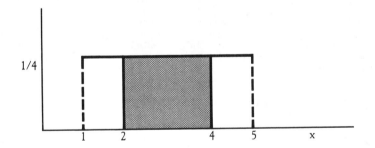

The probability P(2 $\leq$ X $\leq$ 4) is the area of the shaded rectangle
whose base is of length 1 and whose height is $\frac{1}{4}$. Thus

P(2 $\leq$ X $\leq$ 4) = area = base·height = 2($\frac{1}{4}$) = $\frac{1}{2}$.

Example: A probability density curve is given. Find P(X $\geq$ 4).

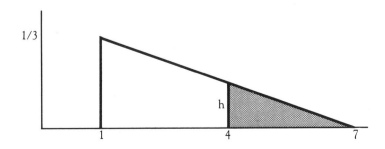

The shaded area is $P(X \geq 4)$.

The shaded triangle is similar to the given triangle. Comparing the ratios of height to base we see

$$\frac{h}{3} = \frac{\frac{1}{3}}{6} \quad \text{or} \quad h = \frac{1}{6} .$$

Thus $P(X \geq 4) = \text{area} = \frac{1}{2} \text{ base} \cdot \text{height} = \frac{1}{2}(3)(\frac{1}{6}) = \frac{1}{4}$.

Notes:

a) The total area under a probability density curve is 1.

b) For any particular value x of a continuous variable X, the probability is 0, i.e. $P(X = x) = 0$. (There is no area above a single point.)

8) <u>Joint Probability Distributions</u>

Occasionally 2 or more random variables are needed to describe the out-comes of a probability experiment. Probabilities are then assigned to pairs of values (x,y) and the variables X and Y have a joint distribution.

Example: Two wheels of fortune are shown. Let X be the outcome of the first, Y the second.

The joint distribution of X and Y is given in the table on the next page.

	y			
	1	2	3	Marginal Totals
1	$\frac{1}{16}$	$\frac{1}{16}$	$\frac{1}{8}$	$\frac{4}{16} = \frac{1}{4}$
2	$\frac{1}{16}$	$\frac{1}{16}$	$\frac{1}{8}$	$\frac{4}{16} = \frac{1}{4}$
x 3	$\frac{1}{16}$	$\frac{1}{16}$	$\frac{1}{8}$	$\frac{4}{16} = \frac{1}{4}$
4	$\frac{1}{16}$	$\frac{1}{16}$	$\frac{1}{8}$	$\frac{4}{16} = \frac{1}{4}$
Marginal Totals	$\frac{4}{16} = \frac{1}{4}$	$\frac{4}{16} = \frac{1}{4}$	$\frac{4}{8} = \frac{1}{2}$	1

For example, the probability of a 2 on spinner I and a 3 on spinner II is found at the intersection of the row $x = 2$ and column $y = 3$. Thus

$$P(X = 2, Y = 3) = \frac{1}{8} .$$

This table was produced using the product rule. For example,

$$P(X = 2 \text{ and } Y = 3) = P(X = 2) \cdot P(Y = 3)$$
$$= \frac{1}{4} \times \frac{1}{2} = \frac{1}{8}$$

$$P(X = 1 \text{ and } Y = 2) = P(X = 1) \cdot P(Y = 2)$$
$$= \frac{1}{4} \cdot \frac{1}{4} = \frac{1}{16} .$$

9) Marginal Distributions

Each of the variables X and Y of a joint distribution has a probability distribution which does not involve the other variable. These are called the marginal distributions of X and Y respectively and are found by summing the rows and columns of the joint distribution table.

Example: The joint distribution of the outcomes of the 2 wheels of fortune of the last example is repeated below.

	y 1	2	3	Marginal Totals
1	$\frac{1}{16}$	$\frac{1}{16}$	$\frac{1}{8}$	$\frac{4}{16} = \frac{1}{4}$
2	$\frac{1}{16}$	$\frac{1}{16}$	$\frac{1}{8}$	$\frac{4}{16} = \frac{1}{4}$
x 3	$\frac{1}{16}$	$\frac{1}{16}$	$\frac{1}{8}$	$\frac{4}{16} = \frac{1}{4}$
4	$\frac{1}{16}$	$\frac{1}{16}$	$\frac{1}{8}$	$\frac{4}{16} = \frac{1}{4}$
Marginal Totals	$\frac{4}{16} = \frac{1}{4}$	$\frac{4}{16} = \frac{1}{4}$	$\frac{4}{8} = \frac{1}{2}$	1

The marginal distribution of X is obtained
from the row totals and is given at the right.
Note this is simply the distribution of outcomes
for spinner I.

The marginal distribution of Y is obtained
from the column totals and is given at the
right. This is simply the distribution of
outcomes for spinner II.

x	P(x)
1	$\frac{1}{4}$
2	$\frac{1}{4}$
3	$\frac{1}{4}$
4	$\frac{1}{4}$

x	P(y)
1	$\frac{1}{4}$
2	$\frac{1}{4}$
3	$\frac{1}{2}$

Example: Two draws without replacement are
made from a box containing 6 balls numbered
1, 1, 2, 3, 3, and 3. Let X be the number
on the first ball drawn and Y the number on
the second. Find the joint distribution of X
and Y and the marginal distributions. We use the product rule
in the form

$$P(X = x, Y = y) = P(X = x) \cdot P(Y = y | X = x)$$

$$P(X = 1, Y = 1) = \frac{2}{6} \cdot \frac{1}{5} = \frac{2}{30}$$

$$P(X = 1, Y = 2) = \frac{2}{6} \cdot \frac{1}{5} = \frac{2}{30}$$

$$P(X = 1, Y = 3) = \frac{2}{6} \cdot \frac{3}{5} = \frac{6}{30}$$

$$P(X = 2, Y = 1) = \frac{1}{6} \cdot \frac{2}{5} = \frac{2}{30}$$

$$P(X = 2, Y = 2) = \frac{1}{6} \cdot \frac{0}{6} = 0$$

Etc. The joint distribution table follows

		y			Marginal Totals
		1	2	3	
	1	$\frac{2}{30}$	$\frac{2}{30}$	$\frac{6}{30}$	$\frac{10}{30} = \frac{1}{3}$
x	2	$\frac{2}{30}$	0	$\frac{3}{30}$	$\frac{5}{30} = \frac{1}{6}$
	3	$\frac{6}{30}$	$\frac{3}{30}$	$\frac{6}{30}$	$\frac{15}{30} = \frac{1}{2}$
Marginal Totals		$\frac{10}{30} = \frac{1}{3}$	$\frac{5}{30} = \frac{1}{6}$	$\frac{15}{30} = \frac{1}{2}$	1

The marginal distributions of X and Y are given below.

x	P(x)
1	$\frac{1}{3}$
2	$\frac{1}{6}$
3	$\frac{1}{2}$

y	P(y)
1	$\frac{1}{3}$
2	$\frac{1}{6}$
3	$\frac{1}{2}$

10) **Independent Random Variables**

The random variables X and Y are independent if and only if their joint probability distribution is the product of their marginal distributions:

$$P(X, Y) = P(X) \cdot P(Y).$$

This definition readily extends to 3 or more variables.

Example: The joint probability distribution of X and Y is given. Are X and Y independent?

		y		
	1	2	3	Marginal Totals
x 1	.08	.12	.20	.4
2	.12	.18	.30	.6
Marginal Totals	.2	.3	.5	.10

We need to check whether $P(X = x, Y = y) = P(X = x) \cdot P(Y = y)$ for every pair (x,y) . This amounts to checking the product of the marginal row and column totals with the table entry at the intersection of the row and column.

$$P(X = 1, Y = 1) = .08 = (.4)(.2) = P(X = 1) \cdot P(Y = 1)$$
$$P(X = 1, Y = 2) = .12 = (.4)(.3) = P(X = 1) \cdot P(Y = 2)$$
$$P(X = 1, Y = 3) = .20 = (.4)(.5) = P(X = 1) \cdot P(Y = 3)$$
$$P(X = 2, Y = 1) = .12 = (.6)(.2) = P(X = 2) \cdot P(Y = 1)$$
$$P(X = 2, Y = 2) = .18 = (.6)(.3) = P(X = 2) \cdot P(Y = 2)$$
$$P(X = 2, Y = 3) = .30 = (.6)(.5) = P(X = 2) \cdot P(Y = 3)$$

All products check. Thus the variables are independent.

Calculator Usage Illustration

Compute the expected value.

Find $\mu = 3 \cdot (.45) + 4 \cdot (.20) + 5 \cdot (.35)$

Method 1) Direct summation of the individual terms

$$3 \; \boxed{\times} \; .45 \; \boxed{+} \; 4 \; \boxed{\times} \; .20 \; \boxed{+} \; 5 \; \boxed{\times} \; .35 \; \boxed{=} \; \underline{3.9}$$

Method 2) Accumulate the sum in memory

$$3 \; \boxed{\times} \; 4.5 \; \boxed{=} \; \boxed{\text{STO}} \; 4 \; \boxed{\times} \; .20 \; \boxed{=} \; \boxed{\text{SUM}} \; 5 \; \boxed{\times} \; .35 \; \boxed{=} \; \boxed{\text{SUM}}\;\boxed{\text{RCL}} \; \underline{3.9}$$

Note: As an alternative, the partial sum of all but the last term can be accumulated in memory, then recalled and added to the last term.

$$3 \; \boxed{\times} \; 4.5 \; \boxed{=} \; \boxed{\text{STO}} \; 4 \; \boxed{\times} \; .20 \; \boxed{=} \; \boxed{\text{SUM}} \; 5 \; \boxed{\times} \; .35 \; \boxed{=} \; \boxed{+} \; \boxed{\text{RCL}} \; \boxed{=} \; \underline{3.9}$$

Note: When summing to memory it is a good practice to store the first term rather than adding it to the memory. This clears out any results that are left over from previous calculations.

Calculator Usage Illustration

Compute the standard deviation of a distribution

Find $\sigma = \sqrt{(3 - 3.9)^2(.45) + (4 - 4.9)^2(.20) + (5 - 3.9)^2(.35)}$

Method 1) Direct summation followed by taking the square root

$$\boxed{(}\; 3 \; \boxed{-} \; 3.9 \; \boxed{)} \; \boxed{x^2} \; \boxed{\times} \; .45 \; \boxed{+} \; \boxed{(} \; 4 \; \boxed{-} \; 3.9 \; \boxed{)} \; \boxed{x^2} \; \boxed{\times} \; .20$$

$$\boxed{+} \; \boxed{(} \; 5 \; \boxed{-} \; 3.9 \; \boxed{)} \; \boxed{x^2} \; \boxed{\times} \; .35 \; \boxed{=} \; \boxed{\sqrt{\;}} \; \underline{.8888194}$$

Method 2) Accumulate the sum in memory, recall it, and take the square root

$$3 \; \boxed{-} \; 3.9 \; \boxed{=} \; \boxed{x^2} \; \boxed{\times} \; .45 \; \boxed{\text{STO}} \; 4 \; \boxed{-} \; 3.9 \; \boxed{=} \; \boxed{x^2} \; \boxed{\times} \; .20$$

$$\boxed{=} \; \text{SUM} \; 5 \; \boxed{-} \; 3.9 \; \boxed{=} \; \boxed{x^2} \; \boxed{\times} \; .35 \; \boxed{=} \; \boxed{\text{SUM}}\;\boxed{\text{RCL}} \; \boxed{\sqrt{\;}} \; \underline{.8888194}$$

Calculator Usage Illustration

Compute the variance of a distribution using $\sigma^2 = \Sigma\, x^2 P(x) - \mu^2$.
Find $\sigma^2 = 3^2 \cdot (.45) + 4^2 \cdot (.20) + 5^2 \cdot (.35) - (3.9)^2$.

Method 1)   Direct evaluation

3 $\boxed{x^2}$ $\boxed{\times}$ $.45$ $\boxed{+}$ 4 $\boxed{x^2}$ $\boxed{\times}$ $.20$ $\boxed{+}$ 5 $\boxed{x^2}$ $\boxed{\times}$ $.35$ $\boxed{-}$ 3.9 $\boxed{x^2}$ $\boxed{=}$ $\underline{.79}$

Method 2)   Accumulate the sum in the memory

3 $\boxed{x^2}$ $\boxed{\times}$ $.45$ $\boxed{=}$ $\boxed{STO}$ 4 $\boxed{x^2}$ $\boxed{\times}$ $.20$ $\boxed{=}$ $\boxed{SUM}$ 5 $\boxed{x^2}$ $\boxed{\times}$ $.35$

$\boxed{=}$ $\boxed{SUM}$ $\boxed{RCL}$ $\boxed{-}$ 3.9 $\boxed{x^2}$ $\boxed{=}$ $\underline{.79}$

Exercise Set V.1

1) The possible values of X are 0, 1, 2, and 3.

A sample space of equally likely outcomes is

S = {(H,H,H),(H,H,T),(H,T,H),(T,H,H),(H,T,T),(T,H,T),(T,T,H),(T,T,T)}.

From this we easily find the following distribution:

x	0	1	2	3
P(x)	$\frac{1}{8}$	$\frac{3}{8}$	$\frac{3}{8}$	$\frac{1}{8}$

3) Let x = 0 ↔ white (2 of the 8 sectors)

x = 1 ↔ red (1 of the 8 sectors)

x = 2 ↔ blue (3 of the 8 sectors)

x = 3 ↔ orange (2 of the 8 sectors)

x	P(x)
0	$\frac{2}{8} = \frac{1}{4}$
1	$\frac{1}{8}$
2	$\frac{3}{8}$
3	$\frac{2}{8} = \frac{1}{4}$

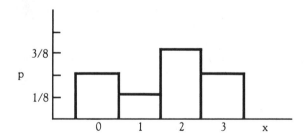

5) <u>NO.</u> The sum of the probabilities is not 1.

7) a) The probability that none of the 40 favor an increase.

b) The probability that 8 or fewer (at most 8) favor an increase.

c) The probability that between 20 and 30 (inclusively) favor an increase.

d) The probability that more than 20 (at least 21) favor an increase.

9) a) P(X < 4) = P(X = 2) + P(X = 3)

$$= \frac{1}{36} + \frac{2}{36} = \frac{3}{36} = \frac{1}{12}$$

b) P(5 ≤ X < 8) = P(X = 5) + P(X = 6) + P(X = 7)

$$= \frac{4}{36} + \frac{5}{36} + \frac{6}{36} = \frac{5}{12}$$

c) $P(X = 10) = \frac{3}{36} = \frac{1}{12}$

d) $P(X > 12) = 0$

11)   a)

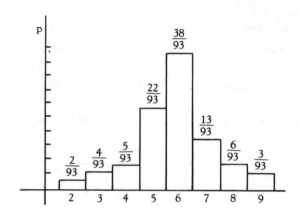

b) $P(X < 5) = P(X = 2) + P(X = 3) + P(X = 4)$

$= \frac{2}{93} + \frac{4}{93} + \frac{5}{93} = \frac{11}{93}$

c) $P(X \geq 7) = P(X = 7) + P(X = 8) + P(X = 9)$

$= \frac{13}{93} + \frac{6}{93} + \frac{3}{93} = \frac{22}{93}$

13) $P(X = 3) = P(D,D,D) = (.02)(.02)(.02) = .000008$

$P(X = 2) = P(D,D,G) + P(D,G,D) + P(G,D,D)$
$= (.02)(.02)(.98) + (.02)(.98)(.02) + (.98)(.02)(.02)$
$= 3(.02)^2(.98) = .001176$

$P(X = 1) = P(D,G,G) + P(G,D,G) + P(G,G,D)$
$= 3(.02)(.98)^2 = .057624$

$P(X = 0) = P(G,G,G) = (.98)(.98)(.98) = (.98)^3 = .941192$

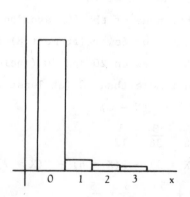

15) $P(X = 1) = \frac{1}{2}$

$P(X = 2) = P(T,H) = \frac{1}{2} \times \frac{1}{2} = (\frac{1}{2})^2 = \frac{1}{4}$

$P(X = 3) = P(T,T,H) = (\frac{1}{2})^3 = \frac{1}{8}$

$P(X = x) = (\frac{1}{2})^x$

$P(X \geq 3) = 1 - P(X \leq 2)$ (using the complement)

$\qquad\qquad = 1 - P(X = 1) - P(X = 2)$

$\qquad\qquad = 1 - \frac{1}{2} - \frac{1}{4} = 1 - \frac{3}{4} = \frac{1}{4}$

17)

Number	1	2	3	4	5	6	7	8	9	10
Divisors	1	1,2	1,3	1,2,4	1,2	1,2,3,6	1,7	1,2,4,8	1,3,9	1,2,5,10
x	1	2	2	3	2	4	2	3	3	4

x	1	2	3	4
$P(X = x)$	$\frac{1}{10}$	$\frac{4}{10} = \frac{2}{5}$	$\frac{2}{10} = \frac{1}{5}$	$\frac{3}{10}$

Exercise Set V.2

1) a) $\mu_X = \Sigma\, xP(x) = 1.60$

x	P(x)	xP(x)	$x^2P(x)$
0	.2	.00	.00
1	.2	.20	.20
2	.4	.80	1.60
3	.2	.60	1.80

$\sigma_X^2 = \Sigma\, x^2P(x) - (\mu_X)^2$

$\quad = 3.6 - (1.6)^2$

$\quad = 1.04$

$\sigma_X = \sqrt{1.04} = 1.0198$

$\quad\quad = 1.02$

$$\Sigma\, xP(x) = 1.60 \quad \Sigma\, x^2P(x) = 3.60$$

b) $\mu \pm \sigma = 1.60 \pm 1.02 = 2.62$ or .58.

The observations $x = 1$ and $x = 2$ are on this interval.

$P(.58 \le X \le 2.62) = P(X = 1) + P(X = 2) = .2 + .4 = .6$

c) $2\sigma_X = 2 \times 1.02 = 2.04$, $\mu \pm 2\sigma = 3.64$ or $-.44$

$P(\mu - 2\sigma \le X \le \mu + 2\sigma) = P(-.44 \le X \le 3.64)$

$\quad\quad\quad\quad\quad\quad\quad\quad = P(X = 0) + P(X = 1) + P(X = 2) + P(X = 3) = 1$

3) Each of the numbers 0, 1, 2, and 3 is equally likely.

a) $P(X = x) = \dfrac{1}{4}$

b) $\mu_X = 0 \times \dfrac{1}{4} + 1 \times \dfrac{1}{4} + 2 \times \dfrac{1}{4} + 3 \times \dfrac{1}{4} = \dfrac{6}{4} = 1.5$

$\sigma_X^2 = (0 = 1.5)^2 \cdot \dfrac{1}{4} + (1 - 1.5)^2 \cdot \dfrac{1}{4} + (2 - 1.5)^2 \cdot \dfrac{1}{4} + (3 - 1.5)^2 \cdot \dfrac{1}{4}$

$\quad = \dfrac{2.25 + .25 + .25 + 2.25}{4} = \dfrac{5}{4} = 1.25$

$\sigma_X = \sqrt{1.25} = 1.12$

5) a) We need the mean μ of the time X spend at the checkout counter. The distribution is at the right.

x	P(x)
1	.15 (15%)
2	.20 (20%)
3	.30 (30%)
4	.25 (25%)
5	.10 (10%)

$\mu = 1 \cdot (.15) + 2 \cdot (.20) + 3 \cdot (.30) + 4 \cdot (.25)$

$\quad\quad\quad\quad\quad\quad\quad\quad\quad\quad + 5 \cdot (.10)$

$\quad = .15 + .40 + .90 + 1.0 + .50$

$\quad = 2.95$ (minutes)

b) The average time spent by one customer is 2.95 minutes. For 900 customers we expect a total time at the checkout counters of around $900 \cdot (2.95) = 2655$ minutes. One clerk working 8 hours has $8 \times 60 = 480$ minutes of checkout time available. Thus we need around

$$\frac{2655}{480} = 5.53 \quad \text{or} \quad 6 \text{ clerks.}$$

7)

g(x)	P(x)
0	$\frac{50}{100} = .50$
.10	$\frac{20}{100} = .20$
.25	$\frac{15}{100} = .15$
.50	$\frac{10}{100} = .10$
1.00	$\frac{2}{100} = .02$
5.00	$\frac{2}{100} = .02$
25.00	$\frac{1}{100} = .01$

$$\mu = 0(.5) + (.10)(.2) + (.25)(.15) + (.50)(.10) + 1.00(.02)$$
$$+ 5.00(.02) + 25.00(.01)$$
$$= .4775 \text{ (about 48¢)}.$$

9) a) Note that the monkey must eventually get the right object since it never
   tries an object a second time.
   The probability it succeeds on the first trial $(x = 1)$ is
   $P(X = 1) = \frac{1}{4}$.
   The probability that it succeeds on the second trial is
   $P(X = 2) = \frac{3}{4} \times \frac{1}{3} = \frac{1}{4}$. $\quad P(F,S) = P(F) \times P(S/F)$
   $P(X = 3) = P(F,F,S) = \frac{3}{4} \times \frac{2}{3} \times \frac{1}{2} = \frac{1}{4}$
   $P(X = 4) = P(F,F,F,S) = \frac{3}{4} \times \frac{2}{3} \times \frac{1}{2} \times \frac{1}{1} = \frac{1}{4}$

   b) $\mu = 1 \cdot (\frac{1}{4}) + 2 \cdot (\frac{1}{4}) + 3 \cdot (\frac{1}{4}) + 4 \cdot (\frac{1}{4}) = \frac{10}{4} = 2.5$

   c) If the monkey is allowed to repeat the experiment on many occasions and
   does not learn from his experiences, then the mean of the number of
   objects he tries (on these occassions) should be somewhere around 2.5.
   If the mean is considerably less than 2.5 then this may be an indication
   of learning. If the mean is much greater than 2.5, then...

11)

x	P(x)	g(x)
0	$\frac{1}{8}$	+5
1	$\frac{3}{8}$	-4
2	$\frac{3}{8}$	-4
3	$\frac{1}{8}$	+5

$$E = 5 \cdot (\tfrac{1}{8}) - 4 \cdot (\tfrac{3}{8}) - 4 \cdot (\tfrac{3}{8}) + 5 \cdot (\tfrac{1}{8})$$

$$= \frac{10}{8} - \frac{24}{8} = -\frac{14}{8} = -1.75$$

On the average you lose $1.75 per game. In 100 plays, your losses will be around $100 \times 1.75 = 175$ (dollars). (Some friend!)

13) **a)** If X has the values 0, 2, 3, then 3X has the values $3 \cdot 0 = 0$, $3 \cdot 2 = 6$, and $3 \cdot 3 = 9$ with the corresponding probabilities

3x	0	6	9
P(3x)	$\frac{1}{6}$	$\frac{1}{2}$	$\frac{1}{3}$

b) $\mu_{3X} = E[3X] = 0 \times \frac{1}{6} + 6 \times \frac{1}{2} + 9 \times \frac{1}{3} = 6$

c) $\mu_X = E[X] = 0 \times \frac{1}{6} + 2 \times \frac{1}{2} + 3 \times \frac{1}{3} = 2$

By Theorem 5.1(2)

$E[3X] = 3E[X] = 3 \cdot 2 = 6$ as found in part b.

15)

	x	P(x)	g(x)
win	0	$\frac{18}{37}$	+1
lose	1	$\frac{19}{37}$	-1

$$E = 1 \cdot (\tfrac{18}{37}) - 1 \cdot (\tfrac{19}{37}) = -\frac{1}{37} = -.027$$

17)

Stock A				Stock B			
	x	P(x)	g(x)		x	P(x)	g(x)
(gain)	1	.5	+20	(gain)	1	.2	+60
(loss)	0	.5	-5	(loss)	0	.8	-5

$$E_A = 20 \cdot (.5) - 5 \cdot (.5) = 7.50 \text{ (dollars)}$$

$$E_B = 60 \cdot (.2) - 5 \cdot (.8) = 8 \text{ dollars}$$

B has the highest expected value. Other factors being equal B would be the best buy.

19) a)

x	P(x)
2	$\frac{1}{5}$
3	$\frac{2}{5}$
4	$\frac{2}{5}$

$$\mu_X = 2 \cdot (\tfrac{1}{5}) + 3 \cdot (\tfrac{2}{5}) + 4 \cdot (\tfrac{2}{5}) = \frac{16}{5} = 3.2$$

b) Use the notation A_1, A_2, B_1, B_2, C for the grades.

Sample	x_1,x_2	$\bar{x}$
A_1,A_2	4,4	4
A_1,B_1	4,3	$\frac{7}{2}$
A_1,B_2	4,3	$\frac{7}{2}$
A_1,C	4,2	3
A_2,B_1	4,3	$\frac{7}{2}$
A_2,B_2	4,3	$\frac{7}{2}$
A_2,C	4,2	3
B_1,B_2	3,3	3
B_1,C	3,2	$\frac{5}{2}$
B_2,C	3,2	$\frac{5}{2}$

$\bar{x}$	$P(\bar{x})$
$\frac{5}{2}$	$\frac{2}{10}$
3	$\frac{3}{10}$
$\frac{7}{2}$	$\frac{4}{10}$
4	$\frac{1}{10}$

c) $\mu_{\bar{X}} = \Sigma \; \bar{x}P(\bar{x}) = (\tfrac{5}{2})(\tfrac{2}{10}) + 3(\tfrac{3}{10}) + (\tfrac{7}{2})(\tfrac{4}{10}) + 4 \cdot (\tfrac{1}{10})$

$$= \frac{64}{20} = 3.2 = \mu_X$$

d)

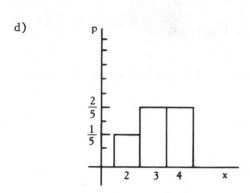
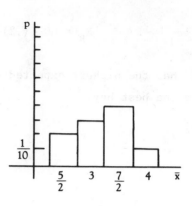

21) a)

x	P(x)	g(x)
(rain) 0	.3	−12000
(no rain) 1	.7	+40000

$$E = (-12000) \cdot (.3) + (40000) \cdot (.07) = 24400$$

b)

x	P(x)	g(x)
(rain) 0	.3	−4800 [12000−12000−4800]
(no rain) 1	.7	35200 [40000−4800]

Don't forget the cost of the insurance in getting g(x).

$$E = -4800(.3) + 35200(.7) = \$23200$$

(Not as good as with no insurance.)

c)

x	P(x)	g(x)
(rain) 0	.3	(−12000 + c(1000) − 400c)
(no rain) 1	.7	40000 − 400c

$$E = (-12000 + 1000c - 400c)(.3) + (40000 - 400c)(.7)$$
$$= 24400 - 100c \quad \text{(reduce 24400 by 100c.)}$$

∴ $E \le 24400$

d) Remember we are talking about expected value. If we play this "game" many times the greatest expected value comes with no insurance. It may be that the promoter cannot stand a $12000 loss. If so he should buy some insurance.

23) a) $E = 0 \cdot (0) + 1 \cdot (.3) + 2 \cdot (.4) + 3 \cdot (.3) = 2.0$

  b) 1 cake in stock

x	P(x)	g(x)	xg(x)
0	0	-4	0.00
1	.3	5	1.50
2	.4	5	2.00
3	.3	5	1.50
			5.00

  $E = 5.00$

  c) 2 cakes in stock

x	P(x)	g(x)	xg(x)
0	0	-8	0.00
1	.3	1	.30
2	.4	10	4.00
3	.3	10	4.00
			8.30

  $E = 8.30$

  d) 3 cakes in stock

x	P(x)	g(x)	xg(x)
0	0	-12	0.00
1	.3	-3	-.90
2	.4	6	2.40
3	.3	15	4.50
			6.00

  $E = 6.00$

Exercise Set V.3

1) a) $P(1.5 \leq X \leq 1.8) = (.3) \cdot 2 = .60$ (Area of the rectangle of base

                                                   $1.8 - 1.5 = .3$ and height 2.0)

    b) $P(X < 1.4) = (.1)(.5) + (.2)(1.0)$ (Sum of the Areas of the two rectangles)

                         $= .05 + .2$

                         $= .25$

    c)) $P(X \leq 1.8) = 1.0$

3) a) $P(X = 3) = 0$ (continuous distribution)

    b) $P(3 < X < 5) = $ length of base $\times$ height $= 2 \cdot (\frac{1}{4}) = \frac{1}{2}$

    c) $P(3 \leq X \leq 5) = \frac{1}{2}$ (same as (b) since $P(X = 3) = P(X = 5) = 0$)

    d) $P(X \leq 6) = 4 \times \frac{1}{4} = 1$ (total area $= 1$)

    e) $P(X > 5.3) = (.7)$ $(\frac{1}{4}) = .175$

5) a) $P(X > 3) = P(X \geq 0) - P(0 \leq X \leq 3) = .5 - .4 = .1$

    b) $P(X = 3) = 0$ (continuous variable)

    c) $P(0 \leq X \leq 5) = P(0 \leq x \leq 3) + P(3 \leq X \leq 5)$

                       $= .4 + .085 = .485$

    d) $P(X > 5) = P(X \geq 0) - P(0 \leq X \leq 5)$

                  $= .5 - .485 = .015$

    e) $P(X \leq 0) = .5$ since $P(X \geq 0) = .5$ (complementary events)

       Since $P(0 \leq x \leq 3) = .4$,

       $P(x \leq 0) + P(0 \leq x \leq 3) = .5 + .4 = .9$

       $P(X \leq 3) = .9$

       $\therefore$ $c = 3$

7) a) $P(2 \leq X \leq 3) = 1 - .45 - .30 = .10$

                   $= .15$ (the total area under the curve is 1)

    b) $P(X \leq 2) = P(0 \leq X \leq 1) + P(1 \leq X \leq 2)$

                 $= .45 + .30 = .75$ (adding the areas)

    c) $P(X \geq 1) = P(1 \leq x \leq 2) + P(2 \leq X \leq 3) + P(3 \leq X \leq 4)$

                    $= .30 + .15 + .10$

                    $= .55$

      or $P(X \geq 1) = 1 - P(0 \leq x \leq 1)$ (using the complement)

                      $= 1 - .45$

                      $= .55$

9) $\mu \approx 85$

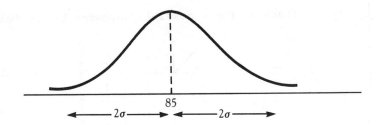

Remember the empirical predictions are very accurate for a bell shaped
distribution.

The probability is around .95 of an outcome being within 2(S.D.) of the
mean.   Thus the range is approximately 4(S.D.).

So $4\sigma \approx 120 - 50 = 70$ and $\sigma \approx \dfrac{70}{4} = 17.5$.

A more pessimistic prediction is obtained using the 6σ estimate.

$$6\sigma \approx 70, \quad \sigma \approx \frac{70}{6} = 11.67$$

σ is probably between 11.67 and 17.5.

11) p is the parameter.   Each choice of p determines a different distribution.

Exercise Set V.4

1) a)

x	y
0	0
0	1
1	0
1	1

(Each of the 4 possible outcomes has probability $\frac{1}{4}$.)

x \ y	0	1	
0	$\frac{1}{4}$	$\frac{1}{4}$	$\frac{1}{2} = P(Y = 0)$
1	$\frac{1}{4}$	$\frac{1}{4}$	$\frac{1}{2} = P(Y = 1)$
	$\frac{1}{2}$	$\frac{1}{2}$	
	P(X=0)	P(X=1)	

b) The marginal distribution of X is obtained from the column sums, that of Y from the row sums.

x	P(x)
0	$\frac{1}{2}$
1	$\frac{1}{2}$

y	P(y)
0	$\frac{1}{2}$
1	$\frac{1}{2}$

3)

x \ y	1	2	3	
0	.1	.1	.05	.25 = P(Y = 0)
1	.05	0	.05	.10 = P(Y = 1)
2	.15	.05	.05	.25 = P(Y = 2)
3	.05	.15	.05	.25 = P(Y = 3)
4	.1	0	.05	.15 = P(Y = 4)
	.45	.30	.25	1.00
	P(X=1)	P(X=2)	P(X=3)	

Marginal X	
x	P(x)
1	.45
2	.30
3	.25

Marginal Y	
x	P(x)
0	.25
1	.10
2	.25
4	.15

If X and Y are independent, then the product of the marginal totals must equal the corresponding table entry

Row 1, column 1: Is $(.45)(.25) = .1$? NO! $\therefore$ Not Independent.

Of course, there are many other entries for which this test fails but only one is needed.

5) a) The distribution of X can be found by listing the sample space:

$$S = \{(H,H),(H,T),(T,H),(T,T)\}.$$

This yields:

x	P(x)
0	$\frac{1}{4}$
1	$\frac{1}{2}$
2	$\frac{1}{4}$

The distribution for the single die is given at the right.

y	P(y)
1	$\frac{1}{6}$
2	$\frac{1}{6}$
3	$\frac{1}{6}$
4	$\frac{1}{6}$
5	$\frac{1}{6}$
6	$\frac{1}{6}$

b) The table entries are the products of the corresponding row and column totals, i.e., the products of the marginal probabilities.

x \ y	0	1	2	
1	$\frac{1}{24}$	$\frac{1}{12}$	$\frac{1}{24}$	$\frac{1}{6}$
2	$\frac{1}{24}$	$\frac{1}{12}$	$\frac{1}{24}$	$\frac{1}{6}$
3	$\frac{1}{24}$	$\frac{1}{12}$	$\frac{1}{24}$	$\frac{1}{6}$
4	$\frac{1}{24}$	$\frac{1}{12}$	$\frac{1}{24}$	$\frac{1}{6}$
5	$\frac{1}{24}$	$\frac{1}{12}$	$\frac{1}{24}$	$\frac{1}{6}$
6	$\frac{1}{24}$	$\frac{1}{12}$	$\frac{1}{24}$	$\frac{1}{6}$
	$\frac{1}{4}$	$\frac{1}{2}$	$\frac{1}{4}$	

row 1 and column 1 entry $= \frac{1}{6} \times \frac{1}{4} = \frac{1}{24}$

row 1 and column 2 entry $= \frac{1}{6} \times \frac{1}{2} = \frac{1}{12}$

etc.

7) a) X + Y = number of fish caught on the first two days.

        E[X + Y] = average number of fish caught on the first 2 days.

        E[X + Y] = E[X] + E[Y]   states that the average 2 day catch is the sum
        of the first and second day averages.

    b) X + Y = total amount won by the husband and wife.

        E[X + Y] = average winning of husband and wife teams.

        E[X + Y] = E[X] + E[Y]   states that the average winning of the husband
        and wife teams is the sum of the averages of the husband's winnings
        and the wife's winnings.

9) For the 2nd test $\mu_Y = 0 \times \frac{1}{5} + 1 \times \frac{4}{5} = \frac{4}{5}$ (same as for X as should be
    clear).

    a) E[X + Y] = E[X] + E[Y]

$$= \mu_X + \mu_Y = \frac{4}{5} + \frac{4}{5} = 2(\frac{4}{5})$$

    b) E[X + Y + Z] = E[X] + E[Y] + E[Z]

$$= \frac{4}{5} + \frac{4}{5} + \frac{4}{5} = 3(\frac{4}{5})$$

    c) The expected number of successes in n trials or tests should be
        $n \cdot (\frac{4}{5})$ = np where p is the probability of success on any trial.

Chapter Test

1) a) P(X = 2) = .20

    b) P(X ≥ 3) = P(X = 3) + P(X = 4) = .30 + .20 = .50

    c) P(2 ≤ X < 5) = P(X = 2) + P(X = 3) + P(X = 4) = .20 + .30 + .20 = .70

    d) Add up the probabilities until .80 is obtained

        P(X ≤ 3) = .10 + .20 + .20 + .30 = .80

        c = 3

2) a) P(X ≥ 5) is the area under the
        density curve to the rigth of
        x = 5.   For this rectangle,
        Area = (base)(height) = $(3)(\frac{1}{6})$
             = $\frac{1}{2}$.
        P(X ≥ 5) = $\frac{1}{2}$

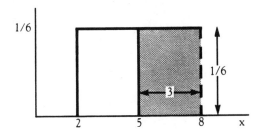

    b) P(2 ≤ X ≤ 3) is the area under
        the density curve above the
        interval 2 ≤ x ≤ 3.   This
        rectangle has
        Area = $(1)(\frac{1}{6})$ = $\frac{1}{6}$.

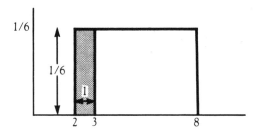

    c) The area under the density
        curve to the right of x = c is
        to be $\frac{1}{3}$.   The corresponding
        rectangle has a base b = 8 − c
        and height h = $\frac{1}{6}$.
        Area = $\frac{1}{3}$ = (8 − c)$(\frac{1}{6})$.
        Solving for c we obtain
        c = 8 − 2 = 6.

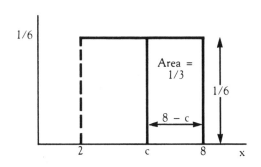

3) a) Let F denote "for" and A "against". When 3 voters are questioned. a sample space of equally likely outcomes is:

$$\{(F,F,F),(F,F,A),(F,A,F),(A,F,F),(F,A,A),(A,F,A),(A,A,F),(A,A,A)\}$$

The event x = 3 can occur in only one way. Thus $P(X = 3) = \frac{1}{8}$.
The event x = 2 can occur in three ways. Thus $P(X = 2) = \frac{3}{8}$.
Similarly we find $P(X = 1) = \frac{3}{8}$ and $P(X = 0) = \frac{1}{8}$.
The distribution follows:

x	P(X = x)
0	$\frac{1}{8}$
1	$\frac{3}{8}$
2	$\frac{3}{8}$
3	$\frac{1}{8}$

b)

x	P(X = x)	xP(X = x)	x^2P(X = x)
0	$\frac{1}{8}$	0	0
1	$\frac{3}{8}$	$\frac{3}{8}$	$\frac{3}{8}$
2	$\frac{3}{8}$	$\frac{6}{8}$	$\frac{12}{8}$
3	$\frac{1}{8}$	$\frac{3}{8}$	$\frac{9}{8}$
		$\frac{12}{8}$	$\frac{24}{8}$

$$\mu = \Sigma \; xP(X = x) = \frac{12}{8} = 1.5$$

$$\sigma^2 = \Sigma \; x^2 P(X = x) - (\Sigma \; xP(X = x))^2$$

$$= \frac{24}{8} - (\frac{12}{8})^2 = 3 = (1.5)^2 = .75$$

$$\sigma = \sqrt{.75} = .866$$

c)

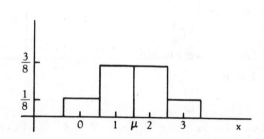

4) $\mu = \Sigma\ xP(X = x) = 4(.20) + 5(.50) + 6(.20) + 7(.10) = 5.2$

5) $\mu = \Sigma\ xP(X = x) = 16(.50) + 17(.30) + 18(.20) = 16.7$

The average of the daily sales is 16.7 dozen.

6) There are only 2 outcomes: The purchaser is alive at the end of the year (1).

The purchaser dies before the end of the year (0).

The probability distribution is as follows:

x	$P(X = x)$	$g(X = x)$	
0	.0018	-992	(Remember to take the paid premium into account in computing the company's loss.)
1	.9982	+8	

a) $\mu = (-992)(.0018) + (8)(.9982) = 6.20$

b) The expected return for $1000 worth of insurance is $6.20.

The expected return for $100,000,000 worth of insurance is

$(100,000)(6.20) = 620,000$ (dollars).

7) a) $P(X = 2, Y = 1) = \dfrac{1}{9}$ (The entry at the intersection of the row headed

$x = 2$ and the column headed $y = 1$.)

b) To find the probability of $x = 1$ by any means, sum the probabilities

of this row. Thus

$$P(X = 1) = 0 + \frac{1}{5} + \frac{2}{15} = \frac{5}{15} = \frac{1}{3}$$

$P(X = 2)$ is found by summing the probabilites of the second row. We find

$$P(X = 2) = \frac{1}{6} + \frac{1}{9} + \frac{1}{4} = \frac{19}{36}$$

Similarly

$$P(X = 3) = \frac{1}{12} + 0 + \frac{1}{18} = \frac{5}{36}\ .$$

x	$P(X = x)$
1	$\dfrac{1}{3}$
2	$\dfrac{19}{36}$
3	$\dfrac{5}{36}$

c) $P(Y = 0)$ is found by summing the probabilities in the column headed

$y = 0$. Thus

$$P(Y = 0) = 0 + \frac{1}{6} + \frac{1}{12} = \frac{3}{12} = \frac{1}{4}\ .$$

Using the columns $Y = 1$ and $Y = 2$ we find

$$P(Y = 1) = \frac{1}{5} + \frac{1}{9} + 0 = \frac{14}{45}$$

$$P(Y = 2) = \frac{2}{15} + \frac{1}{4} + \frac{1}{18} = \frac{79}{180}$$

y	P(Y = y)
0	$\frac{1}{4}$
1	$\frac{14}{45}$
2	$\frac{79}{180}$

$P(X = 2) = \frac{19}{36}$ (from the marginal distribution of X)

Note: Decimals could have been used throughout.

8) We need to check whether $P(X = x) \cdot P(Y = y) = P(X = x, Y = y)$ for each x and y. First find the marginal probabilities.

x \ y	0	1	P(x)
0	0	$\frac{1}{4}$	$\frac{1}{4}$
1	$\frac{1}{3}$	$\frac{5}{12}$	$\frac{9}{12}$
P(y)	$\frac{1}{3}$	$\frac{8}{12}$	

We see that $P(X = 0) \cdot P(Y = 0) = \frac{1}{3} \cdot \frac{1}{3} = \frac{1}{9} \neq 0 = P(X = 0, Y = 0)$.
Thus X and Y are not independent.

9) a) Let X be the number of keys that are tried.
$P(X = 1) = \frac{1}{3}$ since there are 3 keys of which only 1 is correct.
$P(X = 2) = P(\text{incorrect key followed by correct key}) = (\frac{2}{3})(\frac{1}{2}) = \frac{1}{3}$.
$P(X = 3) = P(\text{incorrect key followed by incorrect key followed by correct key}) = (\frac{2}{3})(\frac{1}{2})(\frac{1}{1}) = \frac{1}{3}$.

x	P(X = x)
1	$\frac{1}{3}$
2	$\frac{1}{3}$
3	$\frac{1}{3}$

b) $\mu = (1)(\frac{1}{3}) + 2(\frac{1}{3}) + 3(\frac{1}{3}) = 2$

10) a) The area of the rectangle is
$(2)(\frac{1}{4}) = \frac{1}{2}$. Thus
$P(X \geq 4) = \frac{1}{2}$.

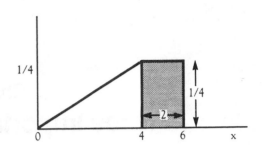

b) The area of the shaded rectangle
is $(1)(\frac{1}{4}) = \frac{1}{4}$. The area of the
triangle is $(\frac{1}{2})(4)(\frac{1}{4}) = \frac{1}{2}$.
$P(X \leq 5) = P(X \leq 4) + P(4 \leq X \leq 5)$
$= \frac{1}{2} + \frac{1}{4} = \frac{3}{4}$

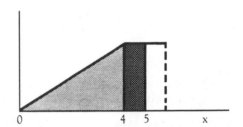

c) The area of the small shaded
triangle is $P(X \leq 2) = (\frac{1}{2})(2)(\frac{1}{8})$
$= \frac{1}{8}$. By the rule for complementary
events,
$P(X \leq 2) + P(2 \leq X \leq 6) = 1$
$\frac{1}{8} + P(2 \leq X \leq 6) = 1$
or $P(2 \leq X \leq 6) = 1 - \frac{1}{8} = \frac{7}{8}$

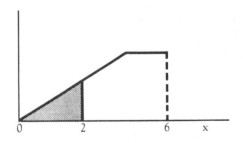

Chapter VI
Three Important Distributions

Chapter 6 introduces three important probability distributions: the binomial, the normal, and the chi-square. The binomial distribution is particularly useful with polling, opinion sampling, and the like. The normal distribution proves to be a useful model for many naturally occurring phenomena as well as having an important role in many inferential processes. The chi-square distribution, which is introduced as a second example of a continuous distribution, has many applications in inferential statistics. Its use with the classical goodness of fit test is illustrated.

Concepts/Techniques

1) Binomial Coefficients

The binomial coefficient $\binom{n}{k}$ is the number of ways in which k objects may be selected from n without regard to the order in which they are selected.

Formula: $\binom{n}{0} = 1$

$$\binom{n}{k} = \frac{n(n - 1)(n - 2) \ \dots \ (n - k + 1)}{k \cdot (k - 1)(k - 2) \ \dots \ 1}$$

Terminology: $\binom{n}{k}$ is read "n choose k".

Example: $\binom{9}{4} = \frac{9 \cdot 8 \cdot 7 \cdot 6}{4 \cdot 3 \cdot 2 \cdot 1} = 126$

Example: $\binom{8}{0} = 1$

Note that $\binom{n}{0}$ is 1 for any positive integer n. Further when k ≠ 0, $\binom{n}{k}$ is given by a fraction whose numerator and denominator each contain k terms. Further the first factor of the numerator is n, the first of the denominators is k and each successive factor is one less than the previous.

Example: Compute $\binom{10}{5}$.

Here n = 10 and k = 5.

There are k = 5 terms in the numerator as well as in the denominator.

The first term of the numerator is 10.

The first term of the denominator is 5.

$$\binom{10}{5} = \frac{10 \cdot 9 \cdot 8 \cdot 7 \cdot 6}{5 \cdot 4 \cdot 3 \cdot 2 \cdot 1} = 252$$

Example: How many 13 card bridge hands can be formed from a standard deck of 52 cards?

The question may be rephrased "How many ways can we choose 13 cards from 52? The answer is

$$\text{"52 choose 13"} = \binom{52}{13}$$

$$\binom{52}{13} = \frac{52 \cdot 51 \cdot 50 \cdot 49 \cdot 48 \cdot 47 \cdot 46 \cdot 45 \cdot 44 \cdot 43 \cdot 42 \cdot 41 \cdot 40}{13 \cdot 12 \cdot 11 \cdot 10 \cdot 9 \cdot 8 \cdot 7 \cdot 6 \cdot 5 \cdot 4 \cdot 3 \cdot 2 \cdot 1}$$

$$= 6.35 \times 10^{11}$$

Example: A population consists of 10 members. How many samples of size 6 has it?

The answer is the number of ways we may choose 6 objects from 10. Thus

$$\binom{10}{6} = \frac{10 \cdot 9 \cdot 8 \cdot 7 \cdot 6 \cdot 5}{6 \cdot 5 \cdot 4 \cdot 3 \cdot 2 \cdot 1} = 210.$$

Example: Compute $\binom{20}{18}$.

Choosing 18 from 20 is equivalent to choosing the 2 from the 20 that will not be used. Thus

$$\binom{20}{18} = \binom{20}{2} = \frac{20 \cdot 19}{2 \cdot 1} = 190.$$

2) Binomial Experiment

A binomial experiment consists of a fixed number n of repeated independent trials with only 2 possible outcomes at each trial, namely
success or failure. Note that the probability p of success on a trial
does not change from trial to trial.

Example: A coin is flipped 10 times. Let success on a trial (flip)
be a head. Then n = 10 and $p = \frac{1}{2}$.

Example: Thirty eight percent of the population smokes. Suppose
we interview a sample of 100 people as to whether or not they smoke.
Let success be a "smoker". Each person interviewed is a trial. Note
that the probability of success changes very slightly from trial
to trial since we would not interview the same person twice. This
change is so slight, however, that we may regard this as a binomial
experiment with n = 100 and p = .38.

3) Binomial Distribution

The probability distribution of the number of successes X in the n
trials of a binomial experiment is called a binomial distribution. A
variable X having this distribution is called a binomial (random)
variable.

Formula: $P(X = x) = \binom{n}{x} p^x q^{n-x}$ where
          n = number of trials
          p = probability of success on any trial
          q = 1 - p = probability of failure on any trial.

Example: In a binomial experiment with n = 3 trials, the probability of success is p = .4. Give the binomial distribution
both as a formula and as a table.

$$n = 3, \ p = .4, \ q = 1 - .4 = .6$$

$$P(X = x) = \binom{3}{x}(.4)^x(.6)^{3-x}$$

Since there are 3 trials, the possible values of X are 0, 1, 2,
and 3.

For x = 0

$$P(X = 0) = \binom{3}{0}(.4)^0(.6)^3 = 1 \cdot 1 \cdot (.6)^3 = .216$$

For x = 1

$$P(X = 1) = \binom{3}{1}(.4)^1(.6)^2 = 3(.4)(.6)^2 = .432$$

For x = 2

$$P(X = 2) = \binom{3}{2}(.4)^2(.6) = 3(.4)^2(.6) = .288$$

For x = 3

$$P(X = 3) = \binom{3}{3}(.4)^3(.6)^0 = 1 \cdot (.4)^3 \cdot 1 = .064$$

The distribution, as a table, follows:

x	P(X = x)
0	.216
1	.432
2	.288
3	.064

Example: A mellon grower estimates that 60% of his crop is ripe. Suppose a random sample of 5 mellons is selected. What is the probability that 2 or fewer of the mellons are ripe?

With n = 5, p = .6, q = 1 - .6 = .4, we need $p(X \le 2)$ where

$$P(X = x) = \binom{5}{x}(.6)^x(.4)^{5-x}$$

$$P(X = 0) = \binom{5}{0}(.6)^0(.4)^5 = .01024$$

$$P(X = 1) = \binom{5}{1}(.6)^1(.4)^4 = .0768$$

$$P(X = 2) = \binom{5}{2}(.6)^2(.4)^3 = .2304$$

$$P(X \le 2) = .01024 + .0768 + .2304 = .31744$$

4) Cumulative Binomial Probabilities

Cumulative sums of binomial probabilities are of the form $P(X \le k) = P(X = 0) + P(X = 1) + \cdots + \cdots + P(X = k)$. For ease of reference, these are tabulated in Table II of the text.

Example: Let $P(X = x) = (\frac{5}{x})(.6)^x(.4)^{5-x}$.

Find $P(X \leq 2)$. (See last example.)

We enter Table A2 in the subtable headed n = 5. At the intersection of the row k = 2 and the column headed p = .6 we find the entry .317.

Thus $P(X \leq 2) = .317$.

Example: Let $P(X = x) = (\frac{10}{x})(.3)^x(.7)^{10-x}$.

Find (a) $P(X \geq 4)$ (b) $P(X = 3)$

(a) Note $P(X \leq 3) + P(X \geq 4) = 1$. Thus

$$P(X \geq 4) = 1 - P(X \leq 3).$$

In Table II and subtable n = 10 we find, at the intersection of the rows k = 3 and p = .3, the entry $P(X \leq 3) = .650$ (rounded). Thus $P(X \geq 4) = 1 - .650 = .350$.

(b) Note $P(X \leq 3) - P(X \leq 2) = P(X = 3)$. From Table II, $P(X \leq 3 = .650$ and $P(X \leq 2) = .383$ (rounded). Thus $P(X = 3) = .650 = .383 = .267$.

Example: Suppose 30% of the students at a large university do not have a declared major. If 20 random student records are examined, what is the probability that 7 or more will show an undeclared major? We seek $P(X \geq 7)$ where n = 20 and p = .3.

$$P(X \geq 7) = 1 - P(X \leq 6) = 1 - .608 = .392 \quad \text{(rounded)}$$

Example: Let $P(X = x) = (\frac{15}{x})(.2)^x(.8)^{15-x}$.

Find k if $P(X \leq k) = .9389$.

Note we are now given a probability which is an entry in the interior of the table for n = 15. Start with the column headed p = .2 and look for this probability. It is found in the row k = 5.

5) The Mean, Variance, and Standard Deviation of the Binomial Distribution

Formula: $\mu_X = np$

$\sigma_X^2 = npq$

$\sigma_X = \sqrt{npq}$

Example: Find the mean, variance, and standard deviation of

$$P(X = x) = \binom{20}{x}(.4)^{x}(.6)^{20-x}$$

Here n = 20 and p = .4. Thus

$$\mu = np = 20(.4) = 8$$

$$\sigma^2 = npq = 20(.4)(.6) = 4.8$$

$$\sigma = \sqrt{npq} = \sqrt{4.8} = 2.19$$

Example: Forty percent of the voters favor a certain issue. If 900 voters are questioned, approximately how many favor the issue? Here n = 900, p = .40 and the expected value is

$$\mu = np = 900(.40) = 360.$$

Note this is just 40% of the "trials".

6) The Proportion $\hat{P}$

The proportion of successes in a binomial experiment is denoted by $\hat{P}$.

Formula: $\hat{P} = \dfrac{X}{n}$ where

X = number of successes

n = number of trials

Example: In 300 throws of a pair of dice, 75 sevens occurred. The proportion of sevens is

$$\hat{p} = \frac{75}{300} = \frac{1}{4} = .25 = 25\%.$$

Note that $\hat{P}$ is a variable. Lower case letters are used for its values.

Example: In a sample of 800 students, 300 had part-time jobs. The proportion of students having a part-time job is

$$\hat{p} = \frac{300}{800} = \frac{3}{8} = .375 = 37.5\%.$$

7) <u>The Mean, Variance, and Standard Deviation of $\hat{P}$</u>

    <u>Formulas</u>: $\mu_{\hat{P}} = p$

$$\sigma_{\hat{P}}^2 = \frac{pq}{n}$$

$$\sigma_{\hat{P}} = \sqrt{\frac{pq}{n}}$$

where

       n = number of trials

       p = probability of success on any trial

       q = 1 - p

    <u>Example</u>: Suppose 60% of all management personnel have a college degree. Let $\hat{P}$ = proportion of 300 managers who have a degree. Find the mean and standard deviation of $\hat{P}$. We regard the sampling experiment as a binomial experiment with n = 300 and p = .60. Then

$$\mu_{\hat{P}} = p = .60$$

$$\sigma_{\hat{P}} = \sqrt{\frac{pq}{n}} = \sqrt{\frac{(.6)(.4)}{300}} = .0283$$

8) <u>The Standard Normal Distribution</u>

The normal distribution is a family of similar distributions. That one having a mean of 0 and a standard deviation of 1 is called the standard normal distribution. The variable Z is traditionally associated with this distribution. Thus

$$\mu_Z = 0$$
$$\sigma_Z = 1$$

The graph of this distribution is shown below:

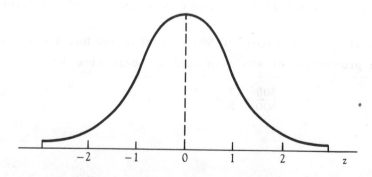

As with any continuous distribution, probabilities are determined by
areas under this curve. And the total area bounded by this curve and the
z axis 1. The distribution of Z is given in Table III (of the text).

Example: Find P($0 \leq Z \leq 1.56$).
In Table III in the left most
column, find z = 1.5. Now
move across this row until
the column 06 is reached.
The number .4406 found is
the desired probability. Thus
P($0 \leq Z \leq 1.56$) = .4406.

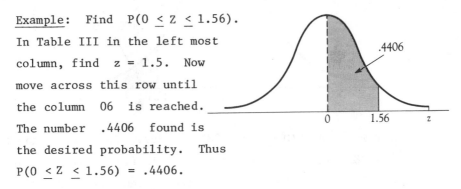

z	.00	.01	.02	.03	.04	.05	.06	.07	.08	.09
1.0	.3413	.3438	.3461	.3485	.3508	.3531	.3554	.3577	.3599	.3621
1.1	.3643	.3665	.3686	.3708	.3729	.3749	.3770	.3790	.3810	.3830
1.2	.3849	.3869	.3888	.3907	.3925	.3944	.3962	.3980	.3997	.4015
1.3	.4032	.4049	.4066	.4082	.4099	.4115	.4131	.4147	.4162	.4177
1.4	.4192	.4207	.4222	.4236	.4251	.4265	.4279	.4292	.4306	.4319
1.5	.4332	.4345	.4357	.4370	.4382	.4394	.4406	.4418	.4429	.4441
1.6	.4452	.4463	.4474	.4484	.4495	.4505	.4515	.4525	.4535	.4545
1.7	.4554	.4564	.4573	.4582	.4591	.4599	.4608	.4616	.4625	.4633
1.8	.4641	.4649	.4656	.4664	.4671	.4678	.4686	.4693	.4699	.4706
1.9	.4713	.4719	.4726	.4732	.4738	.4744	.4750	.4756	.4761	.4767

Note that the area under the right half of the normal density curve is
.5000

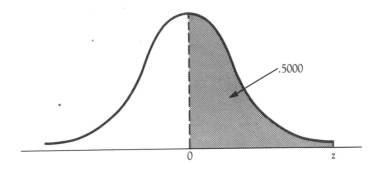

Example: Find P($Z \geq 1.28$).

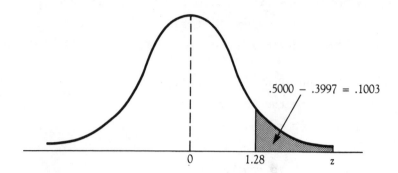

First we find $P(0 \leq Z \leq 1.28) = .3997$.   The area to the right of
$z = 1.28$ is found by subtracting this from   .5000.   Thus
$P(Z \geq 1.28) = .5000 = .3997 = .1003$.   Thus about 10% of the area
is to the right of $z = 1.28$.

Example: Find $P(-1.75 \leq Z \leq 0)$.

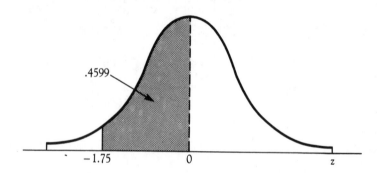

By symmetry this is the area under the curve between   0   and   1.75.

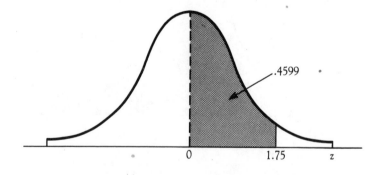

Thus $P(-1.75 \leq Z \leq 0) = P(0 \leq Z \leq 1.75) = .4599$.

Example: Find $P(-2.00 \leq Z \leq 2.00)$.

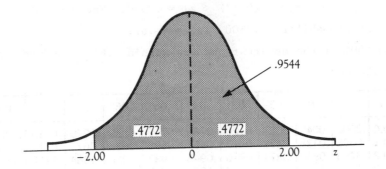

By symmetry this is twice the area between 0 and 2.00.   Now
$P(0 \leq Z \leq 2.00) = .4772.$ Thus $P(-2.00 \leq Z \leq 2.00) = 2(.4772) =$
$.9544 = 95.44\%.$

Note that since $\mu_Z = 0$ and $\sigma_Z = 1$, a z score is simply the number
of standard deviations from the mean.   Thus $P(-2.00 \leq Z \leq 2.00) = 95.44\%$
corresponds to the statement in the empirical rule that about 95% of the
observations of a bell shaped distribution are within 2 standard
deviations of the mean.
Similarly we find

$P(-1.00 \leq Z \leq 1.00) = .3413 + .3413 = .6826 = 68.26\%.$

Thus about 68% of the observations of a normal distribution are within 1
standard deviation of the mean.

Example: Within how many standard deviations of the mean is located
50% of the observations of a normal distribution?
We are now given the probability .5000 and seek the z score
such that $P(-z \leq Z \leq z) = .5000.$ See figure below.

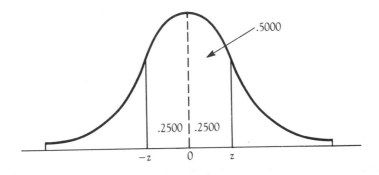

This we find from $P(0 \leq Z \leq z) = .2500.$ The nearest table entry
to the probability .2500 is z = .67.

Thus 50% of the observations are within .67 standard deviations
of the mean.

z	.00	.01	.02	.03	.04	.05	.06	.07	.08	.09
0.0	.0000	.0040	.0080	.0120	.0160	.0199	.0239	.0279	.0319	.0359
0.1	.0398	.0438	.0478	.0517	.0557	.0596	.0636	.0675	.0714	.0753
0.2	.0793	.0832	.0871	.0910	.0948	.0987	.1026	.1064	.1103	.1141
0.3	.1179	.1217	.1255	.1293	.1331	.1368	.1406	.1443	.1480	.1517
0.4	.1554	.1591	.1628	.1664	.1700	.1736	.1772	.1808	.1844	.1879
0.5	.1915	.1950	.1985	.2019	.2054	.2088	.2123	.2157	.2190	.2224
0.6	.2257	.2291	.2324	.2357	.2389	.2422	.2454	.2486	.2517	.2549
0.7	.2580	.2611	.2642	.2673	.2703	.2734	.2764	.2794	.2823	.2852
0.8	.2881	.2910	.2939	.2967	.2995	.3023	.3051	.3078	.3106	.3133
0.9	.3159	.3186	.3212	.3238	.3264	.3289	.3315	.3340	.3365	.3389

9) Percentage Point Notation

Often we must work with the area under the extreme right tail of the
standard normal distribution. If the area to the right of a z score
is α, we denote this score by z_α.

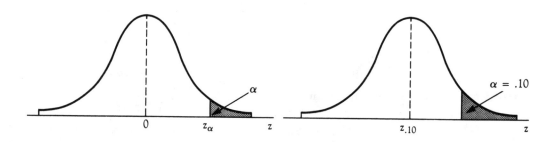

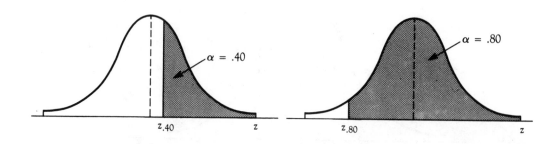

<u>Example</u>: Find $z_{.025}$.

Here $\alpha = .025$ is the given probability. We need that z such

that $P(Z \geq z) = .025$. See figure below.

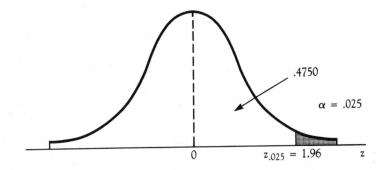

Then $P(0 \leq Z \leq z_{.025}) = .5000 - .025 = .475$. The table entry $.4750$

corresponds to $z = 1.96$.

Thus $z_{.025} = 1.96$.

<u>Example</u>: Find $z_{.18}$.

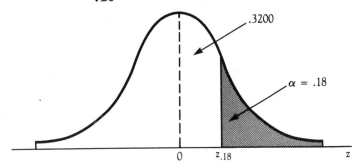

Then $P(0 \leq Z \leq z_{.18}) = .3200$. The nearest table entry is $.3212$.

This corresponds to $z = .92$. This we use as $z_{.18}$.

Thus $z_{.18} = .92$.

<u>Example</u>: Find $z_{.70}$.

Since $\alpha = .70$ is more than $.5000$, the percentage point $z_{.70}$ is

below $z = 0$. See figure below.

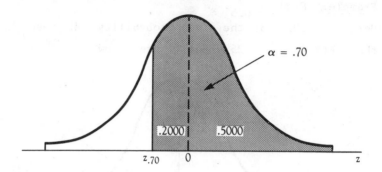

We find $P(0 \leq Z \leq .52) = .1985$ (nearest entry to   .2000).
Thus $z_{.70} = -.52$.

Example: How many standard deviations above the mean is located
the 85th percentile of a normal distribution.

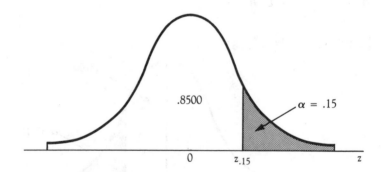

The answer is $z_{.15} = 1.04$ standard deviations (nearest entry to
.3500 is .3508).

10) <u>General Normal Distributions</u>

If X has a normal distribution with mean μ_X and standard deviation σ_X
we may obtain its probabilities from those of the standard normal.

<u>Formula:</u> $P(x_1 \leq X \leq x_2) = P(z_1 \leq Z \leq z_2)$ where

$$z_1 = \frac{x_1 - \mu_X}{\sigma_X} \quad \text{and} \quad z_2 = \frac{x_2 - \mu_X}{\sigma_X}$$

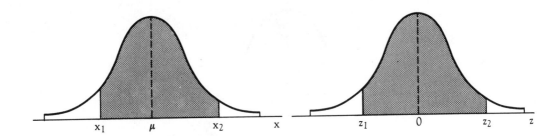

Example: Let X be normally distributed with mean $\mu_X = 70$ and standard deviation $\sigma_X = 12$. Find $P(70 \leq X \leq 86)$.

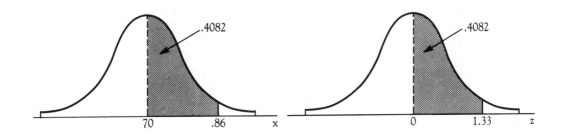

For x = 86,

$$z = \frac{86 - 70}{12} = 1.33.$$

For x = 70,

$$z = \frac{70 - 70}{12} = 0$$

(obvious since x = 70 is located at the mean).

$$P(70 \leq X \leq 86) = P(0 \leq Z \leq 1.33) = .4082.$$

Example: The account balances X of a federation of credit unions are normally distributed with a mean of $1480 and a standard deviation of $430. What percent of the accounts have balances exceeding $2000.00?

First we find the probability that a randomly selected account has a balance exceeding $2000, i.e., $P(X \geq 2000)$

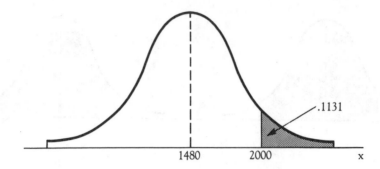

The corresponding z score is computed by

$$z = \frac{x - 1480}{430} = \frac{2000 - 1480}{430} = 1.21$$

$$P(X \geq 2000) = P(Z \geq 1.21) = .5000 - .3869 = .1131 = 11.31\%$$

We interpret this to mean that 11.31% of the balances exceed $2000.

Example: The amount of time students leave their autos in a parking space (at a university during class hours) has been found to be normally distributed with a mean of 204 minutes and a standard deviation of 46 minutes. Find the 90th percentile of this distribution.

The 90th percentile of the standard normal is $z_{.10} = 1.28$. Thus the 90th percentile is 1.28 standard deviations above the mean at

$$x = 204 + (1.28)(46) = 262.9 \text{ minutes.}$$

This could also have been found by solving

$$z = 1.28 = \frac{x - 204}{46} \ .$$

11) Approximating the Binomial Distribution

The normal distribution may be used to approximate the binomial provided both np and nq exceed 5.

Formula: $z = \dfrac{x - np}{\sqrt{npq}}$ where

n = number of trials

p = probability of success on any trial

q = 1 - p

x is the number of successes adjusted by $\pm \frac{1}{2}$ as required.

Example: Find P(X = 20) if n = 60 and p = .3 Both np = 18
and nq = 42 exceed 5. The rectangle of the histogram for x = 20
extends from 19.5 to 20.5. See diagram.

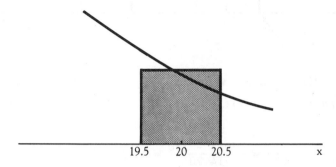

Consequently we need   P(19.5 $\leq$ x $\leq$ 20.5)   where   X   is normally
distributed with

μ = np = (60)(.3) = 18, $\sigma = \sqrt{npq} = \sqrt{60(.3)(.7)}$ = 3.55.

The z scores are obtained from $z = \frac{x - 18}{3.55}$.
For x = 19.5,
$$z = \frac{19.5 - 18}{3.55} = .42.$$

For x = 20.5,
$$z = \frac{20.5 - 18}{3.55} = .70$$

P(X = 20) $\approx$ P(.42 < Z < .70) = .2580 - .1628 = .0952.

Example: Suppose that 40% of all autos are uninsured. If a random
sample of n = 900 autos is examined, what is the probability that
380 or more autos will be uninsured?
We need P(X $\geq$ 380) when p = .40 and n = 900. The interval
for x = 380 of the probability histogram has its left endpoint at
379.5.

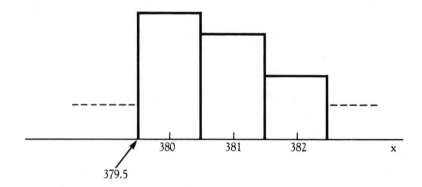

We use $P(X \geq 379.5)$ where $\mu = np = (900)(.4) = 360$,
$\sigma = \sqrt{npq} = \sqrt{(900)(.4)(.6)} = 14.70$, and X is normally distributed.
For $x = 379.5$,

$$z = \frac{379.5 - 360}{14.70} = 1.33.$$

Thus $P(X \geq 380) \approx P(Z \geq 1.33) = .5000 - .4082 = .0918$.

12) <u>Approximating the Distribution of $\hat{P}$ with the Normal</u>

<u>Formula</u>: $z = \dfrac{\hat{P} - p}{\sqrt{\dfrac{pq}{n}}}$ where

   n = number of trials
   p = probability of success on any one trial
   $q = 1 - p$

<u>Example</u>: Suppose that 70% of the students at a large university
have never attended another university. If a random sample of
$n = 500$ are surveyed, what is the probability that between 65%
and 75% will not have attended another university?
Here $n = 500$, $p = .70$ and we need $P(.65 < \hat{p} < .75)$.
The z scores are computed from

$$z = \frac{\hat{p} - .7}{\sqrt{\dfrac{(.7)(.3)}{500}}} = \frac{\hat{p} - .70}{.020}$$

For $\hat{p} = .65$,

$$z = \frac{.65 - .70}{.020} = -2.5.$$

For $\hat{p} = .75$,

$$z = \frac{.75 - .70}{.020} = +2.5.$$

Thus $P(.65 < \hat{p} < .75) = P(-2.5 < Z < 2.5) = .4938 + .4938 = .9876.$

13) The Chi-Square Distribution

The sum of n independent normally distributed random variables has the chi-square (χ^2) distribution with $\nu = n$ degrees of freedom. The distribution is shown below when $\nu = 4$ d.f.

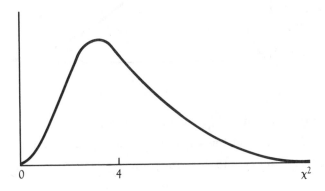

Example: Find $P(0 \leq \chi^2 \leq 13.8)$ when $\nu = 10.$

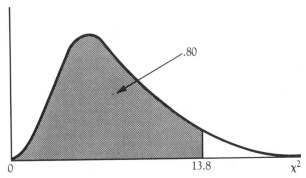

Use Table IV. The nearest entry to 13.4 in the column $\nu = 10$ is 13.4. The row heading is $.80$. Thus

$P(0 \leq \chi^2 \leq 13.8) \approx .80.$

Note that $\chi^2 \geq 0$ so that there is no ambiguity in writing

$P(\chi^2 \leq 13.8) \approx .80.$

14)   <u>Percentage Points (Critical Values) of the χ^2 Distribution</u>

The percentage points χ^2_α are given in Table V. As with the normal distribution

$$P(\chi^2 \geq \chi^2_\alpha) = \alpha$$

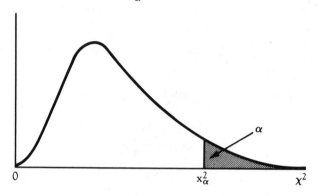

<u>Example</u>: Find $\chi^2_{.05}$ when $\nu = 8$.

At the intersection of the column $\chi^2_{.05}$ and the row $\nu = 8$ we find the entry 15.5. Thus $\chi^2_{.05} = 15.5$.

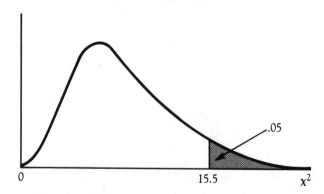

df	0.995	0.990	0.975	0.950	0.900	0.100	0.050	0.025	0.010	0.005
1	0.0000393	0.000157	0.000982	0.00393	0.0158	2.71	3.84	5.02	6.64	7.88
2	0.0100	0.0201	0.0506	0.103	0.211	4.61	6.00	7.38	9.21	10.6
3	0.0717	0.115	0.216	0.352	0.584	6.25	7.82	9.35	11.4	12.9
4	0.207	0.297	0.484	0.711	1.0636	7.78	9.50	11.1	13.3	14.9
5	0.412	0.554	0.831	1.15	1.61	9.24	11.1	12.8	15.1	16.8
6	0.676	0.872	1.24	1.64	2.20	10.6	12.6	14.5	16.8	18.6
7	0.990	1.24	1.69	2.17	2.83	12.0	14.1	16.0	18.5	20.3
8	1.34	1.65	2.18	2.73	3.49	13.4	15.5	17.5	20.1	22.0
9	1.73	2.09	2.70	3.33	4.17	14.7	17.0	19.0	21.7	23.6
10	2.16	2.56	3.25	3.94	4.87	16.0	18.3	20.5	23.2	25.2
11	2.60	3.05	3.82	4.58	5.58	17.2	19.7	21.9	24.7	26.8
12	3.07	3.57	4.40	5.23	6.30	18.6	21.0	23.3	26.2	28.3
13	3.57	4.11	5.01	5.90	7.04	19.8	22.4	24.7	27.7	29.8
14	4.07	4.66	5.63	6.57	7.79	21.1	23.7	26.1	29.1	31.3
15	4.60	5.23	6.26	7.26	8.55	22.3	25.0	27.5	30.6	32.8

Note that $\chi^2_{.05} = 15.5$ in the above example would be considered a large value of χ^2. By the same token $\chi^2_{.95}$ would be considered a small value.

Example: Find $\chi^2_{.95}$ when $\nu = 8$.
$\chi^2_{.95} = 2.73$ is found at the intersection of the column $\chi^2_{.95}$ and the row $\nu = 8$.

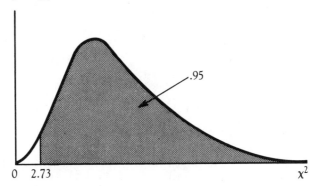

Note that the mean of the Chi-Square Distribution is $\mu = \nu$ and this is located to the right of the "center" of the bell.

15) The Goodness of Fit Test

Procedure. Given observed frequencies $0_1, 0_2, \ldots, 0_n$ and expected frequencies $e_1, e_2, \ldots, e_n$ for n categories, compute

$$\chi^2 = \sum_{i=1}^{n} \frac{(0_i - e_i)^2}{e_i}$$

and compare with χ^2_{α} (α-small) obtained using $\nu = n - 1$ degrees of freedom. If $\chi^2 > \chi^2_{\alpha}$, there is poor agreement between the proposed and observed distributions. We then reject the proposed distribution for the situation at hand.

The selected α (generally .05 or .01) is called the significance level of the test.

The number χ^2_{α} is called the critical value of the test.

Example: We believe that voter registration among students at a university is as follows:

Category	Percent
Democrats	35%
Republicans	20%
Independent	5%
Not Registered	40%

Test that this is the case if a sample of 600 students contained 200 democrats, 130 republicans, 42 independents, and 228 non registered students.

The observed frequencies are

$$O_1 = 200 \text{ (Democrats)}$$
$$O_2 = 130 \text{ (Republicans)}$$
$$O_3 = 42 \text{ (Independent)}$$
$$O_4 = 228 \text{ (Non-Registered)}$$

The expected registrations are

$$e_1 = (.35)(600) = 210$$
$$e_2 = (.20)(600) = 120$$
$$e_3 = (.05)(600) = 30$$
$$e_4 = (.40)(600) = 240$$

This is all summarized in the following table

i	1	2	3	4
O_i	200	130	42	228
e_i	210	120	30	240

First choose the significance level say, $\alpha = .05$. Next find the critical value $\chi^2_{.05} = 7.82$ using Table V with $\nu = 4 - 1 = 3$.
Next we compute the chi-square value:

$$\chi^2 = \frac{(200 - 210)^2}{210} + \frac{(130 - 120)^2}{120} + \frac{(42 - 30)^2}{30} + \frac{(228 - 240)^2}{240}$$

$$= \frac{100}{210} + \frac{100}{120} + \frac{144}{30} + \frac{144}{240} = 6.71$$

Since $x^2 = 6.71$ does not exceed $x^2_{.05} = 7.82$ we do not reject the proposed distribution of voter registrations at the university.

Note. In the x^2 goodness of fit test, the number of degrees of freedom is 1 less than the number of categories.

Note. In order to be able to use the x^2 test, the expected value of each category must be at least 5.

Note. Sometimes we must estimate the expected values from the data (as in the next example).

Example: We have the following data from a survey on accident rates for 4 age groups.

Age	Number Surveyed	Number of Accidents in Last 24 Months
20-30	200	32
30-40	300	37
40-50	400	44

Test the hypothesis that there is no difference in the accident rates for these 3 age groups.

We need an estimate of the common accident rate. This is

$$p = \frac{32 + 37 + 44}{200 + 300 + 400} = \frac{113}{900} = .1256.$$

We will test that this is the rate for each of the age groups. The expected values are

$$e_1 = (.1256)(200) = 25.12$$
$$e_2 = (.1256)(300) = 37.68$$
$$e_3 = (.1256)(400) = 50.24$$

i	1	2	3
0_i	32	37	44
e_i	25.12	37.68	50.24

Choosing $\alpha = .01$ (to be different), we find $x^2_{.01} = 9.21$ when $\nu = 2$.

$$\chi^2 = \frac{(32 - 25.12)^2}{25.12} + \frac{(37 - 37.68)^2}{37.68} + \frac{(44 - 50.24)^2}{50.24} = 2.67$$

Since $\chi^2 = 2.67 < \chi^2_{.01}$ we do not reject the hypothesis that the 3 groups have the same accident rate.

Calculator Usage Illustration

Compute the power of a number

Find $.6^4$.

Method 1) Use the y^x key

$$.6 \; \boxed{y^x} \; 4 \; \boxed{=} \; \underline{.1295}$$

Method 2) Repeated factors

$$.6 \; \boxed{\times} \; .6 \; \boxed{\times} \; .6 \; \boxed{\times} \; .6 \; \boxed{=} \; \underline{.1295}$$

   Note: This method should be used only if your calculator does not have

   the y^x key or its equivalent.

Calculator Usage Illustration

Compute the value of a binomial coefficient $\binom{n}{k}$.

Find $\binom{12}{4}$.

Method 1) Arrange the computation as $\dfrac{12}{4} \cdot \dfrac{11}{3} \cdot \dfrac{10}{2} \cdot 9$

$$12 \; \boxed{\div} \; 4 \; \boxed{\times} \; 11 \; \boxed{\div} \; 3 \; \boxed{\times} \; 10 \; \boxed{\div} \; 2 \; \boxed{\times} \; 9 \; \boxed{=} \; = \underline{495}$$

Method 2) Arrange the computation as $(12 \cdot 11 \cdot 10 \cdot 9) \div (4 \cdot 3 \cdot 2)$

$$12 \; \boxed{\times} \; 11 \; \boxed{\times} \; 10 \; \boxed{\times} \; 9 \; \boxed{=} \; \boxed{\div} \; \boxed{(} \; 4 \; \boxed{\times} \; 3 \; \boxed{\times} \; 2 \; \boxed{)} \; = \underline{495}$$

   Note: If n and k are large this second method produces large numbers

   and a loss of significant digits may occur.

Calculator Usage Illustration

Calculate a probability using the binomial distribution.
Find $\binom{10}{3}(.4)^3(.6)^7$.

Method 1)   Use the memory to store intermediate products

10 $\div$ 3 $\times$ 9 $\div$ 2 $\times$ 8 $=$ [STO] .4 $\boxed{y^x}$ 3 $\times$ [RCL]

$=$ [STO] .6 $\boxed{y^x}$ 7 $\times$ [RCL] $=$ <u>.2149908</u>

Method 2)   Direct evaluation

10 $\div$ 3 $\times$ 9 $\div$ 2 $\times$ 8 $=$ $\times$ .4 $\boxed{y^x}$ 3 $=$ $\times$ .6 $\boxed{y^x}$ 7 $=$ <u>.2149908</u>

Calculator Usage Illustration

Compute the standard deviation of the binomial distribution.
Find $\sigma = \sqrt{(530)(.28)(.62)}$.

Method: Direct evaluation

520 $\times$ .28 $\times$ .62 $=$ $\boxed{\sqrt{}}$ <u>9.59</u>

Calculator Usage Illustration

Compute the standard deviation of the distribution of $\hat{P}$.

Find $\sigma = \sqrt{\dfrac{(.43)(.57)}{650}}$

Method: Direct evaluation

.43 $\times$ .57 $\div$ 650 $=$ $\boxed{\sqrt{}}$ <u>.0194185</u>

Calculator Usage Illustration

Compute the z score corresponding to a proportion $\hat{p}$.

Find $z = \dfrac{.37 - .28}{\sqrt{\dfrac{(.43)(.57)}{650}}}$.

Method 1) Make use of the memory to hold the standard deviation

.43 $\boxed{\times}$ .57 $\boxed{\div}$ 650 $\boxed{=}$ $\boxed{\sqrt{}}$ $\boxed{\text{STO}}$ .37 $\boxed{-}$ .43 $\boxed{=}$ $\boxed{\div}$ $\boxed{\text{RCL}}$ $\boxed{=}$ <u>−3.0898420</u>

Method 2) Multiply the numerator by the reciprocal of the standard deviation

.43 $\boxed{\times}$ .57 $\boxed{\div}$ 650 $\boxed{=}$ $\boxed{\sqrt{}}$ $\boxed{\frac{1}{x}}$ $\boxed{\times}$ $\boxed{(}$ .37 $\boxed{-}$ .43 $\boxed{)}$ $\boxed{=}$ <u>−3.0898420</u>

Method 3) Direct evaluation

.37 $\boxed{-}$ .43 $\boxed{=}$ $\boxed{\div}$ $\boxed{(}$ .43 $\boxed{\times}$ .57 $\boxed{\div}$ 650 $\boxed{)}$ $\boxed{\sqrt{}}$ $\boxed{=}$ <u>−3.0898420</u>

Calculator Usage Illustration

Compute the χ^2 value for a goodness of fit test.

Find $\chi^2 = \dfrac{(12 - 7.2)^2}{7.2} + \dfrac{(10 - 6.3)^2}{6.3} + \dfrac{(9 - 6.5)^2}{6.5}$.

Method 1) Accumulate the separate terms in the memory

12 $\boxed{-}$ 7.2 $\boxed{=}$ $\boxed{x^2}$ $\boxed{\div}$ 7.2 $\boxed{=}$ $\boxed{\text{STO}}$ 10 $\boxed{-}$ 6.3 $\boxed{=}$ $\boxed{x^2}$ $\boxed{\div}$ 6.3

$\boxed{=}$ $\boxed{\text{SUM}}$ 9 $\boxed{-}$ 6.5 $\boxed{=}$ $\boxed{x^2}$ $\boxed{\div}$ 6.5 $\boxed{=}$ $\boxed{+}$ $\boxed{\text{RCL}}$ $\boxed{=}$ <u>6.3345543</u>

Method 2) Direct summation of individual terms

$\boxed{(}$ 12 − 7.2 $\boxed{)}$ $\boxed{x^2}$ $\boxed{\div}$ 7.2 $\boxed{+}$ $\boxed{(}$ 10 $\boxed{-}$ 6.3 $\boxed{)}$ $\boxed{x^2}$ $\boxed{\div}$ 6.3

$\boxed{+}$ $\boxed{(}$ 9 $\boxed{-}$ 6.5 $\boxed{)}$ $\boxed{x^2}$ $\boxed{\div}$ 6.5 $\boxed{=}$ <u>6.3345543</u>

Exercise Set VI.1

1) a) Not a binomial experiment and cannot be approximated as such.
   Reason: The probability of getting a senior varies considerably from
   selection (trial) to selection.

   b) Yes. The probability of getting an answer should be about the same
   on each call. The results on any one trial does not influence the others.

   c) Yes. Replacement is made after each drawing (trial) so one trial does
   not influence the others and the probability of success (your name drawn)
   remains the same on each trial.

   d) May be approximated as such. While it is true we would not question the
   same person twice and thus the probability of success (registered voter)
   changes from trial to trial, the change will be very slight with a large
   population.

3) $P(X = 0) = \binom{4}{0}(.4)^0(.6)^{4-0}$

$\qquad = 1 \cdot 1 \cdot (.6)^4$

$\qquad = .1296$

$P(X = 1) = \binom{4}{1}(.4)^1(.6)^{4-1}$

$\qquad = 4(.4)(.6)^3$

$\qquad = .3456$

$P(X = 2) = \binom{4}{2}(.4)^2(.6)^{4-2}$

$\qquad = 6(.4)^2(.6)^2$

$\qquad = .3456$

$P(X = 3) = \binom{4}{3}(.4)^3(.6)^{4-3}$

$\qquad = 4(.4)^3(.6)$

$\qquad = .1536$

$P(X = 4) = \binom{4}{4}(.4)^4(.6)^{4-4}$

$\qquad = 1(.4)^4 \cdot 1$

$\qquad = .0256$

x	P(x)
0	.1296
1	.3456
2	.3456
3	.1536
4	.0256

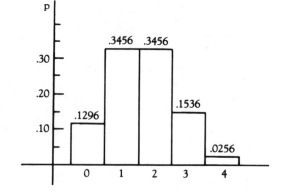

5) $P(X = x) = \binom{5}{x}(.3)^x(.7)^{5-x}$

Each stranger is a trial, n = 5, success is eye contact.

The probability of success is p = .3.

a) $P(X = 3) = \binom{5}{3}(.3)^3(.7)^{5-3}$

$\qquad = 10(.3)^3(.7)^2$

$\qquad = .1323$

b) $P(X \leq 2) = P(X = 2) + P(X = 1) + P(X = 0)$

(since "2 or fewer" occur if the number of eye contacts is "exactly 2"

or "exactly 1" or "exactly 0")

$P(X = 2 = \binom{5}{2}(.3)^2(.7)^{5-2}$

$= 10(.3)^2(.7)^3$

$= .3087$

$P(X = 1) = \binom{5}{1}(.3)^1(.7)^{5-1}$

$= 5(.3)(.7)^4$

$= .36015$

$P(X = 0) = \binom{5}{0}(.3)^0(.7)^{5-0}$

$= 1 \cdot 1 \cdot (.7)^5$

$= .16807$

$P(X \leq 2) = .3087 + .36015 + .16807$

$= .83692$

c)   At least 1 is the complement of $x = 0$ (none)

$P(X \geq 1) = 1 - P(X = 0) = 1 - .16807$ (from (b))

$= .83193$

7)   a) $\mu = np = 100(.7) = 70$ (70% of 100 trials is 70)

$\sigma = \sqrt{npq} = \sqrt{100(.7)(.3)} = \sqrt{21} = 4.58$

b) $\mu = np = 60(\frac{1}{2}) = 30$

$\sigma = \sqrt{npq} = \sqrt{60(\frac{1}{2})(\frac{1}{2})} = \sqrt{15} = 3.87$

c) $\mu = np = 400(.25) = 100$

$\sigma = \sqrt{npq} = \sqrt{400(.25)(.75)} = \sqrt{75} = 8.66$

9)   n = 400 trials, $p = \dfrac{900}{30000} = .03$

$\mu = np = 400 \times (.03) = 12$

If the needles are packaged as produced, then we need the defective needles
to have occurred randomly in the selection process. If the needles are
selected from a completed production run, the selection process is random.

11)   Use Table II with n = 15 and p = .4.

a) $P(X \le 6) = .6098$

b) $P(X \ge 3) = 1 - P(X \le 2)$

$$= 1 - .0271$$

$$= .9729$$

Note: $P(X \le 2)$ and $P(X \ge 3)$ are the probabilities of complementary events.

c) $P(4 \le X \le 7) = P(X \le 7) - P(X < 4)$

$$= P(X \le 7) - P(X \le 3)$$

$.2173 \le Y$ $= .7869 - .0905$

≤ 7869 $= .6964$

13)   n = 15

p = .40

Success corresponds to a busy line.

$P(X \ge 4) = 1 - P(X \le 3)$ (complementary events)

$$= 1 - .0905 \qquad \text{(cumulative table, n = 15, p = .40)}$$

$$= .9095$$

15)   p = .8, success corresponds to a germinating seed

n = 25

x = 16

By poor we interpret 16 or fewer.

$P(X \le 16) = .0468 = 4.68\%$

Only about 4.7% of all samples of size n = 25 should show such poor germination.

17)   Success corresponds to a defective item

n = 15   trials

a)   If p = .05, what is $P(X \ge 3)$?

$P(X \ge 3) = 1 - P(X \le 2)$

$$= 1 - .9638$$

$$= .0362 = 3.62\%$$

Thus about 4.6% get rejected. Remember, if p = .05, the expected number of defective is = .05 × 15 = .75 < 1.

b)   If p = .10, what is $P(X \ge 3)$?

$P(X \ge 3) = 1 - P(X \le 2)$

$$= 1 - .8159$$

$$= .1841 = 18.41\%$$

c) If p = .20, what is $P(X \geq 3)$?

$P(X \geq 3) = 1 - P(X \leq 2)$

$= 1 - .3980$

$= .6020$

This particular lot acceptance plan is not very effective in detecting a lot with even 20% defective. Even then it will fail about 40% of the time.

19) 16 steps forward, 9 backward, difference = 16 - 9 = 7 (forward). Let success correspond to a forward step. The probability of success is $p = \frac{1}{2}$ if he cannot tell forward from backward.

$P(X \geq 16) = 1 - P(X \leq 15)$

$= 1 - .8852$ (using n = 25; p = .5)

$= .1148$

$= 11.48\%$

If he cannot tell forward from backward, there is only about a 11.5% chance of him being 7 or more steps forward of where he started. This would seem to indicate that he has some sense of direction.

21) The probability of 3 red cards in 5 draws is

$p = (\begin{smallmatrix}5\\3\end{smallmatrix})(\frac{1}{2})^3(\frac{1}{2})^2 = \frac{10}{32} = \frac{5}{16}$

The probability of repeating this (success) in 2 of 3 trials is

$P(X = 2) = (\begin{smallmatrix}3\\2\end{smallmatrix})(\frac{5}{16})^2(\frac{3}{16})^1 = .2014$

23) a) At least as many trials as successes are needed. Thus if there are k successes, we will need k or more trials, i.e. $X \geq k$.

b) The probability p of a 7 is $p = \frac{6}{36} = \frac{1}{6}$.

k = 2 sevens are needed. The probability that x trials will be needed is

$P(X = x) = (\begin{smallmatrix}x-1\\2-1\end{smallmatrix})(\frac{1}{6})^2(\frac{5}{6})^{x-2}$,   x = 2,3,4,...

For x = 2 trials to produce 2 sevens

$P(X = 2) = (\begin{smallmatrix}2-1\\2-1\end{smallmatrix})(\frac{1}{6})^2(\frac{5}{6})^0 = \frac{1}{36}$.

For x = 3 trials to produce 2 sevens

$P(X = 3) = (\begin{smallmatrix}3-1\\2-1\end{smallmatrix})(\frac{1}{6})^2(\frac{5}{6})^{3-2}$

$= (\begin{smallmatrix}2\\1\end{smallmatrix})(\frac{1}{6})^2(\frac{5}{6}) = \frac{5}{108}$.

25) The population size is N = 10.

There are n = 3 trials.

Success corresponds to a brown rat.

k = 6

$$P(X = x) = \frac{\binom{6}{x}\binom{10-6}{3-x}}{\binom{10}{3}}$$

$$P(X = 3) = \frac{\binom{6}{3}\binom{4}{3-3}}{\binom{10}{3}} = \frac{\binom{6}{3}\binom{4}{0}}{\binom{10}{3}} = \frac{\left(\frac{6 \cdot 5 \cdot 4}{3 \cdot 2 \cdot 1}\right) \cdot 1}{\frac{10 \cdot 9 \cdot 8}{3 \cdot 2 \cdot 1}} = \frac{1}{6}$$

Exercise Set VI.2

1) a) The table entry corresponding to z = 1.24 is .3925.

    b) desired area = P(0 $\leq$ Z $\leq$ 2.0) - P(0 $\leq$ z $\leq$ 1.0)

                     = .4772 - .3413 = .1359

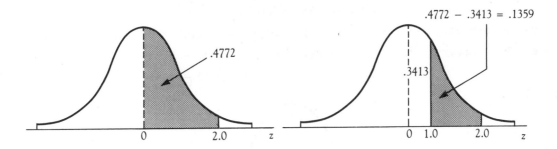

    c) by symmetry

        desired area = 2P(0 $\leq$ Z $\leq$ 1.40)

                     = 2(.4192) = .8384

    d) desired area = .5000 - .4904 = .0096

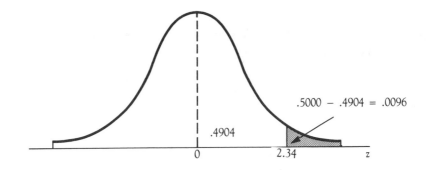

    e) desired area = P(Z $\leq$ 1.50) = P(Z $\leq$ 0) + P(0 $\leq$ Z $\leq$ 1.50)

                     = .5000 + .4332 = .9332

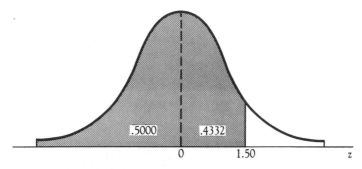

f) by symmetry, desired area = $P(Z \geq 1.68) = .5000 - .4535 = .0465$

3) a) $P(0 \leq -Z \leq 1.43) = .4236$ (table entry)

    b) $P(Z \geq 2.2) = .5000 - P(0 \leq Z < 2.2)$
        $= .5000 - .4861 = .0139$

    c) $P(.7 \leq Z \leq 1.4) = P(0 \leq Z \leq 1.4) - P(0 \leq Z <$
        $= .4192 - .2580 = .1612$

    d) by symmetry $P(-1.6 < Z < 1.6) = 2P(0 \leq Z < 1.6) = 2(.4452) = .8904$

    e) by symmetry $P(Z < -2.4) = P(Z > 2.4) = .5000 - P(Z \leq 2.4)$
        $= .5000 - .4918 = .0082$

5) a)

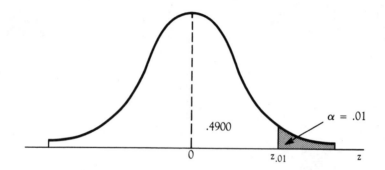

$z_{.01}$ is found using the table entry $.4900 = .5000 - .01$

$z_{.01} = 2.33$ (nearest entry)

    b)

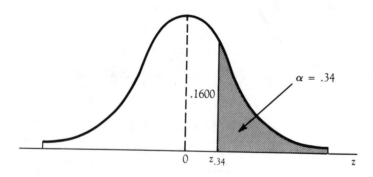

We use the table entry $.5000 - .34 = .1600$

$z_{.34} = .41$ (nearest entry)

c)

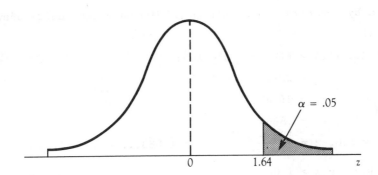

$z_{.05} = 1.64$ (nearest entry to   .4500)

d)

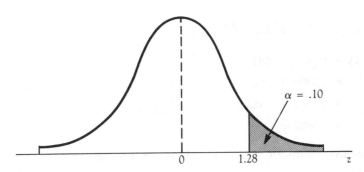

$z_{.10} = 1.28$

e)

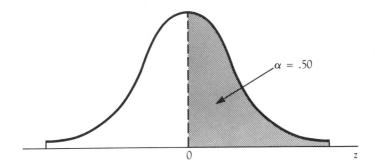

$z_{.50} = 0$

7) $z = \dfrac{x - \mu}{\sigma} = \dfrac{x - 123}{12}$

a)   When $x = 141$, $z = \dfrac{141 - 123}{12} = \dfrac{18}{12} = 1.5.$

When $x = 105$, $z = \dfrac{105 - 123}{12} = \dfrac{-18}{12} = -1.5.$ This could also have been obtained by symmetry since 141 and 105 are 16 units above and below the mean.

$$P(105 \leq X \leq 141) = P(-1.5 \leq z \leq 1.5)$$
$$= 2P(0 \leq z \leq 1.5)$$
$$= 2(.4332)$$
$$= .8664$$

b) When $x = 136$, $z = \dfrac{136 - 123}{12} = \dfrac{13}{12} = 1.083\ldots = 1.08$ (rounded)

$$P(X > 136) = P(z > 1.08)$$
$$= .5000 - P(z < 1.08)$$
$$= .5000 - .3599$$
$$= .1401$$

c) When $x = 148$, $z = \dfrac{148 - 123}{12} = 2.08$

$$P(X \leq 148) = P(z \leq 2.08)$$
$$= .5000 + P(0 \leq z \leq 2.08)$$
$$= .5000 + .4812$$
$$= .9812$$

d) In the normal table we find $z = 1.35$ corresponds to an area or probability of .4115.

$$1.35 = z = \frac{(123 + k) - 123}{12} = \frac{k}{12}$$

$1.35 = \dfrac{k}{12}$ yields $k = 12 \cdot 1.35 = 16.2$.

Thus $k = 16.2$ and $P(106.8 < X \leq 139.2) = .8230$.

e) To the area .4750 corresponds $z = 1.96$. Thus x_1 is 1.96 S.D. above the mean, i.e., $x_1 = 123 + (1.96)(12) = 146.52$.

9) Let X = Daily production

$\mu = 10000$ and $\sigma = 1200$

Find $P(X < 8000)$

$z = \dfrac{x - 10000}{1200}$

When $x = 8000$

$z = \dfrac{8000 - 10000}{1200}$

$= \dfrac{-2000}{1200}$

$= -1.67$

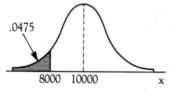

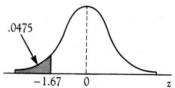

$$P(X < 8000) = P(z < -1.67)$$
$$= P(z > 1.67)$$
$$= .5000 - .4525$$
$$= .0475$$

11) Let X = operating life of a battery

$\mu = 110$ and $\sigma = 14$

$$z = \frac{x - 110}{14}$$

$P(X < 80)$ will give the percentage which fail before 80 hours.

When $x = 80$, $z = \dfrac{80 - 110}{14} = \dfrac{-30}{14}$

$$= -2.14$$

$$P(X < 80) = P(z < -2.14)$$
$$= .01621$$
$$= 1.62\%$$

Thus about 1.62% will fail within the warranty period.

13) Let X = length of time to complete the exam.

$\mu = 86$ and $\sigma = 14$

We seek the time x such that $P(X < x_1) = .90$, i.e. the probability is
.90 that an examinee completes the test in less than x_1 minutes.

Let $z_1 = \dfrac{x_1 - 86}{14}$.

Then $P(Z < z_1) = .90$ and $P(0 \leq z < z_1) = .40$.

Corresponding to the area .4000 we find $z_1 = 1.28$ (approximately).

Solve:

$$1.28 = \frac{x_1 - 86}{14}$$

$$x_1 = 86 + 1.28 \times 14 = 103.92$$

104 minutes would probably be used.

15) Note that some of the oversized rods by machine A qualify as acceptable by
B and some of B's undersized rods are acceptable by the dimensions established
for A.

Machine A

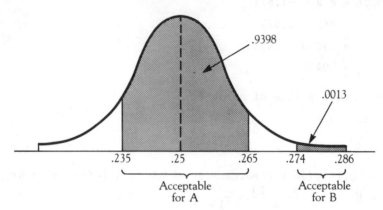

We need   P(.235 ≤ X ≤ .265) + P(.274 ≤ X ≤ .286)

$$z = \frac{x - .25}{.008}$$

When x = .265, z = 1.88 and when x = .235, z = -1.88 and

P(.235 ≤ X ≤ .265) = P(-1.88 ≤ z ≤ 1.88) = 2(.4699) = .9398.

When x = .286, z = 4.5 and when x = .274, z = 3.   Thus

P(.274 ≤ X ≤ .286) = P(3 ≤ z < 4.5) = .0013.

Thus the probability of an acceptable rod from A is

p = .9398 + .0013 = .9411.

Machine B

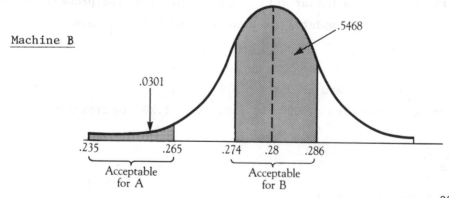

P(.274 < X < .286) = P(.235 < x < .265) is needed. $z = \frac{x - .28}{.008}$

P(.274 < X < .286) = P(-.75 < Z < .75) = 2(.2734) = .5468

P(.235 < X < .265) = P(-5.63 < Z < -1.88) ≈ P(Z < -1.88) = .5000 - .4699

                                                                          = .0301

The probability of an acceptable rod from B is   .5468 + .0301 = .5769

Suppose each produce 10000 rods.

Then from A we get 9411 rods   (.9411 = 94.11%).

And from B we get 5769 rods   (.5769 = 57.69%).

Thus of the 20000 rods we would get = 15180 acceptable rods.

This is $\frac{15180}{20000} \times 100 = 75.90\%$.

Or, more simply, since both machines produce the same number,

$p = \frac{.9411 + .5769}{2} = .7590 = 75.90\%$. Thus 24.10% are unacceptable.

17) Let X represent age.

We are given $P(X \geq 35) = .62$.

Now $P(z \geq .31) = .62$ (approximately)

Also $z = -.31 = \frac{35 - \mu}{10}$.

Thus $\mu = 35 + 10(.31)$

$= 38.1$.

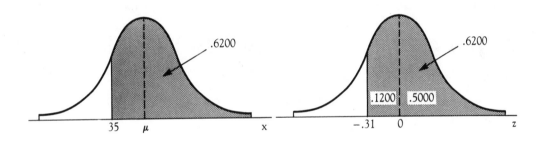

19) Let X represent the service life of a randomly selected thermocouple.

The problem is to find x such that $P(X < x) = .03$.

Now $z = \frac{x - 35}{4}$ and $P(z < -1.88) = .03$.

Thus $-1.88 = \frac{x - 35}{4}$ and $x = 35 - 4(1.88) = 27.48$.

About every 27 1/2 months the thermocouples should be replaced.

21) We need $P(X \geq 300)$.

$z = \frac{x - 266}{16}$

When $x = 300$, $z = \frac{300 - 266}{16} = 2.13$.

$P(X \geq 300) = P(Z \geq 2.13) = .5000 - .4834 = .0166 = 1.66\%$

23) Using the hint

$$z = \frac{x - 125}{20} = \frac{y - 500}{100}$$

Solving for y,

$$y - 500 = 100 \left(\frac{x - 125}{20}\right) = 5(x - 125) = 5x - 625$$

$$y = 5x - 625 + 500 = 5x - 125$$

Thus $y = 5x - 125.$

25) The specification limits are $2200 + 150 = 2350$ and $2200 - 150 = 2050$.
Let   X   represent the resistance. A unit fails if $X > 2350$ or $X < 2050$.
Thus $P(X < 2050) + P(X > 2350)$ is needed.

Using $z = \frac{x - 2200}{80}$ we find $z = \pm 1.88$ correspond to these limits.   Further
$P(Z < -1.88) = P(Z > 1.88) = .0301.$ Thus
$P(X < 2050) + P(X > 2350) = .0301 + .0301 = .0602 = 6.02\%.$

27) Let   X   be the concentration of the pesticide.
$\mu_X = 3.2$ and $\sigma_X = 1.4$ ten days following the application.
We seek $P(X < 6.0)$

$$z = \frac{x - 3.2}{1.4} .$$

When $x = 6.0$

$$z = \frac{6.0 - 3.2}{1.4} = \frac{2.8}{1.4} = 2.0$$

$P(X < 6.0) = P(Z < 2.0) = .5000 + .4772 = .9772.$

Exercise Set VI.3

1) a) $\mu = np = 800(.65) = 520$

$\sigma = \sqrt{npq} = \sqrt{(800)(.65)(.35)}$

$= 13.5$

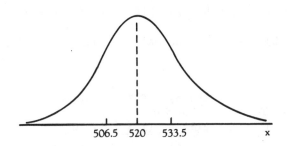

b) $\mu = np = 200(\frac{1}{2}) = 100$

$\sigma = \sqrt{npq} = \sqrt{(200)(\frac{1}{2})(\frac{1}{2})}$

$= \sqrt{50}$

$= 7.07$

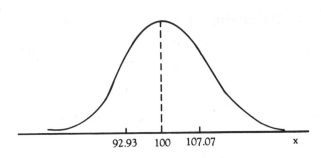

c) $\mu = 400(\frac{1}{4}) = 100$

$\sigma = \sqrt{(400)(\frac{1}{4})(\frac{3}{4})}$

$= \sqrt{75} = 8.66$

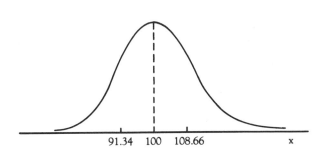

3) a) Using the normal approximation, $P(189.5 \leq X \leq 210.5)$ is needed where

$\mu = np = (500)(.4) = 200$ and $\sigma = \sqrt{(500)(.4)(.6)} = 10.95$.

$z = \dfrac{x - 200}{10.95}$

When $x = 210.5$, $z = \dfrac{210.5 - 200}{10.95} = .96.$

When $x = 189.5$, $z = \dfrac{189.5 - 200}{10.95} = - .96$ (also by symmetry).

$P(189.5 \leq X \leq 210.5) = P(-.96 \leq z \leq .96) = .6630.$

b) Using the normal approximation, $P(X \leq 195.5)$ is needed.

When $x = 195.5$, $z = \dfrac{195.5 - 200}{10.95} = -.41.$

$P(X \leq 195.5) = P(Z \leq -.41) = .3409.$

c) $P(X > 215) = P(X \geq 216)$ when X is binomially distributed. We need $P(X \geq 215.5)$ when the normal approximation is used.

When $x = 215.5$, $z = \dfrac{215.5 - 200}{10.95} = 1.42.$

$P(X \geq 215.5) = P(Z \geq 1.42) = .0793.$

5) a) 38% or more of the sample corresponded to success.

b) $\mu_{\hat{P}} = p = .35$, $\sigma_{\hat{P}} = \sqrt{\dfrac{pq}{n}} = \sqrt{\dfrac{(.35)(.65)}{200}} = .034$

$z = \dfrac{\hat{P} - .35}{.034}$

When $\hat{p} = .38$, $z = \dfrac{.38 - .35}{.034} = .88.$

$\begin{aligned} P(\hat{P} \geq .38) &\simeq P(Z \geq .88) \\ &= .5 - .3106 \\ &= .1894 \end{aligned}$

7) We want the binomial probability $P(X \leq 260)$. This will be approximated by $P(X \leq 260.5)$ using a normal distribution with $\mu = (.70)(400) = 280$ and $\sigma = \sqrt{(.70)(.30)(400)} = 9.17.$

$z = \dfrac{260.5 - 280}{9.17} = -2.13$

$\begin{aligned} P(X \leq 260) &\simeq P(Z \leq -2.13) \\ &= .5000 - .4834 \\ &= .0166 \end{aligned}$

9) $p = .60$

$n = 500$

$P(\hat{p} \geq .58)$ is needed.

$\mu_{\hat{P}} = p = .60$

$\sigma_{\hat{P}} = \sqrt{\dfrac{(.6)(.4)}{500}} = .0219$

$$z = \frac{\hat{p} - .60}{.0219}$$

When $\hat{p} = .58$, $z = \frac{.58 - .60}{.0219} = -.91$

$P(\hat{p} \geq .58) \approx P(Z \geq -.091)$

$\qquad = .3186 + .5000$

$\qquad = .8186.$

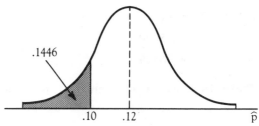

.1446

.10   .12 $\hat{p}$

11) The airline is in trouble if the number of no shows is less than 30, i.e.
$\hat{p} < \frac{30}{300} = .1$. We need $P(\hat{p} < .1)$.

Here $n = 300$, $p = .12$,

$\mu_{\hat{p}} = p = .12$,

$\sigma_{\hat{p}} = \sqrt{\frac{(.12)(.88)}{300}} = .0188$, and

$z = \frac{.1 - .12}{.0188} = -1.06$

$P(\hat{p} < .1) = P(Z < -1.06)$

$\qquad = .5000 - .3554$

$\qquad = .1446$

The probability of having enough seats is thus $1 - .1446 = .8554.$

13) $P(\hat{p} < .15)$ is needed.

$\mu_{\hat{p}} = p = .20$, $\sigma_{\hat{p}} = \sqrt{\frac{(.20)(.80)}{100}} = .04$ and

$z = \frac{\hat{p} - .20}{.04}$

When $\hat{p} = .15$, $z = \frac{.15 - .20}{.04} = -1.25.$

$P(\hat{p} < .15) = P(Z < -1.25) = .5000 - .3944 = .1056.$

15) If 70 are correct and 50 are incorrect
the score is $3(70) - 1(50) = 160$.

We need $P(X \geq 70)$

$n = 120$

$p = \frac{1}{2} = .5$ if each answer is a guess.

With endpoint correction and the normal

distribution we need $P(X \geq 69.5)$.

Using $\mu_X = np = \frac{1}{2}(120) = 60$ and

$\sigma_X = \sqrt{npq} = \sqrt{(120)(.5)(.5)} = 5.47,$

we find $z = \frac{69.5 - 60}{5.47} = 1.74.$

Thus

$\begin{cases} c = \text{correct} \\ w = \text{incorrect} \\ c + w = 120 \text{ (total number of} \\ \qquad\qquad\qquad \text{questions)} \\ 3c - w = 160 \\ 4c = 280 \\ c = 70 \end{cases}$

$P(X \geq 70) = P(Z \geq 1.74)$

$\qquad\qquad\quad = .5000 - .4591$

$\qquad\qquad\quad = .0409$

There is about a 4% chance of passing if the student guesses at every answer.

Exercise Set VI.4

1) a) Entering the x^2 table in the column headed $\nu = 6$, we find the entry
      10.6 corresponds to a probability of .90.
      Thus $P(0 \le x^2 \le 10.6) = .90$ (also note, $x^2_{.10} = 10.6$ when $\nu = 6$).

   b) Entering the x^2 table in the column headed $\nu = 15$, we find the entry
      25.0 corresponds to a probability of .95.
      Thus $P(0 \le x^2 \le 25.0) = .95$ and $P(x^2 > 25.0) = 1 - .95 = .05$.

3) a) $P(x^2 > 14.9) = 1 - P(0 \le x^2 \le 14.9)$

$$= 1 - .995$$

$$= .005$$

   And from Table V, we find $x^2_{.005} = 14.9$.

   b) $P(x^2 > 18.3) = 1 - P(0 \le x^2 \le 18.3)$

$$= 1 - .95$$

$$= .05$$

   And from Table V, we find $x^2_{.05} = 18.3$.

5) If the die is fair, then in 120 tosses there should be

$$e_1 = \frac{1}{6}(120) = 20 \text{ ones,}$$

$$e_2 = \frac{1}{6}(120) = 20 \text{ twos, etc.}$$

Outcomes	1	2	3	4	5	6
0_i	20	25	18	19	23	15
e_i	20	20	20	20	20	20

There are $\nu = 6 - 1 = 5$ d.f.
We reject the die as being fair if the computed x^2 value exceed $x^2_{.05} = 11.1$.

$$x^2 = \frac{(20-20)^2}{20} + \frac{(25-20)^2}{20} + \frac{(18-20)^2}{20} + \frac{(19-20)^2}{20} + \frac{(23-20)^2}{20} + \frac{(15-20)^2}{20}$$

$$= 0 + 1.25 + .20 + .05 + .45 + 1.25$$

$$= 3.20.$$

Since $3.20 \not> 11.1$, do not reject the die as being fair.

7) Mendel's model predicts that of each $6 + 2 + 2 + 1 = 11$ peas, 6 will be round and yellow, 2 round and green, 2 oblong and yellow, and 1 oblong and green.

Thus
$$P(\text{round and yellow}) = \frac{6}{11}$$
$$P(\text{round and green}) = \frac{2}{11}$$
$$P(\text{oblong and yellow}) = \frac{2}{11}$$
$$P(\text{oblong and green}) = \frac{1}{11}$$

In all he observed $315 + 108 + 101 + 32 = 556$ peas.

$$e_1 = \frac{6}{11}(556) = 303.27$$
$$e_2 = \frac{2}{11}(556) = 101.09$$
$$e_3 = \frac{2}{11}(556) = 101.09$$
$$e_4 = \frac{1}{11}(556) = 50.55$$

	Round and Yellow	Round and Green	Oblong and Yellow	Oblong and Green
O_i	315	108	101	32
e_i	303.27	101.09	101.09	50.55

Using $\nu = 4 - 1 = 3$, we find $\chi^2_{.10} = 6.25$

$$\chi^2 = \frac{(315 - 303.27)^2}{303.27} + \frac{(108 - 101.09)^2}{101.09} + \frac{(101 - 101.09)^2}{101.09} + \frac{(32 - 50.55)^2}{50.55} = 7.733.$$

Since $\chi^2 = 7.733 > \chi^2_{.10}$ the model should be rejected.

9) Each dial spins 300 times. Thus there are 900 trials in all. The total number of bells was $39 + 28 + 30 = 97$ and the probability estimate of a bell on any one of the dials is $\frac{97}{900} = .1078$ (assuming the same probability for each dial). Then the expected frequency for each dial in 300 trials is

$$(.1078)(300) = 32.34$$

	Dial 1	Dial 2	Dial 3
O_i	39	28	30
e_i	32.34	32.34	32.34

$$\chi^2 = \frac{(39 - 32.34)^2}{32.34} + \frac{(28 - 32.34)^2}{32.34} + \frac{(30 - 32.34)^2}{32.34} = 2.123.$$

When $\nu = 3 - 1 = 2$, $\chi^2_{.05} = 5.99$.

$\chi^2 = 2.123$ is considerably less than $\chi^2_{.05}$.

Thus, do not reject the hypothesis of equal frequencies of bells on the 3 dials.

11)

	Independents	Democrats	Republicans
0_i	33	151	116
e_i	30	150	120

$$\chi^2 = \frac{(33 - 30)^2}{30} + \frac{(151 - 150)^2}{150} + \frac{(116 - 120)^2}{120}$$

$$= .300 + .007 + .130$$

$$= .437$$

From Table IV with $\nu = 2$

$P(\chi^2 < .45) = .20$ (nearest entry)

Thus the probability of getting as good agreement as reported is less than .20.

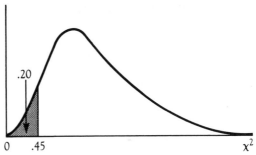

13) Let X = number of boys in a 3 child family. We have the following distribution

x	$P(x)$
0	$\frac{1}{8}$
1	$\frac{3}{8}$
2	$\frac{3}{8}$
3	$\frac{1}{8}$

$e_1 = \frac{1}{8}(400) = 50$

$e_2 = \frac{3}{8}(400) = 150$

$e_3 = \frac{3}{8}(400) = 150$

$e_4 = \frac{1}{8}(400) = 50$

Number of Boys	0	1	2	3
0_i	40	180	120	60
e_i	50	150	150	50

$$\chi^2 = \frac{(40 - 50)^2}{50} + \frac{(180 - 150)^2}{150} + \frac{(120 - 150)^2}{150} + \frac{(60 - 50)^2}{50} = 16$$

Using $\nu = 4 - 1 = 3$ d.f., $\chi^2_{.05} = 7.82$.

Since $\chi^2 = 16 > 7.82$ we reject the model that boys and girls are equally likely in 3 children families.

15) $\mu = \nu = 50$

$\sigma = \sqrt{2\nu} = \sqrt{2(50)} = 10$

$z_{.05} = 1.65$

$1.65 = \dfrac{\chi^2 - 50}{10}$. Solving for χ^2

$\qquad \chi^2 = 50 + (1.65)(10)$

$\qquad\quad = 66.5$

The table entry when $\nu = 50$ is

$\qquad \chi^2_{.05} = 67.5.$

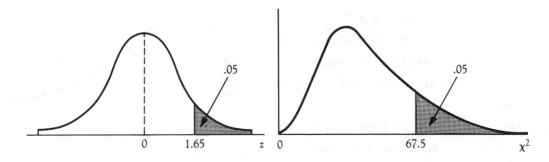

Chapter Test

1) a) May be approximated as such. Each executive surveyed represents a trial.
      Success represents the belief that the next president will be a democrat.

   b) Binomial experiment. Each applicant represents a trial. Success
      corresponds to a pass.

   c) No. The probability of a defective watch (success) changes as watches
      are withdrawn from the shipment.

   d) May be approximated as such. Each family represents a trial. Success
      corresponds to having a dental plan.

2) $n = 3$, $p = .4$, and $q = .6$

   a) $\mu = np = (3)(.4) = 1.2$

      $\sigma = \sqrt{npq} = \sqrt{(3)(.4)(.6)} = .8485$

   b) $P(X = 0) = \binom{3}{0}(.4)^0(.6)^3 = (1)(1)(.6)^3 = .216$

      $P(X = 1) = \binom{3}{1}(.4)^1(.6)^{3-1} = (3)(.4)(.6)^2 = .432$

      $P(X = 2) = \binom{3}{2}(.4)^2(.6)^{3-2} = (3)(.4)^2(.6) = .288$

      $P(X = 3) = \binom{3}{3}(.4)^3(.6)^{3-3} = 1(.4)^3(1) = .064$

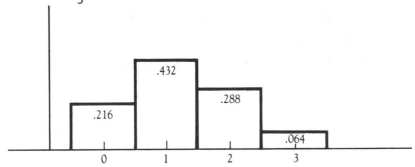

3) a) $P(X \geq 10) = 1 - P(X \leq 9) = 1 - .7553 = .2447$

   b) $P(X \geq 10 \,|\, X \text{ binomial}) \approx P(X \geq 9.5 \,|\, X \text{ normal})$

      $= P(Z \geq .68) = .5000 - .2517 = .2483$

      Here $\quad z = .68 = \dfrac{9.5 - (20)(4)}{\sqrt{(20)(.4)(.6)}}$

4) $P(X = x) = \binom{10}{x}(.3)^x(.7)^{10-x}$

   $n = 10$, $p = .4$

   Using the cumulative tables, we find

   $P(X \geq 4) = 1 - P(X \leq 3) = 1 - .6496 = .3504.$

5) $n = 50$, $p = .75$

$P(\hat{P} < .70)$ is needed.

$$z = \frac{.70 - .75}{\sqrt{\dfrac{(.75)(.25)}{50}}} = -.82$$

$P(\hat{P} < .70) = P(Z < -.82) = .5000 - .2939 = .2061$

6) a) $z_{.05}$ is needed.

The area to the right of $z_{.05}$ is .05.

Look for the z score corresponding to the table entry .5000 – .05 = .4500.

The nearest entry is .4495 (or .4505) corresponding to z = 1.64 (or 1.65).

Thus $z_{.05} = 1.64$ (or 1.65).

b) $z_{.01}$ is needed.

Find the z score corresponding to the table entry .5000 – .01 = .4900.

The nearest table entry is .4901 which corresponds to z = 2.33.

Thus $z_{.01} = 2.33$.

c) $z_{.75}$ is needed.

The area to the right of the z score is .75.

Thus $z_{.75}$ is less than 0.

The table entry is 1 – .75 = .2500 and to this corresponds z = .67 (nearest entry).

Thus $z_{.75} = -.67$.

7) a) $P(X > 500)$ is needed.

$$z = \frac{500 - 340}{80} = 2.00$$

$P(X > 500) = P(Z > 2.00) = .5000 - .4772 = .0228$

b) Find x such that $P(X \le x) = .90$ or equivalently $P(X > x) = .10$.

The corresponding Z score is $z_{.10} = 1.28$.

Find x by solving

$$1.28 = \frac{x - 340}{80} .$$

$x = 340 + 80(1.28) = 442.40$.

8) If they are equally preferred, then each technique has an expected frequency of $\frac{1}{3}$ (90) = 30.

	A	B	C
0_i	24	34	32
e_i	30	30	30

$$x^2 = \frac{(24 - 30)^2}{30} + \frac{(34 - 30)^2}{30} + \frac{(32 - 30)^2}{30} = 1.8667$$

A large value of x^2 is $x^2_{.05} = 5.99$ for $\nu = 2$ d.f.
Since $x^2 = 1.8667 < 5.99$ we cannot conclude that there is any one procedure that is preferred over the others.

9) The estimate for the proportion who are satisfied with their job is

$$\frac{96 + 80 + 92}{400} = .670.$$

There are 140 workers in the professional, 144 in the white collar, and 116 in the blue collar categories. If the degree of job satisfaction is the same for each category, then the expected frequencies are:

$$e_1 = (.670)(140) = 93.90$$

$$e_2 = (.670)(144) = 96.48$$

$$e_3 = (.670)(116) = 77.72$$

	Professional	White Collar	Blue Collar
0_i	96	80	92
e_i	93.90	86.48	77.72

$$x^2 = \frac{(96 - 93.90)^2}{93.90} + \frac{(80 - 96.48)^2}{96.48} + \frac{(92 - 77.72)^2}{77.72} = 5.49$$

For $\nu = 2$ d.f., $x^2_{.05} = 5.99$.
Since $x^2 = 5.49 < 5.99$ we cannot conclude that the degree of job satisfaction is different for these three categories.

10)   Note the 90th percentile is being used. The Z score corresponding to
      this is $z_{.10}$ = 1.28. We find the 90th percentile by solving

$$1.28 = \frac{x - 100}{15}$$

$$x = 100 + (1.28)(15)$$

$$= 119.2.$$

Chapter VII
Sampling Distributions

Introduction

An understanding of the relationship of sample means to the population mean, sample variances to the population variance, and sample proportions to the population proportion is essential to the processes of inferential statistics. The distribution of these sample statistics is the subject of this chapter.

Concepts/Techniques

1) Population Parameters

A population parameter is a number which identifies some important characteristic of the population distribution.

Example: The mean μ, the variance σ^2, and the standard deviation are population parameters.

Example: The proportion p of a population having some property of interest is a population parameter.

2) Sample Statistics

A random variable which derives its values from all samples of a certain size, n, is called a sample statistic.

Example: The mean $\overline{X}$ is a sample statistic or variable whose values are the means of samples of a given fixed size, say n.

Example: The variance S^2 is a sample statistic whose values are the variances s^2 of samples of size n.

Example: The standard deviation $S = \sqrt{S^2}$ is a sample statistic.

Example: The proportion $\hat{P}$ is a sample statistic whose values are the proportions of samples of size n which have a certain property or characteristic.

3) Sampling Distributions

A sampling distribution is the probability distribution of a sample statistic.

4) Unbiased Estimator

A sample statistic is said to be an unbiased estimator of a population parameter if its mean or expected value is equal to the population parameter. The (sample) mean $\overline{X}$ can be shown to be an unbiased estimator of the population mean, that is

$$E(\overline{X}) = \mu_{\overline{X}} = \mu_X = \mu.$$

The variance S^2 can be shown to be an unbiased estimator of the population variance σ^2.

$$E(S^2) = \mu_{S^2} = \sigma^2$$

Example: Let $P = \{1,2,3\}$ represent a population. Find the probability distribution of the mean $\overline{X}$ of samples of size n = 2 and verify that

$$\mu_{\overline{X}} = \mu_X$$

First we see that the population mean is

$$\mu = \frac{1 + 2 + 3}{3} = 2$$

The $\binom{3}{2} = 6$ samples of size n = 2 and their means are tabulated below along with the probability distribution of $\overline{X}$

Sample	$\overline{x}$
1,2	$\frac{3}{2}$ = 1.5
1,3	$\frac{4}{2}$ = 2
2,3	$\frac{5}{2}$ = 2.5

$\overline{x}$	$P(\overline{x})$
1.5	$\frac{1}{3}$
2	$\frac{1}{3}$
2.5	$\frac{1}{3}$

Probability Distribution of $\overline{X}$

The mean of the statistic $\overline{X}$ is

$$\mu_{\overline{X}} = \Sigma\ \overline{x}P(\overline{x}) = (1.5)\left(\frac{1}{3}\right) + 2\left(\frac{1}{3}\right) + 2.5\left(\frac{1}{3}\right) = \frac{1.5 + 2 + 2.5}{3} = 2$$

Thus $\mu_{\overline{X}} = \mu_X = 2$.

Thus the mean of the sample means is the population mean and $\overline{X}$ is an unbiased estimator of μ.

5) **The Distribution of $\overline{X}$ (Large Samples Case)**

The Central Limit theorem states that when n is large, generally $n \geq 30$, the distribution of the mean $\overline{X}$ is approximately normal with mean and standard deviation

$$\mu_{\overline{X}} = \mu = \text{Population Mean}$$

$$\sigma_{\overline{X}} = \frac{\sigma}{\sqrt{n}} = \frac{\text{Population Standard Deviation}}{\sqrt{\text{Sample Size}}}$$

Example: Suppose the IQ's X of elementary school children have mean $\mu = 100$ and standard deviation $\sigma = 10$. Describe the distribution of $\overline{X}$ for samples of size $n = 64$.

Note we are to describe the distribution of all sample means where each sample is of size $n = 64$. Since $n = 64 \geq 30$, the distribution of the sample means is approximately normal. The mean and standard deviation of this normal dsstribution are

$$\mu_{\overline{X}} = \mu = 100$$

$$\sigma_{\overline{X}} = \frac{\sigma}{\sqrt{n}} = \frac{10}{\sqrt{64}} = 1.25.$$

The distributions of both X and $\overline{X}$ are shown on the next page.

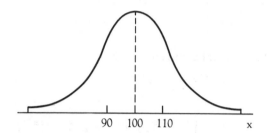 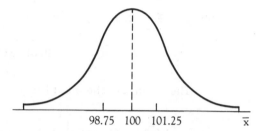

Notice that as the sample size n gets larger, the standard
deviations of the distributions of $\overline{X}$ get smaller due to the n
in the denominator of the expression $\sigma_{\overline{X}} = \dfrac{\sigma}{\sqrt{n}}$. This means that
for large samples, the deviation of the sample means from the
population mean tends to be small.

6) <u>Probabilities for $\overline{X}$ (Large Sample Case)</u>
z scores are obtained from values of $\overline{X}$ by

$$z = \frac{\overline{x} - \mu}{\sigma_{\overline{X}}} = \frac{\overline{x} - \mu}{\dfrac{\sigma}{\sqrt{n}}}.$$

<u>Example</u>: Reaction times of healthy adults have a mean $\mu = 42$
seconds and standard deviation $\sigma = .18$ seconds. What is the
probability that the mean reaction time of a sample of n = 36
healthy adults will exceed 0.46 seconds?
The problem is to find $P(\overline{X} > .46)$. The needed area is located
under the distribution curve of $\overline{X}$ to the right of $\overline{x} = .46$.
The corresponding z score is found using the formula

$$z = \frac{\overline{x} - .42}{\dfrac{.18}{\sqrt{36}}} = \frac{\overline{x} - .42}{.03}$$

Using $\overline{x} = .46$, we find

$$z = \frac{.46 - .42}{.03} = 1.33.$$

Thus $P(\overline{X} > .46) = P(Z > 1.33) = .5000 - .4082 = .0918.$

We interpret this to mean that about 9.18% of all samples of

size n = 36 from this population will have means exceeding .46

seconds.

Note that when finding probabilities related to $\overline{X}$ it is the distri-

bution of $\overline{X}$ that is used – not that of X.

Example: Suppose heart rates X (beats per minute) of adults

are normally distributed with mean $\mu = 80$ and standard deviation

$\sigma = 9.$

a) What percent of all adults have rates exceeding 83 beats

per minute?

b) What percent of all samples of size n = 36 have mean rates

exceeding 83 beats per minute?

Answer (a). Here we are to find $P(X > 83)$ where $\mu = 80$ and

$\sigma = 9.$ Thus

$$z = \frac{x - 80}{9} = \frac{83 - 80}{9} = .33 \quad \text{and}$$

$$P(X > 83) > P(.33) = .5000 - .1293 = .3707$$

Thus about 37.07% of all adults have heart rates exceeding 83

beats per minute.

Answer (b). The problem here is to find $P(\overline{X} > 83)$ where

$\mu_{\overline{X}} = 80$ and $\sigma_{\overline{X}} = \dfrac{9}{\sqrt{36}} = 1.5.$ Note this is a probability

involving $\overline{X}.$ Thus the sampling distribution of $\overline{X}$ is used.

$$z = \frac{\overline{x} - 80}{1.5} = \frac{83 - 80}{1.5} = 2$$

Thus

$$P(\overline{X} > 83) = P(Z > 2) = .0228.$$

Thus about 2.28% of all samples of n = 36 adults will have a mean

rate exceeding 83 beats per minute.

7)   The Standard Error of the Mean

The standard deviation of the mean $\overline{X}$ is called the standard error of

the mean. Notation: S.E.

Thus

$$\text{S.E.} = \sigma_{\overline{X}} = \frac{\sigma}{\sqrt{n}}$$

Notes

1) The formula for obtaining z scores from sample means $\bar{x}$ can now be expressed as

$$z = \frac{\bar{x} - \mu}{S.E.} \; .$$

2) The S.E. (being the standard deviation of $\bar{X}$) is the unit of measure that is used by the statistician. We speak, for example, of a mean $\bar{x}$ being a certain number of standard errors from the population mean.

3) For large samples the empirical rule may now be restated as follows: Within 1 standard error of the population mean is located approximately 68% of the sample means.

Within 2 standard errors (actually 1.96) of the population mean is located approximately 95% of the sample means.

8) <u>The Distribution of $\bar{X}$ (Normal Population Case)</u>

If the population is normally distributed then so is the mean $\bar{X}$ (regardless of whether n is small or large).

Example: A sample of size $n = 9$ is selected from a normal population with $\mu = 150$ and $\sigma = 21$. What is the probability that the sample mean is less than 140? We seek $P(\bar{X} < 140)$ and know that $\bar{X}$ is normally distributed since this is true of the population.

The standard error is $S.E. = \dfrac{21}{\sqrt{9}} = 7.$

Thus $z = \dfrac{\bar{x} - \mu}{S.E.} - \dfrac{\bar{x} - 150}{7} = \dfrac{140 - 150}{7} = -1.43$

$P(\bar{X} < 140) = P(Z < -1.43) = .5000 - .4236 = .0764$

9) <u>The Distribution of $\bar{X}$ (Normal Population, σ Unknown)</u>

When σ is unknown and must be approximated by the sample standard deviation s, the t distribution with $\nu = n - 1$ degrees of freedom is used with

$$t = \frac{\bar{x} - \mu}{\dfrac{s}{\sqrt{n}}}$$

in place of the normal.

Notes

a) When the approximation of σ by s is necessary, we then use

$$S.E. = \frac{s}{\sqrt{n}}.$$

b) When the sample size n is 30 or more the normal distribution may be used with

$$z = \frac{\bar{x} - \mu}{\frac{s}{\sqrt{n}}}.$$

10) <u>The t Distribution</u>

The t distribution is similar in shape to the normal. Only percentage points or right tail critical values of this distribution are given as these are all that are needed in inference work.

<u>Example</u>: Using $\nu = 8$ d.f., find $t_{.05}$ and interpret.
In the column headed .05 and in the row $\nu = 8$ we find $t_{.05} = 1.860$. Thus the probability of a t value exceeding 1.860 is .05, i.e.

$$P(T > 1.860) = .05.$$

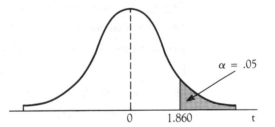

<u>Example</u>: A sample of size $n = 4$ from a normal population with a mean $\mu = 500$ had mean $\bar{x} = 660$ and standard deviation $s = 100$. Estimate the probability of obtaining a sample mean of this size or larger.

We seek $P(\bar{X} \geq 660)$.

The t distribution is applicable since the population is normal.

$$S.E. = \frac{s}{\sqrt{n}} = \frac{100}{\sqrt{4}} = 50$$

$$t = \frac{\bar{x} - \mu}{S.E.} = \frac{\bar{x} - 500}{50} = \frac{660 - 500}{50} = 3.20$$

For $\nu = 4 - 1 = 3$, we see that

$$t_{.025} = 3.182 \quad \text{and} \quad t_{.01} = 4.541.$$

Thus $P(T \geq 3.20)$ is somewhere between .025 and .01.

Thus the probability of obtaining a sample mean of at least 660 is between .025 and .01.

11) The Distribution of $\hat{P}$ (Large Sample Case)

When n is large, the proportions $\hat{p}$ of samples of size n have a distribution which is approximately normal with mean and standard deviation

$$\mu_{\hat{P}} = p = \text{population proportion}$$

$$\sigma_{\hat{P}} = \sqrt{\frac{pq}{n}}$$

Notes:

a) The sample size must be such that both np and nq are at least 5.

b) The standard deviation $\sigma_{\hat{P}}$ of $\hat{P}$ is called the standard error of the proportion $\hat{p}$. Thus we write

$$\text{S.E.} = \sqrt{\frac{pq}{n}}$$

Example: Suppose p = .18 = 18% of all freshman college students receive financial aid. Each sample of n = 400 students has a proportion $\hat{p}$ that receive financial aid. Describe the distribution of these proportions. This is simply the distribution of the sample statistic $\hat{p}$. Since np = 400(.18) = 72 and nq = 400(.82) = 328 both exceed 5, the distribution of $\hat{p}$ is approximately normal. The mean and standard error are

$$\mu_{\hat{P}} = p = .18$$

$$\text{S.E.} = \sqrt{\frac{pq}{n}} = \sqrt{\frac{(.18)(.82)}{400}} = .0192 = 1.92\%$$

The distribution is shown below.

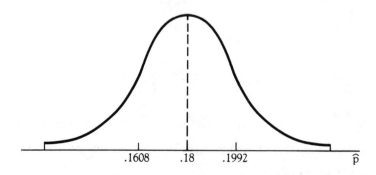

.1608   .18   .1992 $\hat{p}$

Thus about 68% of all sample proportions are within 1.92% of the population proportion $p = 18\%$, i.e., 68% of all sample proportions are between .1608 and .1992.

12) Computing Probabilities for $\hat{P}$ (Large Sample Case)

Z scores are computed using

$$z = \frac{\hat{p} - \mu_{\hat{p}}}{S.E.} = \frac{\hat{p} - p}{\sqrt{\dfrac{pq}{n}}}$$

Example: Suppose that 70% of the population is opposed to the 55 m.p.h. speed limit. What is the probability that less than 65% of a sample of $n = 600$ people will oppose this speed limit? We are to find $P(\hat{P} < .65)$. The distribution of $\hat{p}$ is approximately normal with mean and standard error:

$$\mu = p = .70$$

$$S.E. = \sqrt{\frac{pq}{n}} = \sqrt{\frac{(.70)(.30)}{600}} = .0187$$

The required area is shown below.

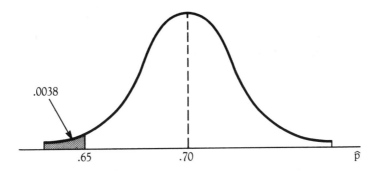

The Z score corresponding to $\hat{p} = .65$ is

$$z = \frac{\hat{p} - .70}{.0187} = \frac{.65 - .70}{.0187} = -2.67$$

Thus $P(\hat{P} < .65) = P(Z < -2.67) = .5000 - .4962 = .0038$.
The chances of getting a sample with a proportion this small are extremely slight.

In inference applications involving the proportion $\hat{p}$, large samples on the order of 1000 to 2500 subjects are often used in order to keep the standard error small. Small samples are rarely used.

13) <u>The Distribution of the Variance s^2 (Normal Population)</u>

Let s^2 be the variance associated with samples of size n from a normal population with variance σ^2. Then

$$\chi^2 = \frac{(n - 1)s^2}{\sigma^2}$$

has the chi-square distribution with $\nu = n - 1$ degree of freedom.

> Example: A normal population has mean $\mu = 500$ and variance $\sigma^2 = (100)^2 = 10000$. Estimate the probability that a random sample of size n = 16 will have a variance exceeding $(140)^2 = 19600$.
>
> We seek $P(s^2 > 19600)$.
>
> The χ^2 value corresponding to $s^2 = 19600$ is
>
> $$\chi^2 = \frac{(n - 1)s^2}{\sigma^2} = \frac{(16 - 1)19600}{10000} = 29.4$$
>
> From the χ^2 tables with $\nu = n - 1 = 16 - 1 = 15$ d.f. we find $\chi^2_{.025} = 27.5$ and $\chi^2_{.01} = 30.6$. Thus $P(s^2 > 19600)$ is somewhere between .025 and .01.

14) <u>The F Distribution</u>

The F distribution shown below is similar in shape to the Chi square distribution. It has only <u>non negative values</u> f and is skewed to the right.

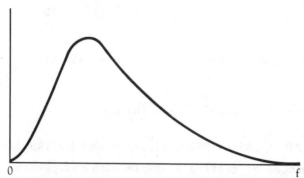

0   f

The F distribution has 2 associated degrees of freedom ν_1 and ν_2 which later will be identified with numerators and denominators, respectively, of ratios.

The mean of the F distribution is $\mu_F = \dfrac{\nu_2}{\nu_2 - 2}$. This mean is quite close to 1 even for moderate values of ν_2. Critical values of F are represented by f_α or $f_\alpha(\nu_1, \nu_2)$ if we wish to emphasize the degrees of freedom.

As with the χ^2 distribution only the right tail critical values or percentage points are tabulated for F. Note that due to the 2 separate degrees of freedom one page or table is needed for each choice of α.

Example: Find $f_{.05}(8,10)$.

The first mentioned degree of freedom is $\nu_1 = 8$ and is found as a column heading. The second, $\nu_2 = 10$ is a row heading. At the intersection of this column and row in the table headed $\alpha = .05$ we find

$$f_{.05}(8,10) = 3.07.$$

This area is shaded in the figure below.

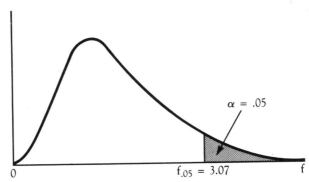

The F distribution having $\nu_1 = 8$ d.f. and $\nu_2 = 10$ d.f.

Note that the mean of this distribution is

$$\mu_F = \frac{\nu_2}{\nu_2 - 2} = \frac{10}{10 - 2} = \frac{10}{8} = 1.125.$$

Critical values f_α for large α are generally not tabulated. These values which are located under the left "tail" of the distribution may be found using the formula

$$f_\alpha(\nu_1, \nu_2) = \frac{1}{f_{1-\alpha}(\nu_2, \nu_1)} \, .$$

<u>Example</u>: Find $f_{.99}(6,10)$.

$$f_{.99}(6,10) = \frac{1}{f_{.01}(10,6)}$$ (Note: interchange the degrees of freedom.)

$$= \frac{1}{7.87} = .13$$

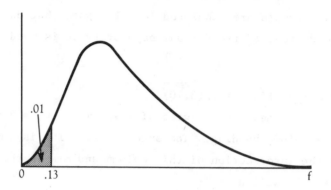

15) <u>The Distribution of the Ratio of Two Sample Variances from a Normal Population</u>

Let s_1^2 and s_2^2 be the variances of two samples of size n_1 and n_2 respectively from a normal population. Then the ratio

$$f = \frac{s_1^2}{s_2^2}$$

is the value of a variable F having the F distribution with $\nu_1 = n_1 - 1$ d.f. and $\nu_2 = n_2 - 1$ d.f.

Notes:

a) Since the expected value of F is approximately 1 for moderate size ν_2, the values of the ratios s_1^2/s_2^2 should cluster close to 1. A large deviation from this would be an indication that possibly the two samples are not from the same population.

b) The numbering of the samples is generally chosen so that s_1^2 is the larger of the 2 sample variances. In that way $f = s_1^2/s_2^2 > 1$ and we only need be concerned with f values greater than 1.

<u>Example</u>: Two independent samples of overweight adults are selected

and treated with separate weight reduction therapies.   The sample
sizes and the variances of the weight losses are given below.
Estimate the probability of obtaining such different variances
if in fact there is no difference in the therapies.

Therapy A	Therapy B
n = 12	n = 10
s^2 = 196	s^2 = 620

If there is no difference in the therapies, then both samples are
from the same population which we will assume to be approximately
normal.   Since the B sample has the larger variance we choose it
as population 1.   Thus n_1 = 10, s_1^2 = 620, n_2 = 12, s_2^2 = 196   and
$f = \frac{620}{196} = 3.163$.
We find $f_{.05}(9,11) = 2.90$ and $f_{.025}(9,11) = 3.59$.
Thus the probability of such different sample variances is between
.05   and   .025.   This is quite small and would probably be taken
as an indication that the therapies produce different results.

Calculator Usage Ilustration

Compute the standard error of $\overline{X}$.

Find S.E. $= \dfrac{84}{\sqrt{42}}$.

Method 1) Direct evaluation: $84 \div \sqrt{42}$.

$$84 \;\boxed{\div}\; 42 \;\boxed{\sqrt{}}\; \boxed{=} \;\underline{\underline{12.961481}}$$

Method 2) Multiply the numerator by the reciprocal of the square root of the sample size.

$$42 \;\boxed{\sqrt{}}\; \boxed{\dfrac{1}{x}}\; \boxed{\times}\; 84 \;\boxed{=}\; \underline{\underline{12.961481}}$$

Calculator Usage Illustration

Compute the standard error of $\hat{P}$.

Find S.E. $= \sqrt{\dfrac{(.27)(.73)}{570}}$

Method: Direct evaluation.

$$.27 \;\boxed{\times}\; .73 \;\boxed{\div}\; 570 \;\boxed{=}\; \boxed{\sqrt{}}\; \underline{\underline{.018595}}$$

Calculator Usage Illustrations

Compute the Z score corresponding to a sample mean.

Find $z = \dfrac{38.5 - 31.6}{\dfrac{7}{\sqrt{40}}}$.

Method 1) Direct evaluation. Arrange the computations as
$(38.5 - 31.6) \div (7 \div \sqrt{40})$.

38.5 ⊟ 31.6 ⊟ ⊡ (7 ⊡ 40 √) ⊟ 6.2342045

Method 2) Arrange the computation as $(38.5 - 31.6) \div \left(\dfrac{1}{\sqrt{40}} \times 7 \right)$

38.5 ⊟ 31.6 ⊟ ⊡ (40 $\sqrt{x}$ $\frac{1}{x}$ ⊠ 7) ⊟ 6.2342045

or equivalently

(38.5 ⊟ 31.6) ⊡ (40 $\sqrt{x}$ $\frac{1}{x}$ ⊠ 7) ⊟ 6.2342045

Method 3) First compute the standard error and store in the memory. Then
call it back for the division after the numerator is computed.

40 $\sqrt{x}$ $\frac{1}{x}$ ⊠ 7 ⊟ [STO] 38.5 ⊟ 31.6 ⊟ ÷ RCL ⊟ 6.2342045

or equivalently

40 $\sqrt{x}$ $\frac{1}{x}$ ⊠ 7 ⊟ [STO] (38.5 ⊟ 31.6) ⊡ RCL ⊟ 6.2342045

Calculator Usage Illustrations

Compute the Z score corresponding to a sample proportion.

$$\text{Find} \quad z = \frac{.381 - .304}{\sqrt{\dfrac{(.381)(.619)}{450}}}$$

Method 1) Arrange the computations as $(.381 - .304) \div \sqrt{.381 \times .619 \div 450}$

.381 $\boxed{-}$ .304 $\boxed{=}$ $\div$ $\boxed{(}$ .381 $\boxed{\times}$ .619 $\boxed{\div}$ 450 $\boxed{=}$ $\boxed{\sqrt{x}}$ $\boxed{)}$ $\boxed{=}$ <u>3.3634825</u>

Note: Some calculators will not permit the number of operations and symbols within a set of parentheses used above. If this should be the case, try either of the following.

Method 2) Compute the standard error first, then invert it, and multiply by the numerator

.381 $\boxed{\times}$ .619 $\boxed{\div}$ 450 $\boxed{=}$ $\boxed{\sqrt{x}}$ $\boxed{\frac{1}{x}}$ $\boxed{\times}$ $\boxed{(}$ .381 $\boxed{-}$ .304 $\boxed{)}$ $\boxed{=}$ <u>3.3634825</u>

Method 3) Compute the standard error and store it in the memory. Then call it out for the division after the numerator is computed.

.381 $\boxed{\times}$ .619 $\boxed{\div}$ 450 $\boxed{=}$ $\boxed{\sqrt{x}}$ $\boxed{STO}$ .381 $\boxed{-}$ .304 $\boxed{=}$ $\boxed{\div}$ $\boxed{RCL}$ $\boxed{=}$ <u>3.3634825</u>

Calculator Usage Illustration

Compute the ratio of two sample variances to obtain an f score.

$$\text{Compute} \quad f = \frac{(38.4)^2}{(35.6)^2}$$

Method 1) Arrange the computation as $(38.4)^2 \div (35.6)^2$.

38.4 $\boxed{x^2}$ $\boxed{\div}$ 35.6 $\boxed{x^2}$ $\boxed{=}$ $\underline{1.1634895}$

or equivalently

38.4 $\boxed{x^2}$ $\boxed{\div}$ $\boxed{(}$ 35.6 $\boxed{x^2}$ $\boxed{)}$ $\boxed{=}$ $\underline{1.1634895}$

Method 2) Arrange the computation as $(38.4 \div 35.6)^2$. (This saves one step over the first of the above arrangements.)

38.4 $\boxed{\div}$ 35.6 $\boxed{=}$ $\boxed{x^2}$ $\underline{1.1634895}$

Exercise Set VII.1

1) a) $\mu_{\overline{X}} = \mu_X$ = Population Mean = 8.4

    b) $\sigma_{\overline{X}} = \dfrac{\sigma_X}{\sqrt{n}} = \dfrac{\text{Population S.D.}}{\sqrt{\text{Sample Size}}} = \dfrac{2.4}{\sqrt{64}} = .3$

    c) The Standard Error of the Mean

    d) $\overline{X}$ is normally distributed

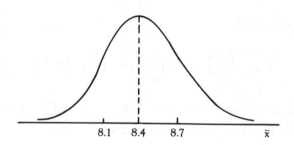

8.1   8.4   8.7 $\overline{x}$

3) Since X is normal, the same is true of $\overline{X}$ regardless of the sample size
    and $\mu_{\overline{X}} = \mu = 84.$

    a) $\mu = 84$ and   S.E. $= \dfrac{28}{\sqrt{16}} = 7$

        $\overline{x} = 91$ is seen to be 1 (S.E.) above the mean.

        Of course, $z = \dfrac{91 - 84}{7} = 1.0.$ Shows this.

    b) $\mu = 84$ and   S.E. $= \dfrac{28}{\sqrt{196}} = 2$

        $z = \dfrac{79.5 - 84}{2} = -2.25.$

        Thus $\overline{x} = 79.5$ is 2.25 S.E. below the mean

    c) $\mu = 84$ and   S.E. $= \dfrac{28}{\sqrt{1600}} = .7$

        $z = \dfrac{86.5 - 84}{.7} = 3.57$

        Thus $\overline{x} = 86.5$ is 3.57 (S.E.) above the mean.

    d) $\mu = 84$ and   S.E. $= \dfrac{28}{\sqrt{6}} = 11.43.$

        $z = \dfrac{100 - 84}{11.43} = 1.40$

        Thus $\overline{x} = 100$ is 1.40 (S.E.) above the mean.

5) S.E. $= \dfrac{12}{\sqrt{64}} \doteq 1.5$

a) $P(\overline{X} > 132)$ is needed.

$z = \dfrac{\overline{x} - 128}{1.5} = \dfrac{132 - 128}{1.5} = 2.67$

$P(\overline{X} > 132) = P(Z > 2.67) = .5000 - .4962 = .0038$

b) $P(126 < \overline{X} < 130)$ is needed

When $\overline{x} = 130$, $z = \dfrac{130 - 128}{1.5} = 1.33$

When $\overline{x} = 126$, $z = \dfrac{126 - 128}{1.5} = -1.33$

$P(126 < \overline{X} < 130) = P(-1.33 < Z < 1.33) = 2(.4082) = .8164$

7) a) This problem pertains to the distribution of X.

We need $P(X < 25000)$.

$z = \dfrac{x - \mu}{\sigma} = \dfrac{x - 25600}{1840}$

When $x = 25000$

$z = \dfrac{25000 - 25600}{1840} = \dfrac{-600}{1840} = -.33$

$P(X < 25000) = P(Z < -.33)$

$= .5000 - .1293$

$= .3707$

b) This problem pertains to the distribution of $\overline{X}$.

We use $Z = \dfrac{\overline{x} - 25600}{S.E.}$.

S.E. $= \dfrac{1840}{\sqrt{16}} = \dfrac{1840}{4} = 460$

When $\overline{x} = 25000$,

$z = \dfrac{25000 - 25600}{460} = \dfrac{-600}{460} = -1.30.$

$P(\overline{X} < 25000) = P(Z < -1.30$

$= .5000 - 4032$

$= .0968$

c) 37.07% of all students have a starting salary below $25000. 9.68% of all samples of n = 16 students have a mean starting salary below $25000.

9) $P(\overline{X} > 205)$ is needed.

$\mu_{\overline{X}} = 202$ and S.E. $= \dfrac{12.5}{\sqrt{36}} = 2.08$

When $\overline{x} = 205$, $z = \dfrac{205 - 202}{2.08} = 1.44$

$$P(\bar{X} > 205) = P(Z > 1.44)$$

$$= .5000 - .4251 = .0749$$

11) $P(\bar{X} < 5.74)$ is needed when $\mu_{\bar{X}} = \mu_X = 6.89$

When $\bar{x} = 5.74$, $z = \dfrac{5.74 - 6.89}{\dfrac{3.20}{\sqrt{68}}} = -2.96$

$$P(\bar{X} < 5.74) = P(Z < -2.96)$$

$$= .5000 - .4985 = .0015$$

13) Larger samples should strongly resemble the population. Thus their means should be close to the population mean and, as a result, these means should have a small standard deviation, i.e. standard error.

Recall $\sigma_{\bar{X}} = \text{S.E.} = \dfrac{\sigma}{\sqrt{n}}$.

As the sample size n becomes large, the denominator becomes large and the ratio $\dfrac{\sigma}{\sqrt{n}}$ becomes small.

15) $\bar{X}$ is continuous also

17) The mean of the medians of all samples of a particular size n is equal to the population mean, i.e.

$$E(\tilde{X}) = \mu.$$

The symmetry of the distribution is needed.

19) a) $\mu_X = $ Population Mean $= 100$

   b) $\mu_{\bar{X}} = $ Mean of the Sample Means $= $ Population Mean $= 100$

   c) $\sigma_X = $ Standard Deviation of the Population $= 30$

   d) $\sigma_{\bar{X}} = $ Standard Deviation of the Sample Means $= \dfrac{\sigma}{\sqrt{n}} = \dfrac{30}{\sqrt{100}} = 3.0$

   e) S.E. $= $ Standard Error of the Mean $= \sigma_{\bar{X}} = 3.0$ (from part (d))

Exercise Set VII.2

1) The distribution of X is normal and the approximation $\sigma \simeq s$ is being used.

3) a) $t_{.05} = 1.721$ (The entry at the intersection of the .05 column and the $\nu = 21$ d.f. row.)

   b) $t_{.10} = 1.337$ (The entry at the intersection of the .10 column and the $\nu = 16$ d.f. row.)

   c) $t_{.005} = 3.169$ when $\nu = 10$ d.f.

5) $P(\overline{X} > 15.2)$ is needed.

$$S.E. = \frac{3.4}{\sqrt{16}} = .85$$

When $\overline{x} = 15.2$, $t = \frac{15.2 - 13}{.85} = 2.59$

$P(\overline{X} > 15.2) = P(T > 2.59)$

For $\nu = 16 - 1 = 15$ d.f., $t_{.025} = 2.131$ and $t_{.01} = 2.60$

Thus $P(\overline{X} \geq 15.2)$ is about .01.

The event is extremely unlikely.

7) In approximating the distribution of $\overline{X}$ using s, there is a great deal more variability with which to contend then when using σ.

9) $P(\mu - 1.75(S.E.) < \overline{X} < \mu + 1.75(S.E.)) = P(-1.75 < t < 1.75)$

$$\approx 1 - .10 = .90$$

Note: For $\nu = 16 - 1 = 15$ d.f., we find $t_{.05} = 1.753$

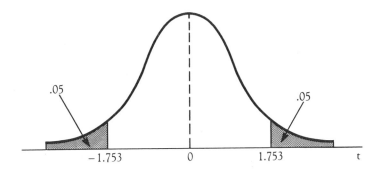

Exercise Set VII.3

1) $\hat{P}$ is approximately normal with $\mu_{\hat{P}} = p = .6$ and

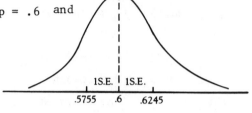

$$S.E. = \sqrt{\frac{pq}{n}} = \sqrt{\frac{(.6)(.4)}{400}} = .0245$$

.5755   .6   .6245
1 S.E. | 1 S.E.

3) The distribution of $\hat{P}$ is approximately normal with mean

$$\mu = .4 \quad \text{and} \quad S.E. = \sqrt{\frac{(.4)(.6)}{625}} = .0196$$

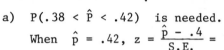

a) $P(.38 < \hat{P} < .42)$ is needed.
   When $\hat{p} = .42$, $z = \dfrac{\hat{p} - .4}{S.E.} =$

   $$\frac{.42 - .40}{.0196} = 1.02$$

   When $\hat{p} = .38$, $z = \dfrac{.38 - .40}{.0196} = -1.02$

   $P(.38 < \hat{P} < .42) = P(-1.02 < Z < 1.02)$
   $\qquad\qquad\qquad\qquad = 2(.3461) = .6922$

.3461   .3461

.38 .40 .42 $\hat{p}$

b) $P(\hat{P} > .44)$ is needed.

   When $\hat{p} = .44$, $z = \dfrac{.44 - .40}{.0196} = 2.04$

   $P(\hat{P} > .44) = P(Z > 2.04)$
   $\qquad\qquad\qquad = .5000 - .4793 = .0207$

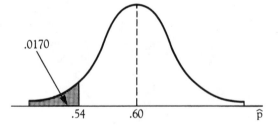

.0207

.40   .44 $\hat{p}$

5) $P(\hat{P} < .54)$ is needed.

   $$z = \frac{\hat{p} - .60}{S.E.} = \frac{.54 - .60}{\sqrt{\dfrac{(.60)(.40)}{300}}} = -2.12$$

   $P(\hat{P} < .54) = P(z < -2.12)$
   $\qquad\qquad\qquad = .5000 - .4830 = .0170$

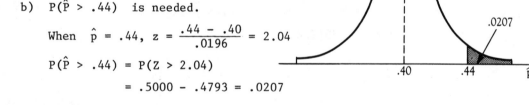

.0170

.54   .60 $\hat{p}$

7) $p = .45$
   $n = 500$
   $P(\hat{P} > .50)$ is needed.

   $$S.E. = \sqrt{\frac{(.45)(.55)}{500}} = .0222$$

   When $\hat{p} = .50$,

   $$z = \frac{.50 - .45}{.0222} = 2.25$$

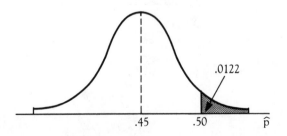

.0122

.45   .50 $\hat{p}$

$P(\hat{P} > .50) = P(z > 2.25)$

$= .5000 - .4878$

$= .0122$

9) $p = .67$ and $n = 300$

$P(.64 < \hat{P} < .70)$ is needed

S.E. $= \sqrt{\dfrac{(.67)(.33)}{300}} = .0271$

When $\hat{p} = .70$, $z = \dfrac{.70 - .67}{.0271} = 1.11$

By symmetry, $z = -1.11$ when $\hat{p} = .64$.

$P(.64 < \hat{P} < .70) = P(-1.11 < Z < 1.11)$

$= 2(.3665) = .7330$

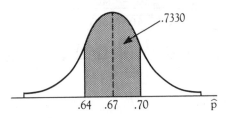

11) $p = .61$, $n = 124$

$P(\hat{P} < .55)$ is needed

S.E. $= \sqrt{\dfrac{(.61)(.39)}{124}} = .0438$

$z = \dfrac{.55 - .61}{.0438} = -1.37$

$P(\hat{P} < .55) = P(Z < -1.37)$

$= .5000 - .4147$

$= .0853$

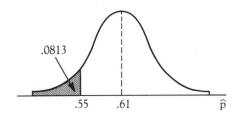

13) $p = .50$ and $n = 900$

$P(\hat{P} > .54)$ is needed

When $\hat{p} = .54$, $z = \dfrac{.54 - .50}{\sqrt{\dfrac{(.50)(.50)}{900}}} = 2.4$

$P(\hat{P} > .54) = P(Z > 2.4)$

$= .5000 - .4918 = .0082$

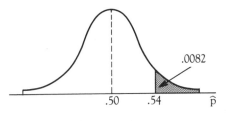

15) $p = \dfrac{1}{4} = .25$ and $n = 400$

$P(\hat{P} > .265)$ is needed

When $\hat{p} = .265$, $z = \dfrac{.265 - .250}{\sqrt{\dfrac{(.25)(.75)}{400}}} = .69$

$P(\hat{P} > .265) = P(Z > .69)$

$= .5000 - .2549 = .2451$

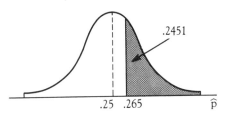

17) $P(.37 < \hat{p} < .43) = P(-z_1 < z < z_1) = .99$

Thus $z_1 = z_{.005} = 2.58$

$$2.58 = \frac{.43 - .40}{\sqrt{\frac{(.40)(.60)}{n}}}$$

$$n = \frac{(2.58)^2(.40)(.60)}{(.03)^2} = 1775.04$$

Take $n \geq 1776$

Exercise Set VII.4

1) $\sigma^2 = 23$, $n = 16$

$P(s^2 > 30)$ is needed

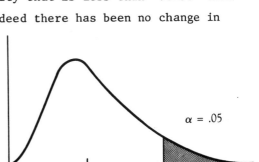

$\chi^2 = \dfrac{(n-1)s^2}{\sigma^2} = \dfrac{(16-1)(30)}{23} = 19.56$

For $\nu = 15$ d.f., $P(\chi^2 > 19.3) =$

$1.0000 - .80 = .20$

Thus the nearest estimate is

$P(s^2 > 30) \approx .20$.

3) $S > 20$ implies $s^2 > 400$

$P(s^2 > 400)$ is needed

$\chi^2 = \dfrac{(n-1)s^2}{\sigma^2} = \dfrac{(15-1)(400)}{225} = 24.89$

We note that for $\nu = 14$ d.f.,

$\chi^2_{.05} = 23.7$ (nearest entry to 24.88)

Thus $P(\chi^2 > 24.88) \approx .05$.

5) a) In the $\alpha = .05$ table at the intersection of the column headed 7
   and the row headed 3 is found $f_{.05}(5,7) = 3.14$.

   b) In the $\alpha = .01$ table at the intersection of the column headed 20
   and the row headed 17 is found $f_{.01}(20,17) = 3.16$.

7) $S_1 = 18.9$ $s_1^2 = 357.21$ $\nu_1 = 15$

   $S_2 = 11.2$ $s_2^2 = 125.44$ $\nu_2 = 20$

   $f = \dfrac{357.21}{125.44} = 2.8476$

   $f_{.025}(15,20) = 2.57$

   $f > f_{.025}$.

This observed increase has a probability that is less than .025. This
is a highly unlikely occurrence if indeed there has been no change in
the variance of the emissions.

9) $\mu_F = \dfrac{\nu_2}{\nu_2 - 2}$

   a) $\mu_F = \dfrac{3}{3-2} = 3$

   Large values of f exceed, say

   $f_{.05} = 9.01$.

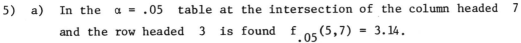

b) $\mu_F = \dfrac{6}{6-2} = 1.5$

Large values of f exceed,
say, $f_{.05} = 4.39$.

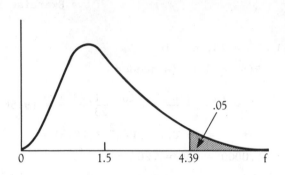

c) $\mu_F = \dfrac{40}{40-2} = \dfrac{40}{38} = 1.053$

Large values of f exceed,
say, $f_{.05} = 2.45$.

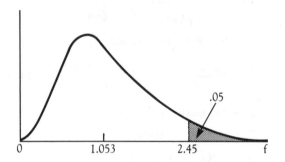

11) Look in the column $\nu_1 = 10$. As we move down the column, $f(10,\nu_2)$ decreases.

Look in the row $\nu_2 = 10$. As we move across the row, $f(\nu_1,10)$ becomes smaller once ν_1 exceeds 2.

As we move across and down the table, $f(\nu_1,\nu_2)$ decreases ultimately to 1.

Since $f = \dfrac{s_1^2}{s_2^2}$, the probability must be very high that $\dfrac{s_1^2}{s_2^2}$ is near 1 when both ν_1 and ν_2 are large, i.e., the sample variances are almost equal. Or, if we are considering $f = \dfrac{s_1^2/\sigma_1^2}{s_2^2/\sigma_2^2}$, the ratios s_1^2/σ_1^2 and s_2^2/σ_2^2 must be almost equal.

13) a) $f_{.95}(4,6) = \dfrac{1}{f_{.05}(6,4)}$

$= \dfrac{1}{6.16} = .1623$

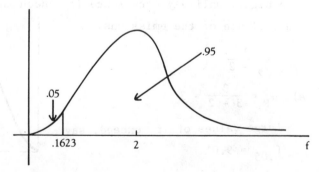

b) $f_{.99}(10,12) = \dfrac{1}{f_{.01}(12,10)}$

                $= \dfrac{1}{4.71} = .2123$

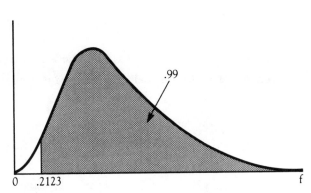

Exercise Set VII.5

1) Thirteen of the 16 specialists underestimated the size of sphere number 1.
    Seven of the 16 specialists underestimated the size of sphere number 12.

2) 1) May not be totally localized.

    2) May not be spherical.

    3) May move around somewhat as it is measured. The largest cross sectional
        measurement might be used.

3) The average bias is $\bar{b} = \frac{-4.63}{12} = -.3858$.
    The average diameter of the spheres is

    $$\bar{x} = \frac{66.8}{12} = 5.567$$

    $$\frac{.3858}{5.567} \times 100 = 6.93\%   \text{(Relative bias)}$$

4) Specialists 2,3, 7, 8, 12, and 13 all the time.
    All rounded to the nearest centimeter for the larger spheres.

5) For the 25% criterion, $\frac{12}{64} = .1875 = 18.75\%$.
    For the 50% criterion, $\frac{5}{64} = .0781 = 7.81\%$.

Chapter Test

1) a) $t_{.05} = 1.753$

   b) $f_{.05}(8,10) = 3.07$

   c) $f_{.95}(6,4) - \dfrac{1}{f_{.05}(4,6)} = \dfrac{1}{4.53} = .2207$

2) The distribution is approximately normal with $\mu = 78$ and $\sigma = S.E. = \dfrac{18}{\sqrt{36}} = 3.$

3) a) The distribution of $\hat{P}$ is approximately normal.

   $$\mu = p = .65 \quad \text{and} \quad \sigma = S.E. = \sqrt{\dfrac{pq}{n}} = \sqrt{\dfrac{(.65)(.35)}{200}} = .0337$$

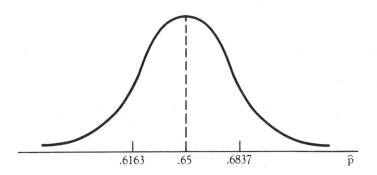

.6163   .65   .6837 $\hat{p}$

   b) $P(\hat{p} < .55)$ is needed

   $z = \dfrac{.55 - .65}{.0337} = -2.96$

   $P(\hat{p} < .55) = P(Z < -2.96) = .5000 - .4985 = .0015$

4) Note that since the population is normal, the sample means have a normal distribution.

   $P(-.5 < z < .5)$ is needed since $z = \dfrac{x - \mu}{S.E.} = \pm.5$

   $P(-.5 < z < .5) = .1915 + .1915 = .3830$

5) $p = .68,\ n = 100$

   $P(\hat{P} < .50)$ is needed.

   $z = \dfrac{.50 - .68}{\sqrt{\dfrac{(.68)(.32)}{100}}} = -3.86$

   $P(\hat{P} < .50) = P(Z < -3.86) = .0000$ (to 4 decimal places)

6) $P(\overline{X} > 760)$ is needed

$$S.E. = \frac{250}{\sqrt{400}} = 12.50$$

$$z = \frac{760 - 743}{12.50} = 1.36$$

$P(\overline{X} > 760) = P(z > 1.36) = .5000 = .4131 = .0869$

7) $P(\overline{X} \geq 108)$ is needed

$$z = \frac{108 - 100}{\dfrac{10}{\sqrt{10}}} = 2.53$$

$P(\overline{X} \geq 108) = P(z \geq 2.53) = .5000 - .4943 = .0057$

8) a) Note: This involves the distribution of X.

   $P(X > 840)$ is needed

   $$z = \frac{840 - 800}{100} = .4$$

   $P(X > 840) = P(Z > .4) = .5000 - .1554 = .3446 = 34.46\%$

   b) Note: This involves the distribution of $\overline{X}$.

   $P(\overline{X} > 840)$ is needed

   $$z = \frac{840 - 800}{\dfrac{100}{\sqrt{36}}}$$

   $P(\overline{X} > 840) = P(Z > 2.4) = .5000 - .4918 = .0082 = .82\%$

9) $\chi^2 = \dfrac{(n - 1)s^2}{\sigma^2}$

$$\chi^2 = \frac{(n - 1)s^2}{\sigma^2} = \frac{(23 - 1)60}{43} = 30.70$$

For $\nu = 22$ d.f. we note $\chi^2_{.10} = 30.8$.

Thus the probability of obtaining such a sample variance is approximately .10.

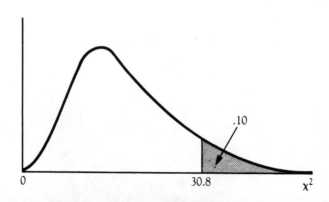

10)   a) $f = \dfrac{s_2^2}{s_1^2} = \dfrac{2.63}{.49} = 5.367$

$f_{.05}(12,11) = 2.7876$

Since $f > f_{.05}$, we conclude that there must have been some effect due to the treatment.

b)   Normality assumptions for the growths under both sets of conditions.

11) $\mu = \dfrac{\nu_2}{\nu_2 - 2} = \dfrac{10}{8} = 1.25$

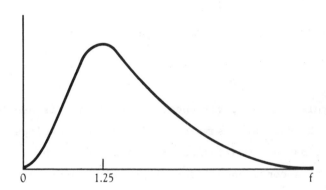

12)

samples	$\overline{x}$
1,1	1.0
1,3	2.0
1,4	2.5
3,1	2.0
3,3	3.0
3,4	3.5
4,1	2.5
4,3	3.5
4,4	4.0

$\overline{x}$	$P(\overline{x})$
1.0	$\frac{1}{9}$
2.0	$\frac{2}{9}$
2.5	$\frac{2}{9}$
3.0	$\frac{1}{9}$
3.5	$\frac{2}{9}$
4.0	$\frac{1}{9}$

Chapter VIII
Estimation

Often a population mean, variance, or proportion is needed and must be estimated from the data of a sample. This chapter develops techniques for producing these estimates. Confidence intervals are at the center of the discussion. The end points of a confidence interval provide bounds for the quantity of interest. The confidence level is a measure of the reliability of this interval estimate.

Concepts/Techniques

1) Point Estimates

   A point estimate is a single numerical estimate of a population parameter. Generally point estimates are derived from samples.

   Example: A sample mean $\bar{x}$ is a point estimate of the population mean μ.

   Example: A sample proportion $\hat{p}$ is a point estimate of the population proportion p.

2) Confidence Intervals

   A population parameter is rarely equal to its point estimate. Still it should be located on some interval near the point estimate. The longer this interval, the more confident we are that the interval contains the parameter.

In this text, a confidence interval for the population mean μ is an interval of the form

$$\bar{x} - E < \mu < \bar{x} + E.$$

Similarly, a confidence interval for the population proportion p is of the form

$$\hat{p} - E < p < \hat{p} + E.$$

The number E is called the maximum error of the estimate. The numbers $\bar{x} \pm E$ are called the upper and lower limits of the confidence interval.

Example: Suppose a sample furnishes the mean $\bar{x} = 175$ and on the basis of this, we construct the confidence interval

$$175 - 10 < \mu < 175 + 10 \quad \text{or simply}$$
$$165 < \mu < 185$$

The maximum error of the estimate is $E = 10$ and the upper and lower limits of the confidence interval are 185 and 165 respectively.

Example: Suppose a sample has produced the proportion $\hat{p} = .60$ (60%). If the maximum error of the estimate is $E = .05$ (5%), then the confidence interval for the population proportion p is

$$.60 - .05 < p < .60 + .05 \quad \text{or simply}$$
$$.55 < p < .65.$$

The upper and lower limits are .65 and .55 respectively.

3) The Confidence Level (Confidence Coefficient)

The level of confidence associated with an interval $\bar{x} - E < \mu < \bar{x} + E$ is the probability that the sample mean will differ from the population mean by at most E. This may be interpreted as the proportion of times that the constructed confidence intervals "capture" the mean μ if the sampling experiment is performed repeatedly. It is thus a measure of the reliability of the procedure used in constructing the confidence interval.

Similar statements apply to confidence intervals for the proportion p.

   Example: Suppose 150 - 10 < μ < 150 + 10 is a 95% confidence
   interval for μ. This means first that there is a 95% chance that
   a randomly selected sample will have a mean that is within E = 10
   units of μ. Moreover, the procedure used in constructing this
   interval (explained below) will produce an interval which contains
   μ about 95% of the time.

4) Computing the Error of the Estimate E (Large Sample Case)
   For a confidence level c such as .90, .95, or .99, let $\alpha = 1 - c$.
   Then a large sample confidence interval for μ has the error of estimate

$$E = z_{\alpha/2}(S.E.) = z_{\alpha/2} \frac{\sigma}{\sqrt{n}} \; .$$

Similarly for the proportion p,

$$E = z_{\alpha/2}(S.E.) = z_{\alpha/2} \sqrt{\frac{pq}{n}}$$

Obviously we rarely ever know σ and it must be approximated by the
standard deviation s of the sample. Thus for $n \geq 30$

$$E = z_{\alpha/2} \frac{s}{\sqrt{n}} \; .$$

Similarly, the product pq must be approximated by $\hat{p}\hat{q}$ from the sample.
Then

$$E = z_{\alpha/2} \sqrt{\frac{\hat{p}\hat{q}}{n}} \; .$$

   Example: A sample of n = 100 hospital records were surveyed as
   to the length of stay of non-terminally ill patients. These times
   had mean $\bar{x}$ = 9.8 (days) and standard deviation s = 3.2 days.
   Find a 95% confidence interval for the mean length of stay μ of
   hospital patients.
      1) c = .95
      2) $\alpha = 1 - .95 = .05, \frac{\alpha}{2} = .025$
      3) $z_{.025} = 1.96$
      4) $E = z_{\alpha/2} \frac{s}{\sqrt{n}} = (1.96) \frac{3.2}{\sqrt{100}} = .63$ days

5)   The limits of the confidence interval are $\overline{x} \pm E = 9.8 \pm .63$
     or simply   9.17   and   10.43.

6)   The 95% confidence interval is

$$9.17 < \mu < 10.43.$$

Example: Of   1600   graduating college seniors,   x = 400   indicated
that they wished they had chosen a different major.   Find a 99%
confidence interval for the proportion of graduating seniors who are
so disposed.

Note first that the sample proportion is $\hat{p} = \dfrac{400}{1600} = .25.$

1)   c = .99

2) $\alpha = 1 - .99 = .01, \frac{\alpha}{2} = .005$

3) $z_{.005} = 2.58$

4) $E = z_{\alpha/2} \sqrt{\dfrac{\hat{p}\hat{q}}{n}} = 2.58\sqrt{\dfrac{(.25)(.75)}{1600}} = .028$

5) $\hat{p} \pm E = .25 \pm .028.$ These limits are   .222   and   .278

6)   The 99% confidence interval is

$$.222 < p < .278.$$

This we are 99% certain that somewhere between 22.2% and 27.8%
of the graduating college seniors wished they had chosen a
different major.

5)   Estimating the Sample Size

Suppose we plan to estimate the population mean μ with the mean $\overline{x}$
of a sample.   How large a sample is needed if the error of this estimate
E   is not to exceed a given number with a certainty $c = 1 - \alpha$.
If an estimate of the population proportion σ is available, then the
minimal sample size to accomplish this is obtained by solving for   n
in the equation:

$$E = z_{\alpha/2} \frac{\sigma}{\sqrt{n}} .$$

This yields $n = \left(\dfrac{z_{\alpha/2}\sigma}{E} \right)^2.$

If a similar type of estimate is needed for the population proportion   p,
the sample size is obtained by solving

$$E = z_{\alpha/2} \sqrt{\frac{pq}{n}} \; .$$

This yields $n = \left(\dfrac{z_{\alpha/2}}{E} \right)^2 pq.$

Here also we need an estimate of the product pq. If this is not available, use

$$pq = \frac{1}{4} \; .$$

Example: An air force agency needs an estimate of the mean refueling time for a particular type of fighter aircraft that will be in error by at most 1.5 minutes with a 95% certainty. How many aircrafts should be tested if they know the standard deviation of such refueling times to be around $\sigma = 6$ minutes?

Note that $\bar{x} - 1.5 < \mu < \bar{x} + 1.5$ is to be a 95% confidence interval. Thus E = 1.5.

For c = .95, α = .05, $\dfrac{\alpha}{2}$ = .025 and $z_{.025}$ = 1.96.

Substituting into $E = z_{\alpha/2} \dfrac{\sigma}{\sqrt{n}}$, we obtain

$$1.5 = (1.96) \frac{6}{\sqrt{n}} \; .$$

Solving for n yields

$$n = \left(\frac{(1.96)6}{1.5} \right)^2 = 61.47.$$

Thus the sample size should be at least 62.

Example: A medical center needs an estimate of the proportion of positive reactions to a tubercular skin test that is accurate to within 3% with a 99% certainty. What size sample should be used if they believe that the proportion of positive reactions is around p = .20?

In this case E = .03 = 3% and we want $\hat{p} - .03 < p < \hat{p} + .03$ to be a 99% confidence interval.

For c = .99, $\dfrac{\alpha}{2}$ = .005 and $z_{.005}$ = 2.58.

Substituting into $E - z_{\alpha/2} \sqrt{\dfrac{pq}{n}}$ using p = .2 and q = .8 we obtain

$$.03 = 2.58 \sqrt{\frac{(.2)(.8)}{n}} \; .$$

Solving for n yields

$$n = \left(\frac{2.58}{.03} \right)^2 (.2)(.8) = 1183.36.$$

Thus at least 1184 people should be tested.

Example: Suppose a 95% confidence interval is needed for the proportion p of newlyweds who plan on a family with exactly 2 children. What size sample is needed if the maximum error of the estimate E is to be at most .04 (4%)?

The 95% confidence interval is to be of the form

$$\hat{p} - .04 < p < \hat{p} + .04.$$

Thus $E = .04$, $\alpha = .05$, $\frac{\alpha}{2} = .025$, $z_{\alpha/2} = 1.96$.
Since no estimate of pq is available, we use $pq = \frac{1}{4} = .25$.
Thus we obtain n by solving

$$.04 = 1.96 \sqrt{\frac{.25}{n}}.$$

We find

$$n = \left(\frac{1.96}{.04} \right)^2 (.25) = 600.25.$$

Thus the sample size should be at least 601.

6) The Maximum Error of the Estimate E (Small Sample Case)

When the sample size is small, the approximation of the population standard deviation σ by that of the sample s introduces so much error that the normal distribution may no longer be used with $\overline{X}$. If the population being sampled is normal, we can, however, use the t distribution.

The maximum error of the estimate E is for the confidence interval then computed by

$$E = t_{\alpha/2} \frac{s}{\sqrt{n}}.$$

The number ν of degrees, freedom to be used with the z distribution is $\nu = n - 1$.

Example: A sample of n = 9 guinea pigs are sedated with a certain dosage of a new drug. The times X until the pigs are fully awake

are recorded and found to have mean $\bar{x}$ = 96.3 minutes and standard
deviation s = 36.0 minutes. Find a 90% confidence interval
for the mean time μ for such guinea pigs to recover from this
sedative.

   1) c = .90

   2) α = 1 - .90 = .10 and $\frac{\alpha}{2}$ = .05

   3) Using ν = 9 - 1 = 8, we find $t_{.05}$ = 1.86

   4) The maximum error of the estimate is

$$E = t_{.05} \frac{s}{\sqrt{n}} = (1.86) \frac{36.0}{\sqrt{9}} = 22.32.$$

   5) The upper and lower limits are $\bar{x} \pm E$ = 96.3 ± 22.32.
       These are 73.98 and 118.62.

   6) The 90% confidence interval is

$$73.98 < \mu < 118.62.$$

Additional Examples

Example: Give the formula for computing the maximum error of the
estimate E for a 92% confidence
interval for the mean μ. Assume
the sample size is large.
Since c = .92, α = 1 - .92 = .08,
and $\frac{\alpha}{2}$ = .04.
Thus $E = z_{.04} \frac{\sigma}{\sqrt{n}} \approx 1.75 \frac{s}{\sqrt{n}}$.

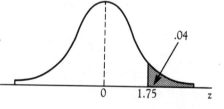

Example: The following are 95%
confidence intervals for the
mean birth weight (in ounces) of a certain breed of dog. With which
would the point estimate μ = 13 be more reliable?

   a) 12.3 < μ < 13.7   b) 11 < μ < 15

The interval 12.3 < μ < 13.7 has the smallest maximum
error of the estimate E, namely E = .7. Thus the estimate
μ = 13 from this study would be more reliable.

Example: A report stated that the mean water temperature for a
lake is 18 ± 2 degrees C. If this is a confidence interval
derived from n = 49 observations having $\bar{x}$ = 18 and s = 7,

find the level of confidence.

The interval is $18 - 2 < \mu < 18 + 2$.   Thus $E = 2$.

Substituting into $E = z_{\alpha/2} \dfrac{s}{\sqrt{n}}$ we obtain

$$2 = z_{\alpha/2} \frac{7}{\sqrt{49}} = z_{\alpha/2}.$$

Thus $z_{\alpha/2} = 2.0$, $\dfrac{\alpha}{2} = .025$, $\alpha \approx .05$,   and $c = .95 = 95\%$.

Example: We wish to design a study for estimating the mean checking account balances of families to within \$100 with a 95% certainty. How large a sample of account balances is needed if the standard deviation of such account balances is thought to be around \$450? The confidence interval will be of the form

$$\overline{x} - 100 < \mu < \overline{x} + 100.$$

Thus $E = 100$. For a 95% confidence interval $z_{.025} = 1.96$. Substituting these and the estimate $\sigma = 450$ into $E = z_{\alpha/2} \dfrac{\sigma}{\sqrt{n}}$, we obtain

$$100 = 1.96 \left(\frac{450}{\sqrt{n}} \right).$$

Solving for   n   yields

$$n = \left(\frac{(1.96)(450)}{100} \right)^2 = 77.8.$$

Thus   n   should be at least   78.

Example: An auto parts manufacturer needs an estimate of the mean assembly time μ for a certain component. Suppose a sample of $n = 16$ assembly times had mean $\overline{x} = 34$ minutes and standard deviation $s = 8$ minutes. Find a 99% confidence interval for μ if such assembly times are approximately normal.

1) $c = .99$

2) $\alpha = 1 - .99 = .01$ and $\dfrac{\alpha}{2} = .005$

3)   For $\nu = 16 - 1 = 15$, we find $t_{.005} = 2.947$

4) $E = t_{\alpha/2} \dfrac{s}{\sqrt{n}} = \dfrac{(2.947)(8)}{\sqrt{16}} = 5.9$

5)   The upper and lower limits are $34 + 5.9 = 39.9$ and $34 - 5.9 = 28.1$ respectively.

6)   The 99% confidence interval is

$$28.1 < \mu < 39.9.$$

Example: Assuming a sample size of n = 50, which would be longer, a 95% confidence interval constructed using $z_{.025}$ or one using $t_{.025}$?

Since $t_{.025} > z_{.025}$, the confidence interval constructed using the t distribution would be longer.

Example: Suppose that of 700 patients who were declared in need of surgery, 280 sought a second opinion. Find a 95% confidence interval for the proportion of surgery candidates who seek a second opinion.

1) c = .95

2) $\alpha = .05, \frac{\alpha}{2} = .025$

3) $z_{.025} = 1.96$ and $\hat{p} = \frac{280}{700} = .40$ (40%)

4) $E = z_{\alpha/2}\sqrt{\frac{\hat{p}\hat{q}}{n}} = 1.96\sqrt{\frac{(.40)(.60)}{700}} = .036$

5) The upper limit of the confidence interval is
    $\hat{p} + E = .40 + .036 = .436.$ The lower limit is
    $\hat{p} - E = .40 = .036 = .364.$

6) The 95% confidence interval is

$$.364 < p < .436.$$

Example: A direct mail advertiser needs an estimate of the proportion of people who read advertisements which are received by mail. How many people would need be sampled to estimate this proportion to within .02 (2%) with a 95% certainty?

The desired sample size is associated with the 95% confidence interval

$$\hat{p} - .02 < p < \hat{p} + .02.$$

The sample size is found from $E = z_{.025}\sqrt{\frac{pq}{n}}$.

Using E = .02, $z_{.025} = 1.96$ and estimating pq by $pq = \frac{1}{4} = .25$ (since there is no available estimate of p), we obtain

$$.02 = 1.96\sqrt{\frac{.25}{n}}.$$

Solving for n yields

$$n = \left(\frac{1.96}{.02}\right)(.25) = 2401.$$

Thus at least 2401 people should be used.

Calculator Usage Illustration

Calculate the maximum error of the estimate of a large sample confidence
interval for μ.

Find $E = (1.96) \dfrac{(38)}{\sqrt{40}}$.

Method 1)   Direct evaluation.

$$1.96 \; \boxed{\times} \; 38 \; \boxed{\div} \; 40 \; \boxed{\sqrt{}} \; \boxed{=} \; \underline{11.7763220}$$

Method 2)   Multiply the numerator by the reciprocal of the denominator.

$$40 \; \boxed{\sqrt{}} \; \boxed{\tfrac{1}{x}} \; \boxed{\times} \; 1.96 \; \boxed{\times} \; 38 \; \boxed{=} \; \underline{11.7763220}$$

Calculator Usage Illustration

Calculate the limits of a confidence interval for μ.

Find 17.46 ± 11.78.

Method: Use the fact that lower limit = upper limit $-$ 2E.   When the maximum
             error E is entered, store it for use in finding the lower limit.

$$17.46 \; \boxed{+} \; 11.78 \; \boxed{\text{STO}} \; \boxed{=} \; \underline{29.24} \; \boxed{-} \; \boxed{\text{RCL}} \; \boxed{\times} \; 2 \; \boxed{=} \; \underline{5.68}$$

Calculator Usage Illustration

Calculate the maximum error of the estimate of a large sample confidence
interval for p.

Find $E = 1.64 \sqrt{\dfrac{(.44)(.56)}{300}}$.

Method: Evaluate the radical first, then multiply by 1.64.

$$.44 \; \boxed{\times} \; .56 \; \boxed{\div} \; 300 \; \boxed{=} \; \boxed{\sqrt{}} \; \boxed{\times} \; 1.64 \; \boxed{=} \; \underline{.0470}$$

Exercise Set VIII.1

1) a) 95% The limits are of the form $\bar{x} \pm z_{.025}$ (S.E.).

Thus $\alpha = 2(.025) = .05$ and $c = 1 - .05 = .95 = 95\%$.

  b) 98% The limits are of the form $\bar{x} \pm z_{.01}$ (S.E.).

Thus $\alpha = 2(.01) = .02$ and $c = 1 - .02 = .98 = 98\%$.

  c) 90% The limits are of the form $\bar{x} \pm z_{.05}$ (S.E.).

Thus $\alpha = 2(.05) = .10$ and $c = 1 - .10 = .90 = 90\%$.

  d) 68.26% The limits are of the form $\bar{x} \pm 1$ (S.E.) $= \mu \pm z_{.1587}$ (S.E.).

Thus $\alpha = 2(.1587) = .3174$ and $c = 1 - .3174 = .6826 = 68.26\%$.

3) a) $\frac{\alpha}{2} = .01$ and $\alpha = .02$. Thus $c = 1 - .02 = .98 = 98\%$

  b) $\frac{\alpha}{2} = .05$ and $\alpha = .10$. Thus $c = 1 - .10 = .90 = 90\%$

  c) $\frac{\alpha}{2} = .02$ and $\alpha = .04$. Thus $c = 1 - .04 = .96 = 96\%$

5) S.E. $= \dfrac{18}{\sqrt{100}} = 1.8$

  a) $c = .90$, $\alpha = 1 - .90 = .10$, and $\frac{\alpha}{2} = .05$.

$E = z_{.05}$(S.E.) $= (1.64)(1.8) = 2.95$

The confidence interval is:

$$70 - 2.95 < \mu < 70 + 2.95 \quad \text{or equivalently}$$
$$67.05 < \mu < 72.95$$

  b) $c = .95$, $\alpha = 1 - .95 = .05$, and $\frac{\alpha}{2} = .025$

$E = z_{.025}$(S.E.) $= (1.96)(1.8) = 3.53$

The confidence interval is:

$$70 - 3.53 < \mu < 70 + 3.53 \quad \text{or equivalently}$$
$$66.27 < \mu < 73.53$$

  c) $c = .99$, $\alpha = 1 - .99 = .01$, and $\frac{\alpha}{2} = .005$

$E = z_{.005}$(S.E.) $= (2.57)(1.8) = 4.63$

The confidence interval is:

$$70 - 4.63 < \mu < 70 + 4.63 \quad \text{or equivalently}$$
$$65.37 < \mu < 74.63$$

  d) $c = .80$, $\alpha = 1 - .80 = .20$, and $\frac{\alpha}{2} = .10$

$E = z_{.10}$(S.E.) $= (1.38)(1.8) = 2.30$

The confidence interval is:

$$70 - 2.30 < \mu < 70 + 2.30 \quad \text{or equivalently}$$
$$67.70 < \mu < 72.30$$

7) $c = .90$, $\alpha = 1 - .90 = .10$, $\frac{\alpha}{2} = .05$, and $z_{.05} = 1.64$

S.E. $= \dfrac{12}{\sqrt{140}} = 1.01$

$E = (z_{.05})(\text{S.E.}) = (1.64)(1.01) = 1.66$

The confidence interval is

$$38.50 - 1.66 < \mu < 38.50 + 1.66 \quad \text{or equivalently}$$
$$36.84 < \mu < 40.16$$

9) Note that the interval is of the form $72 - 3 < \mu < 72 + 3$. Thus

$$E = 3 = (z_{\alpha/2})(\text{S.E.}) = (z_{\alpha/2})\left(\frac{8.40}{\sqrt{49}}\right) = (z_{\alpha/2})(1.20)$$

Solving for $z_{\alpha/2}$ yields

$$z_{\alpha/2} = \frac{3}{1.20} = 2.5.$$

The area to the right of $z_{\alpha/2} = 2.5$ is $\frac{\alpha}{2} = .0062$.
Thus $\alpha = .0124$ and $c = 1 - .0124 = .9876 = 98.76\%$.

11) Note that the maximum error of the estimate is $E = 5$.
Further $c = .997$, $\alpha = 1 - .997 = .003$, $\frac{\alpha}{2} = .0015$, and $z_{.0015} = 2.97$.
$E = (z_{.0015})(\text{S.E.}) = (2.97)\left(\dfrac{20}{\sqrt{n}}\right) = \dfrac{59.4}{\sqrt{n}}$

Equating these values of E, we obtain:

$$E = 5 = \frac{59.4}{\sqrt{n}}.$$

Solving: $\sqrt{n} = \dfrac{59.4}{5} = 11.88$

$$n = (11.88)^2 = 141.13$$

$$\text{Choose} \quad n \geq 142.$$

13) $c = .99$, $\alpha = 1 - .99 = .01$, $\frac{\alpha}{2} = .005$, and $z_{.005} = 2.58$

S.E. $= \dfrac{.53}{\sqrt{36}} = .09$

$E = (z_{.005})(\text{S.E.}) = (2.58)(.09) = .23$

The upper and lower limits are $3.40 + .23 = 3.63$ and $3.40 - .23 = 3.17$.

The 99% confidence interval is

$$3.17 < \mu < 3.63.$$

For 100 of these castings, the confidence interval is

$$317 < \mu' < 363.$$

15) $E = (z_{.025})(S.E.) = (1.96) \dfrac{54}{\sqrt{40}} = 16.73$

The limits are $1845 + 16.73 = 1861.73$ and $1845 - 16.73 = 1828.27$.
The confidence interval is

$$1828.27 < \mu < 1861.73.$$

17) White Group

$E = (1.96) \dfrac{18}{\sqrt{524}} = 1.54$

$148.9 - 1.54 < \mu < 148.9 + 1.54$ or

$147.36 < \mu < 150.44$

Black Group

$E = (1.96) \dfrac{18}{\sqrt{308}} = 2.01$

$165.0 - 2.01 < \mu < 165.0 + 2.01$ or

$162.99 < \mu < 167.01$

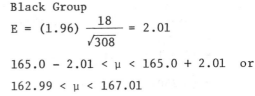

19)

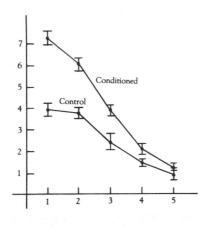

Exercise Set VIII.2

1) $\nu = 15 - 1 = 14$ d.f. and $t_{.025} = 2.145$

$E = (t_{.025})(S.E.) = (2.145)\dfrac{5.62}{\sqrt{15}} = 3.11$

$41.2 - 3.11 < \mu < 4.12 + 3.11$ or equivalently

$38.09 < \mu < 44.31$

3) $\nu = 9 - 1 = 8$ and $t_{.025} = 2.306$

$E = (t_{.025})(S.E.) = (2.306)\dfrac{2.3}{\sqrt{9}} = 1.77$

Upper Limit $= 17.1 + 1.77 = 18.87$

Lower Limit $= 17.1 = 1.77 = 15.33$

$15.33 < \mu < 18.87$

5) $\nu = 10 - 1 = 9$ d.f. and $t_{.005} = 3.250$

$E = (3.250)\dfrac{.5}{\sqrt{10}} = .51$

Upper Limit $= 7.9 + .51 = 8.41$

Lower Limit $= 7.9 = .51 = 7.39$

$7.39 < \mu < 8.41$

7) $\nu = 18 - 1 = 17$ d.f. and $t_{.05} = 1.740$

$E = (1.740)\dfrac{145}{\sqrt{18}} = 59.5$

Upper Limit $= 1246 + 59.5 = 1305.5$

Lower Limit $= 1246 - 59.5 = 1186.5$

$1186.5 < \mu < 1305.5$

9) The confidence interval constructed with $t_{\alpha/2}$ is longer than that with $z_{\alpha/2}$ since $t_{\alpha/2} > z_{\alpha/2}$.

11) Before Group (I)

$\nu = 38 - 1 = 37$ d.f. and $t_{.025} = 2.021$ (nearest entry)

$E = (2.021)\dfrac{2.4}{\sqrt{38}} = .79$

$$\text{Upper Limit} = 3.8 + .79 = 4.59$$
$$\text{Lower Limit} = 3.8 - .79 = 3.01$$

$$3.01 < \mu < 4.59$$

0-.5 Group (II)

$\nu = 21 - 1 = 20$ d.f. and $t_{.025} = 2.086$

$E = (2.086) \dfrac{1.9}{\sqrt{21}} = .86$

$$\text{Upper Limit} = 3.6 + .86 = 4.46$$
$$\text{Lower Limit} = 3.6 - .86 = 2.74$$

$$2.74 < \mu < 4.46$$

.51-1.00 Group (III)

$\nu = 12 - 1 = 11$ d.f. and

$t_{.025} = 2.201$

$E = (2.201) \dfrac{1.2}{\sqrt{12}} = .76$

$$1.74 < \mu < 3.26$$

1.0-2.0 Group (IV)

$\nu = 12 - 1 = 11$ d.f. and

$t_{.025} = 2.201$

$E = (2.201) \dfrac{1.6}{\sqrt{12}} = 1.02$

$$1.38 < \mu < 3.42$$

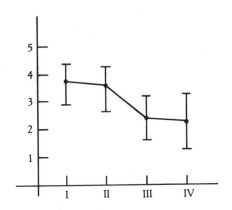

13) Cafteayl Analysis

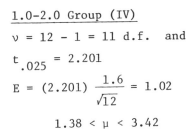

x	x^2
26	676
16	256
20	400
26	676
34	1156
30	900
19	361
16	256
75	5625
24	576
286	10882

$\bar{x} = \dfrac{286}{10} = 28.6$

$s = \sqrt{\dfrac{10882 - \dfrac{(286)^2}{10}}{9}} = 17.32$

$E = (2.262) \dfrac{17.32}{\sqrt{10}} = 12.39$

$$\text{Upper Limit} = 28.6 + 12.39 = 40.99$$
$$\text{Lower Limit} = 28.6 - 12.39 = 16.21$$

$$16.21 < \mu < 40.99$$

Coumaroyl Analysis

x	x^2
7	49
4	16
7	49
6	36
10	100
8	64
6	36
6	36
21	144
26	676
101	1503

$$\bar{x} = \frac{101}{10} = 10.1$$

$$s = \sqrt{\frac{1503 - \frac{(101)^2}{10}}{9}} = 7.32$$

$$E = (2.262)\frac{7.32}{\sqrt{10}} = 5.24$$

Upper Limit = 10.1 + 5.24 = 15.34

Lower Limit = 10.1 − 5.24 = 4.87

$$4.87 < \mu < 15.34$$

Exercise Set VIII.3

1) $E = (z_{.025})(S.E.) = 1.96\sqrt{\dfrac{(.575)(.425)}{400}} = .0484$

Upper Limit = .575 + .0484 = .6234

Lower Limit = .575 − .0484 = .5266

.5266 < p < .6234

3) $\hat{p} = .315$

$E = (z_{.005})(S.E.) = (2.57)\sqrt{\dfrac{(.315)(.685)}{435}} = .0572$

Upper Limit = .315 + .0572 = .3722

Lower Limit = .315 − .0572 = .2578

.2578 < p < .3722

5) $E = (z_{.025})(S.E.) = 1.96\sqrt{\dfrac{(.22)(.78)}{450}} = .0383$

Upper Limit = .22 + .0383 = .2583

Lower Limit = .22 − .0383 = .1817

.1817 < p < .2583

18.17% of 120 = (.1817)(120) = 21.8

25.83% of 120 = (.2583)(120) = 31.0

Thus the expected number of empty seats due to drops is between 22 and 32 inclusively (with a 95% certainty).

7) $\hat{p} = \dfrac{1400}{2000} = .70$

$E = (z_{.025})(S.E.) = 1.96\sqrt{\dfrac{(.70)(.30)}{2000}} = .0201$

Upper Limit = .70 + .0201 = .7201

Lower Limit = .70 − .0201 = .6799

.6799 < p < .7201

9) $E = (z_{.05})(S.E.) = 1.64\sqrt{\dfrac{(.65)(.35)}{400}} = .0391$

Upper Limit = .65 + .0391 = .6891

Lower Limit = .65 − .0391 = .6109

$$.6109 < p < .6891$$

11) $E = .01 = (z_{.025})(S.E.) = 1.96\sqrt{\dfrac{(.5)(.5)}{n}} = \dfrac{(1.96)(.5)}{\sqrt{n}} = \dfrac{.98}{\sqrt{n}}$

Solve: $.01 = \dfrac{.98}{\sqrt{n}}$

$$\sqrt{n} = \dfrac{.98}{.01} = 98$$

$$n = 9604.$$

Take $n \ge 9604$.

Note that we have used $p = \dfrac{1}{2} = .5$ in computing the standard error since no estimate of p is available.

13) $E = .02 = 1.96\sqrt{\dfrac{(.6)(.4)}{n}} = \dfrac{.9602}{\sqrt{n}}$

Solve: $.02 = \dfrac{.9602}{\sqrt{n}}$

$$\sqrt{n} = \dfrac{.9602}{.02} = 48.01$$

$$n = 2304.96$$

Take $n \ge 2305$.

15) <u>No Reminder Group (I)</u>

$E = 1.96\sqrt{\dfrac{(.55)(.45)}{109}} = .0934$

Upper Limit = .55 + .0934 = .6434
Lower Limit = .55 − .0934 = .4566

$$.4566 < p < .6434$$

<u>Letter Reminder Group (II)</u>

$E = 1.96\sqrt{\dfrac{(.837)(.163)}{92}} = .0755$

Upper Limit = .837 + .0755 = .9125
Lower Limit = .837 − .0755 = .7615

$$.7615 < p < .9125$$

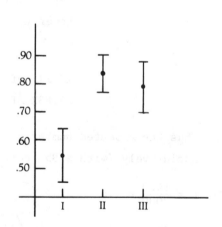

Telephone Reminder Group (III)

$$E = 1.96\sqrt{\frac{(.80)(.20)}{80}} = .0877$$

Upper Limit = .80 + .0877 = .8877

Lower Limit = .80 = .0877 = .7123

.7123 < p < .887

17) $E = (z_{.05})(S.E.) = 1.64\sqrt{\frac{(.173)(.827)}{2943}} = .0114$

Upper Limit = .173 + .0114 = .1844

Lower Limit = .173 − .0114 = .1616

.1616 < p < .1844

Chapter Test

1) $c = .90$, $\alpha = .10$, $\frac{\alpha}{2} = .05$, $z_{.05} = 1.64$

   S.E. $= \dfrac{14.3}{\sqrt{400}} = .715$

   $E = (1.64)(.715) = 1.17$

   $82 - 1.17 < \mu < 82 + 1.17$ or

   $80.83 < \mu < 83.17$

2) $c = .95$, $\alpha = .05$, $\frac{\alpha}{2} = .025$, $\nu = 14 - 1 = 13$ d.f., $t_{.025} = 2.160$

   S.E. $= \dfrac{5.7}{\sqrt{14}} = 1.52$

   $E = (2.160)(1.52) = 3.28$

   $18.3 - 3.28 < \mu < 18.3 + 3.28$ or

   $15.02 < \mu < 21.58$

   The survival times should be normally distributed.

3) $c = .99$, $\alpha = .01$, $\frac{\alpha}{2} = .005$, $z_{.005} = 2.58$

   S.E. $= \sqrt{\dfrac{(.34)(.66)}{250}} = .0300$

   $E = (2.58)(.0300) = .0774$

   $.34 - .0774 < p < .34 + .0774$ or

   $.2626 < p < .4174$

4) $E = 20 = z_{\alpha/2}(\text{S.E.}) = z_{\alpha/2}\left(\dfrac{114.75}{\sqrt{144}}\right) = (z_{\alpha/2})(9.5625)$

   Solving

   $z_{\alpha/2} = \dfrac{20}{9.5625} = 2.09$ (rounded)

   $P(Z > 2.09) = .5000 - .4817 = .0183$

   Thus $\frac{\alpha}{2} = .0183$ and $\alpha = .0366$.

   The confidence level is $1 - .0366 = .9634 = 96.34\%$

5) $n < 30$ so use the t distribution.

   $c = .95$, $\alpha = .05$, $\frac{\alpha}{2} = .025$, $\nu = 20 - 1 = 19$, $t_{.025} = 2.093$

   S.E. $= \dfrac{65}{\sqrt{20}} = 14.53$

   $E = (2.093)(14.53) = 30.42$

   $138 - 30.42 < \mu < 138 + 30.42$

   $107.58 < \mu < 168.42$

6) $E = .03$ is given. Use the worst case estimate $pq = (.5)(.5)$.

$$E = .03 = z_{\alpha/2}\sqrt{\frac{pq}{n}} \approx z_{\alpha/2}\sqrt{\frac{(.5)(.5)}{n}} = (z_{\alpha/2})\frac{(.5)}{\sqrt{n}}$$

$$E = .03 = z_{.025}\sqrt{\frac{pq}{n}} = (1.96)\sqrt{\frac{(.5)(.5)}{n}} = \frac{(1.96)(.5)}{\sqrt{n}} = \frac{.98}{\sqrt{n}}$$

$$n = \left(\frac{.98}{.03}\right)^2 = 1067.11$$

Take $n \geq 1068$.

7) $E = 3 = z_{.005}\frac{\sigma}{\sqrt{n}} = (2.58)\frac{(12)}{\sqrt{n}} = \frac{30.96}{\sqrt{n}}$

$\sqrt{n} = \frac{30.96}{3} = 10.32$

$n = (10.32)^2 = 106.50$

Take $n \geq 107$.

8) (Note: Carry as many decimal places as possible because of the large numbers.)

$S.E. = \frac{16.20}{\sqrt{1144}} = .478963$

$E = (.478963)(1.96) = .9387676$ (about 94¢)

For 1,430,000 adults, the expected amount is

$$(1430000)(38.40) = 54,912,000,$$

and the maximum error is

$$(1430000)(.9387676) = 1,342,437.60.$$

The limits for the city are 53,569,562 and 56,254,438.

9) <u>18-35 Age Group</u>

$\hat{p} = \frac{210}{400} = .525$

$E = 1.96\sqrt{\frac{(.525)(.475)}{400}} = .0489$

$.525 - .0489 < p < .525 + .0489$ or

$.4761 < p < .5739$

35-55 Age Group

$\hat{p} = \dfrac{289}{500} = .578$

$E = (1.96)\sqrt{\dfrac{(.578)(.422)}{500}} = .0433$

$.578 - .0433 < p < .578 + .0433$

$.5347 < p < .6213$

Over 55 Group

$\hat{p} = \dfrac{380}{500} = .76$

$E = 1.96\sqrt{\dfrac{(.76)(.24)}{500}} = .0374$

$.76 - .0374 < p < .76 + .0374$ or

$.7226 < p < .7974$

Combined

$\hat{p} = \dfrac{879}{1400} = .6279$

$E = 1.96\sqrt{\dfrac{(.6279)(.3721)}{1400}} = .0253$

$.6279 - .0253 < p < .6279 + .0253$ or

$.6026 < p < .6532$

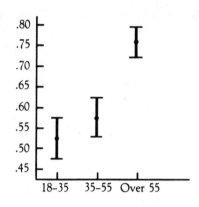

The 3 confidence intervals do not overlap. It does not appear that the proportion of coffee drinkers is the same for each group.

10)

x	x^2
8.2	67.24
8.0	64.00
8.4	70.56
8.5	72.25
8.3	68.89
8.2	67.24
49.6	410.18

$\bar{x} = \dfrac{49.6}{6} = 8.27$

$s = \sqrt{\dfrac{410.18 - \dfrac{(49.6)^2}{6}}{5}} = .175$

$\nu = 6 - 1 = 5$ d.f.   and $t_{.005} = 4.032$

$E = (4.032)\dfrac{(.175)}{\sqrt{6}} = .2881 \approx .29$

$8.27 - .29 < \mu < 8.27 + .29$ or

$7.98 < \mu < 8.56$

Chapter IX
Simple Hypothesis Testing

Elementary hypothesis testing is introduced in this chapter. The topics include: types of hypotheses, decision rules for deciding between two competing hypotheses, the types of errors which can occur, the probability of such errors, the significance level of the test, and the p value of the test.

Concepts/Techniques

1) Hypotheses

An hypothesis is a statement about a population. Hypothesis testing is concerned with deciding between two hypotheses H_0 and H_1 called the null and alternate hypotheses respectively.

The null hypothesis is the hypothesis being tested.

The alternate hypothesis is an hypothesis which we are willing to accept if the null hypothesis is rejected.

Example: Ten years ago the proportion of university applicants electing a busness major was 28%. We believe that there has been a possible increase in interest in this major. To study this we would test

H_0: p = .28 (The proportion is unchanged.)

H_1: p > .28 (The proportion has increased.)

241

The null hypothesis generally reflects the position of no change and
thus will always be worded as an equality.

2) Test Statistics

The null hypothesis is an accepted position. We do not try to prove it.
Rather we test to see if it should be rejected. Such tests are based on
the value of sample statistics such as the mean $\overline{X}$, the proportion $\hat{P}$,
or the corresponding z or t scores. These are called test statistics.

3) Critical Values and the Critical Region

Those values of the test statistic for which the null hypothesis is
rejected form what is called the critical region of the test. Typically
the critical region is composed of one or two semi-infinite intervals.
The boundary or boundaries of these are called the critical value(s)
of the test.

Example: In testing

$$H_0: \quad p = .7$$
$$H_1: \quad p \neq .7$$

we would reject H_0 if the sample proportion is too far in either
direction from p = .7. Suppose the decision rule is to reject H_0
if $\hat{p} < .64$ or $\hat{p} > .76$. The boundaries of these two intervals,
$\hat{p} = .64$ and $\hat{p} = .76$ are the critical values

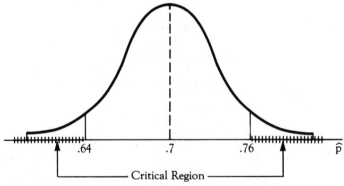

Example: Suppose in testing

$$H_0: \quad \mu = 50$$
$$H_1: \quad \mu > 50$$

the decision rule is to reject H_0 if the sample mean $\bar{x}$ is more than 2 standard errors greater than $\mu = 50$. This is equivalent to saying that the corresponding z score is greater than 2.0 where $z = \dfrac{\bar{x} - 50}{S.E.}$. The critical region is $z > 2.0$ and $z = 2.0$ is the critical value of the test.

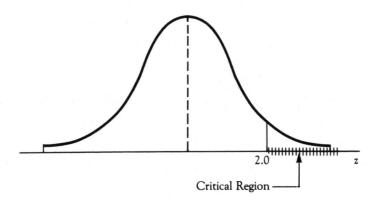

4) <u>Two Tailed vs. Single Tailed Tests</u>

In testing

$$H_0: \quad \mu = 200$$
$$H_1: \quad \mu \neq 200$$

we reject H_0 if the sample mean $\bar{x}$ is too far from $\mu = 200$ in either direction. The critical region thus consists of 2 intervals lying under the tails of the distribution of $\bar{X}$ or of Z or T. This test would accordingly be called a two tailed test.

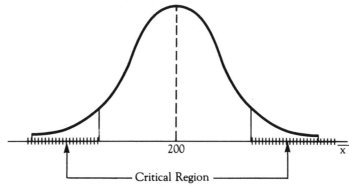

In testing, say

$$H_0: \quad p = .5$$
$$H_1: \quad p > .5$$

we reject H_0 only if $\hat{p}$ is too far to the right of $p = .5$. Thus the critical region will be under the extreme right of the distribution of $\hat{p}$ or Z. Accordingly this test would be called a single tailed or one sided test to the right.

Note that the type of test is determined by the alternate hypothesis. If H_1 is of the form $\mu \neq \mu_0$ or $p \neq p_0$, then a two tail test is used.

If H_1 is of the form $\mu > \mu_0$ or $p > p_0$, then a single tail or one sided test to the right is used.

If H_1 is of the form $\mu < \mu_0$ or $p < p_0$, then a single tail or one sided test to the left is required.

5) Type I and Type II Errors

Decision making always involves the risk of an error. A type I error is made if the null hypothesis is rejected when in fact it is true. A type II error is made if the null hypothesis is not rejected when in fact it is false, i.e., we should have accepted the alternate hypothesis but did not.

   Example: Consider the following test of

$$H_0: \quad p = .50$$
$$H_1: \quad p = .62$$

Suppose the decision rule is to reject H_0 and accept H_1 if the proportion $\hat{p}$ of a random sample of size $n = 400$ exceeds .57, i.e. $\hat{p} > .57$. Find the probabilities α and β of a type I and type II error respectively.

$\alpha = P(H_0 \text{ is rejected} | H_0 \text{ is true})$
   $= P(\hat{p} > .57 | p = .50) = P(z > 2.8) = .5000 - .4974 = .0026.$

   Here we used $z = \dfrac{.57 - .50}{\sqrt{\dfrac{(.50)(.50)}{400}}} = 2.8.$

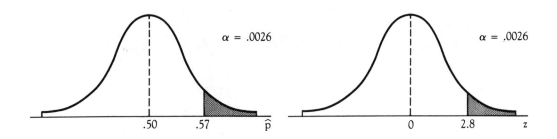

$\beta = P(H_0 \text{ is not rejected} | H_0 \text{ is false})$

$= P(\hat{p} < .57 | p = .62) = P(z < -2.06) = .5000 - .4803 = .0197$

Here the z score is computed assuming $p = .62$, i.e.,

$$z = \frac{.57 - .62}{\sqrt{\dfrac{(.62)(.38)}{400}}} = -2.06$$

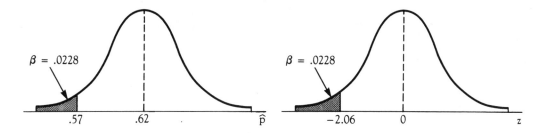

Example: Suppose we are testing

H_0: $p = .60$ (60% of all students are in lower division)

H_1: $p > .60$ (More than 60% of all students are in lower division)

For purposes of the discussion, assume $p = .60$ is correct and
suppose we get a sample where $\hat{p} = .93 = 93\%$ of the students
are in lower division. This would lead us to conclude that $p = .60$
is too small. Thus we would reject H_0 and accept H_1. A type I
error has been made.

Example: Suppose we are testing

H_0:   The drug is effective in curing motion sickness

H_1:   The drug is not effective

Assume for the discussion that H_1 is true, that is the drug is not effective. Suppose, however, due to say the placebo effect nearly all the test subjects report that the drug cured their motion sickness. We would then conclude the drug is effective and not reject H_0. A type II error has been made.

The probability of a type I error is denoted by α, that of a type II by β. Thus

$$\alpha = P(H_0 \text{ is rejected} | H_0 \text{ is true})$$
$$\beta = P(H_0 \text{ is not rejected} | H_0 \text{ is false})$$

6)   <u>The Significance Level of a Test</u>

The significance level of a test is the probability α of a type I error expressed as a percent. Thus

$$\text{Significance Level} = \alpha = P(H_0 \text{ is rejected} | H_0 \text{ is true}).$$

The significance level is chosen to be a small number such as $\alpha = .05$, $\alpha = .025$, $\alpha = .01$, etc. The significance level determines the critical values of a test.

<u>Example</u>: Suppose we plan to test

$$H_0: \quad \mu = 180 \quad (\sigma = 18)$$
$$H_1: \quad \mu \neq 180$$

using the mean $\bar{x}$ of a random sample of size $n = 50$. Find the critical values if the significance level is $\alpha = .05$.

Note that this is a two sided test. Thus the area under each tail of the distribution of $\bar{X}$ and of Z is $\frac{\alpha}{2} = .025$.

The critical values are located $1.96(S.E.)$ above and below the mean $\mu = 180$. Thus the critical values of $\bar{X}$ are

$$\bar{x}_1 = 180 - 1.96 \, \frac{18}{\sqrt{50}} = 175.0 \quad \text{and}$$
$$\bar{x}_2 = 180 + 1.96 \, \frac{18}{\sqrt{50}} = 185.0.$$

If Z is used as the test statistic, then the critical values are
simply ± 1.96.

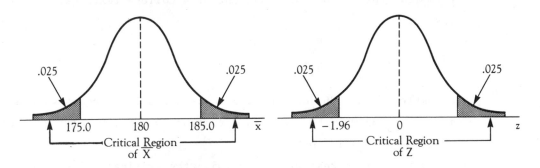

Example: A 1% significance level test based on a sample of n = 500
will be used to test

$$H_0: \quad p = .35$$
$$H_1: \quad p > .35.$$

Find the critical values of both Z and $\hat{P}$ where

$$z = \frac{\hat{P} - .35}{\sqrt{\dfrac{(.35)(.65)}{500}}}.$$

Since this is a one sided test to the right, the area under the
right tail of the normal distribution to the right of the critical
z values is $\alpha = .01$.
Thus the critical value is $z_{.01} = 2.33$.
For $\hat{P}$ the critical value is 2.58(S.E.) above the asserted
p = .35. Thus the critical value is

$$\hat{p} = .35 + z_{.01}(S.E.) = .35 + 2.33 \sqrt{\frac{(.35)(.65)}{500}} = .3997.$$

Obviously it is easier to work with Z as the test statistic than with
either $\overline{X}$ or $\hat{P}$.

7)   Test Procedures (Large Sample Case)
First determine the significance level and then the critical values of
Z based on whether the test is two tailed or single tailed. Then compute
the z value using the data of the sample. If this is more extreme than

a critical value, then reject H_0.

> Example: Let μ be the mean cost of a college textbook. In
> testing
>
> $$H_0: \quad \mu = 34.50$$
> $$H_1: \quad \mu \neq 34.50$$
>
> a sample of $n = 36$ current textbooks had selling costs with a
> mean $\bar{x} = \$32.00$ and standard deviation $s = \$6.30$. If $\alpha = .05$,
> what is the conclusion?
>
> For a two tailed test with $\alpha = .05$, the critical values are
> $\pm z_{.025} = \pm 1.96$.
>
> The value of the test statistic is
>
> $$z = \frac{32.00 - 34.50}{\frac{6.30}{\sqrt{36}}} = -2.38$$
>
> Since $z = -2.38 < -1.96$ we reject H_0 and conclude that the mean
> cost is not $34.50.

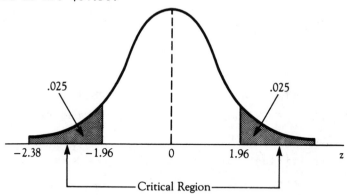

Note in this last example that we approximated the population standard
deviation σ by that of the sample, i.e., by $s = 6.50$.

> Example: Let p be the proportion of new car loans having a 48
> month time period. In 1983 $p = .74$. Suppose we believe this has
> declined (more 60 month loans) and accordingly wish to test
>
> $$H_0: \quad p = .74$$
> $$H_1: \quad p < .74$$

using a significance level of $\alpha = .02$. What is the conclusion
if 350 of a sample of 500 auto loans have a time period of 48
months.

Here $\hat{p} = \dfrac{350}{500} = .70$.

This is a one tailed test to the left with $\alpha = .02$. Thus the
critical value of z is $-z_{.02} = -2.05$.

The value of the test statistic is

$$z = \frac{\hat{p} - .74}{\sqrt{\dfrac{(.74)(.26)}{500}}} = \frac{.70 - .74}{.01962} = -2.04.$$

Since $z = -2.04 \not< -2.05$, we do not reject H_0 at this significance
level.

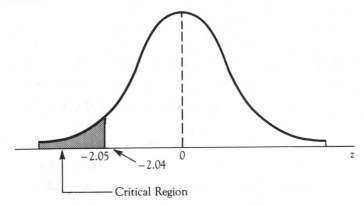

Critical Region

8) The p value of a Test

The p value of a test is the probability of getting a value of the
sample statistic at least as extreme (and in the same direction) as
was actually produced by the sample.

Example: In testing

$$H_0: \quad \mu = 400$$
$$H_1: \quad \mu \neq 400$$

a sample of size $n = 80$ yielded $\bar{x} = 415$ and $s = 65$. Find
the p value of the test.

$\bar{x} = 415$ is under the right tail of the curve. Values that are at
least this extreme are those for which $\bar{x} \geq 415$.

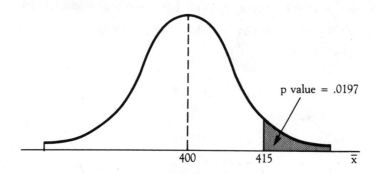

Thus p value = $P(\overline{X} \geq 415 | \mu = 400)$.

The corresponding z score is

$$z = \frac{415 - 400}{\frac{65}{\sqrt{80}}} = 2.06.$$

We find p value = $P(Z > 2.06)$ = .5000 − .4803 = .0197.

Thus the chance of getting this extreme a sample mean is quite

small if indeed $\mu = 400$ is the true population mean.

Note that for a single tailed test, the p value is the smallest

significance level at which the null hypothesis can be rejected.

Since the p value always corresponds to an area under a single tail,

it follows that for a two tailed test, the smallest significance level

at which H_0 can be rejected is twice the p value.

Example: A sample of n = 50 starting salaries for marketing

majors produced $\overline{x}$ = \$1890 and s = \$400. What is the smallest

significance for which the data would justify rejecting H_0 in a

test of

$$H_0: \quad \mu = 1800$$
$$H_1: \quad \mu > 1800$$

p value = $P(\overline{X} \geq 1890 | \mu = 1800)$ = $p(Z > 1.59)$ = .0559

since

$$z = \frac{1890 - 1800}{\frac{400}{\sqrt{50}}} = 1.59.$$

Since this is a single tail test, the smallest significance level
at which H_0 can be rejected is

$$\alpha = p \text{ value} = .0559.$$

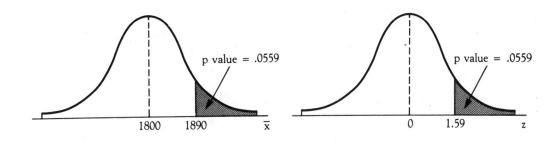

p value = .0559   p value = .0559

1800   1890 $\overline{x}$ 0   1.59   z

Example: Let p be the proportion of accidents in which excess
speed is a factor. In testing

$$H_0: \quad p = .60$$
$$H_1: \quad p \neq .60$$

a study showed excess speed to be a factor in 64% of a sample of
400 accidents. What is the smallest significance level for which
H_0 can be rejected?
p value = $P(\hat{P} > .64 | p = .60) = P(Z > 1.63) = .0516$
Since this is a two tailed test, the smallest significance level
for which the null hypothesis can be rejected is twice the p
value, i.e.,

$$\alpha = 2(.0516) = (.1032) = 10.32\%$$

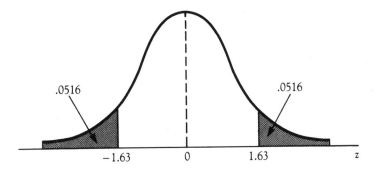

.0516   .0516

-1.63 0   1.63   z

The p value of a test should always be reported.

9) <u>Test Procedures (Small Sample Case)</u>

The t distribution is used in place of the normal when the sample size is below 30 and the population is normal. (This is very similar to the case of small sample confidence intervals.)

Example: Using $\alpha = .01$, test

$$H_0: \quad \mu = 34$$
$$H_1: \quad \mu > 34$$

Assume the population in question is normal and that a sample of n = 8 yielded $\overline{x} = 42$ and s = 6. This is a single tailed test to the right. The critical value $t_{.01} = 2.998$ is obtained using $\nu = 8 - 1 = 7$ d.f.

The value of the test statistic is

$$t = \frac{42 - 34}{\frac{6}{\sqrt{8}}} = 3.77.$$

Since $t = 3.77 > t_{.01}$, reject H_0 and conclude that the mean is greater than 34. Noting that $3.77 > t_{.005}$, we would report the p value as $p < .005$.

10) <u>The Power of a Test</u>

The probability α of a type I error is kept small by choosing α to be .05, .01, .005, etc. The probability β of a type II error is another matter entirely. It can be large. It follows that the smallness of β is a measure of how powerful the test is. Now the smaller β is, the larger is the quantity $1 - \beta$. This number is called the power of the test. Thus

$$\text{Power} = 1 - \beta = 1 - P(\text{Type II Error}).$$

Example: For $\beta = .30$, the power of the test is

$$\text{Power} = 1 - .30 = .70.$$

For $\beta = .10$, the power of the test is

$$\text{Power} = 1 - .10 = .90.$$

Note that $1 - P(\text{Type II Error})$ is the probability of the complementary event "no Type II Error." And this is just the event "H_0 is rejected when H_0 is false" or equivalently "H_1 is accepted when H_1 is true."

<u>Example</u>: Suppose we have a 5% significance level test for which the power is .80 = 80%. Then if H_0 is true, there is only a 5% chance that H_0 will be rejected.
And if H_0 is false, there is an 80% chance that H_0 will be rejected.

<u>Example</u>: Consider a 1% significance level test of

$$H_0: \quad \mu = 140 \quad (\sigma = 30)$$
$$H_1: \quad \mu = 155$$

based on a sample of size $n = 60$. Find the power of the test. The test is carried out as a single tailed test to the right. Thus the critical value of z is $z_{.01} = 2.33$ and for $\overline{X}$ it is
$$\overline{x} = 140 + 2.33\left(\frac{30}{\sqrt{60}}\right) = 149.02.$$

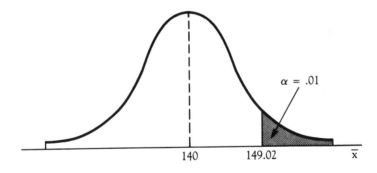

A type II error will result if in fact $\mu = 155$ is the true state of affairs and yet the sample mean is below 149.02. Thus
$$\beta = P(\overline{X} < 149.02 \,|\, \mu = 155) = P(z < -1.54) = .5000 - .4382 = .0618.$$
Here z was computed by

$$z = \frac{149.02 - 155}{\dfrac{30}{\sqrt{60}}}$$

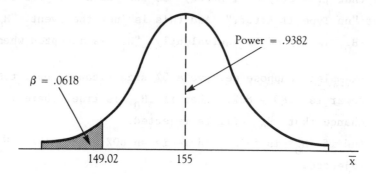

Thus the power is

$$\text{Power} = 1 - \beta = 1 = .0618 = .9382.$$

Note we could also have computed this by

$$\text{power} = P(H_1 \text{ accepted}|H_1 \text{ true}) = P(\overline{X} > 149.02|\mu = 155)$$
$$= .4382 + .5000 = .9382.$$

11) Additional Examples

Example: A large sample test of

$$H_0: \quad \mu = \mu_0$$
$$H_1: \quad \mu \neq \mu_0$$

is as follows. Reject H_0 if the sample mean $\overline{x}$ is more than 2.4
standard errors from μ_0. What is the significance level of the
test?

The critical values are $\pm z_{\alpha/2} = \pm 2.4$.

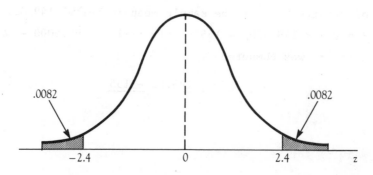

The area to the right of $z = 2.4$ is $\frac{\alpha}{2}$.

$\frac{\alpha}{2} = P(Z > 2.4) = .5000 - .4918 = .0082$

Thus $\alpha = .0164 = 1.64\%$.

This is a 1.64% significance level test.

Example: Suppose the proportion of accountants who actually pass the C.P.A. licensing requirements is estimated to be $p = .40 = 40\%$. A consultant believes this is too low and plans to test

$$H_0: \quad p = .40$$
$$H_1: \quad p > .40$$

and will reject H_0 if the proportion of C.P.A. licensees in a sample of 600 accountants exceeds $.48 = 48\%$. Find α.

First we convert $\hat{p} = .45$ to a z score.

$$z = \frac{.45 - .40}{\sqrt{\dfrac{(.40)(.60)}{600}}} = 2.5$$

Thus the critical region $\hat{P} > .45$ corresponds to $Z > 2.5$. The significance level is

$$\alpha = P(Z > 2.5) = .5000 - .4938 = .0062.$$

Example: Consider a test of

$H_0: \quad \mu = 510$ (California S.A.T. scores have a mean of 510)

$H_1: \quad \mu > 510$ (California S.A.T. scores have a mean greater than 510)

Suppose the decision rule is to reject H_0 if the mean $\bar{x}$ of a random sample of size n exceeds 525. If $\sigma = 100$, for what size sample would this be a 1% significance level test?

This is a one sided test to the right. The critical value $\bar{x} = 515$ corresponds to $z_{.01} = 2.33$. Thus

$$z = \frac{525 - 510}{\dfrac{100}{\sqrt{n}}} = 2.33.$$

Solving for n, we find

$$n = \left(\frac{(2.33)(100)}{15} \right)^2 = 241.28.$$

Thus the sample size must be at least 242.

Exercise Set IX.1

1) a) H_0: $\mu = 57.5$ (The average speed is 57.5 m.p.h.)

        H_1: $\mu > 57.5$ (The average speed exceeds 57.5 m.p.h.)

    b) H_0: $p = .37$ (37% of the teenage population smoke)

        H_1: $p > .37$ (More than 37% of the teenage population smoke)

    c) H_0: $p = .67$ (67% of all students have never skiied)

        H_1: $p \neq .67$ (The proportion of all students who have never skiied is

                not .67 (67%))

    d) H_0: $p = .67$ (67% of all licensed drivers have never had a ticket nor

                been in an accident)

        H_1: $p \neq .67$ (The proportion of all licensed drivers who have never had

                a ticket nor a major accident is not equal to .67 (67%))

    e) H_0: $p = .635$ (63.5% of the present inventory of left over jeans are

                in small sizes)

        H_1: $p < .635$ (Less than 63.5% of the present inventory of left over

                jeans are in small sizes)

    f) H_0: $\mu = 125$ (The mean grade of students having a poor nutritional

                program is 125)

        H_1: $\mu < 125$ (The mean grade of students having a poor nutritional

                program is below 125)

3) A lack of a specific direction in the alternate hypothesis produces a two

    tail test.

    a) H_0: $\mu = 560$ (Rejection region is under both tails of the distribution

        H_1: $\mu \neq 560$ of $\overline{X}$ or Z)

    b) H_0: $\mu = 560$ (Rejection region is under the left tail of the distribution

        H_1: $\mu < 560$ of $\overline{X}$ or Z)

    c) H_0: $\mu = 560$ (Rejection region is under the right tail of the distribution

        H_1: $\mu > 560$ of $\overline{X}$ or Z)

5) H_0: $\mu = 45$ (The average age is 45 days)

    H_1: $\mu > 45$ (The average age exceeds 45 days)

7) S.E. $= \dfrac{4.30}{\sqrt{40}} = .68$

    Reject H_0 if $\overline{x} < 18.70 - 2(.68) = 17.34$.

9) Let p be the proportion of patients with extreme hypertension whose blood
    pressure is reduced by this new medication.

    a) H_0: p = .60   (The proportion remains unchanged)

        H_1: p > .60   (The proportion for which there is a reduction has
                          increased)

        Decision Rule: Reject H_0 if z > 3.0 where

        $$z = \frac{\hat{p} - .60}{\sqrt{\frac{(.60)(.40)}{576}}} = \frac{\hat{p} - .60}{.0204}$$

        or equivalently, reject H_0 if

        $$\hat{p} > .60 + 3(S.E.) = (.60) + 3(.0204) = .6612.$$

    b) H_0: p = .60   (The proportion remains unchanged)

        H_1: p < .60   (The proportion for which there is a reduction in blood
                          pressure has decreased)

        Decision Rule: Reject H_0 if z < -3.0 or equivalently reject H_0 if

        $$\hat{p} < .60 - 3(.0204) = .5388.$$

    c) H_0: p = .60   (The proportion remains unchanged)

        H_1: p ≠ .60   (The proportion for which there is a reduction in blood
                          pressure has changed)

        Decision Rule: Reject H_0 if z > 3.0 or z < -3.0 or equivalently,
        reject H_0 if

        $$\hat{p} > .60 + 3(.0204) = .6612 \quad \text{or}$$
        $$\hat{p} < .60 + 3(.0204) = .5388.$$

11) a) H_0: μ = 112

        H_1: μ > 112

    b) $z = \dfrac{125 - 112}{\frac{28}{\sqrt{30}}} = 2.54$

    c) P(z > 2.54) = .0055

13) S.E. $= \dfrac{3350}{\sqrt{100}} = 335$

    a) H_0: $\mu = 24580$

       H_1: $\mu > 24580$

       The critical region for Z is $Z \geq 2.0$.

       For $\overline{X}$, the critical region is $\overline{X} > 24580 + 2(335) = 25250$.

    b) H_0: $\mu = 24580$

       H_1: $\mu < 24580$

       The critical region for Z is $Z \leq -2.0$.

       The critical region for $\overline{X}$ is $\overline{X} \leq 24580 - 2(335) = 23910$.

15) The critical value of Z is $z = 2.4$

$$z = \frac{\overline{x} - 180}{S.E.} = \frac{188 - 180}{\dfrac{21}{\sqrt{64}}} = 3.05.$$

    Reject H_0 since $z > 2.4$.

17) Yes. Either $z > 2.0$ or $z < -2.0$ and these are also the critical values
of the two tailed test.

19) Reconsider the alternate hypothesis.

    A sample mean exceeding $\mu = 120$ was expected. Instead a sample was
obtained which is $\left(z = \dfrac{112 - 120}{\dfrac{8}{\sqrt{144}}} = -12 \right)$ 12 standard errors below the

    expected $\mu = 120$.

Exercise Set IX.2

1) a) H_0:   The school should retain its certification.

H_1:   The school should lose its certification.

A type I error occurs if the school is decertified when in fact it should
have remained certified. This occurs if $\bar{x} \leq 65$ for the sample.
A type II error occurs if the school retains its certification when in
fact it should have lost its certification. This occurs if $\bar{x} > 65$.
From the schools point of view, the probability α of a type I error
should be small. From the student and general public point of view,
the probability β of a type II error should be small.

   b) H_0:   The driver is not speeding.

H_1:   The driver is speeding.

A type I error results if the driver is cited when, in fact, he or she
is not speeding. This occurs if $\bar{x} > 57.5$.
A type II error results if the driver is speeding but this is not detected.
This occurs if $\bar{x} \leq 57.5$. From the individual drivers point of view
the probability α of a type I error should be small.

   c) H_0:   The patient is healthy (no hypertension).

H_1:   The patient has hypertension.

A type I error occurs if a healthy patient is diagnosed as having
hypertension. This occurs if $\bar{x} > 147$.
A type II error occurs if a patient with hypertension is diagnosed as
healthy. This occurs if $\bar{x} \leq 147$. The probability β of a type II
error should be small.

   d) H_0:   The instructor is effective.

H_1:   The instructor is ineffective.
A type I error results if an effective teacher is judged ineffective.
This occurs if the mean of his evaluations is appreciably below the
department mean.
A type II error results if an ineffective teacher is judged effective.
This occurs if the mean of the sample of instructors grades is not
appreciably below the department mean.
From the instructors point of view, a small probability for a type I
error is desirable. From the students point of view the probability

of a type II error should also be small.

e) H_0: The area is normal (not depressed).

H_1: The area is economically depressed.

A type I error results if a normal area is classified as depressed.
This occurs if $\hat{p} > .15$.
A type II error results if a depressed area is not found to be so. This
occurs if $\hat{p} \leq .15$.
The probability of a type II error should be small.

f) H_0: The child has normal intelligence.

H_1: The child is mentally deficient.

A type I error results if a normal child is classified mentally deficient.
This occurs if $\bar{x} < 80$.
A type II error results if a mentally deficient child is classified as
normal. This occurs if $\bar{x} \geq 80$.

3) S.E. $= \sqrt{\dfrac{(.44)(.56)}{900}} = .0165$

a) $z = \dfrac{.48 - .44}{.0165} = 2.42$

Since $z > 2$, H_0 is rejected and (since H_0 is true) a type I error
results.

b) $z = \dfrac{.45 - .44}{.0165} = .61$

Since $z < .61$, H_0 is not rejected and no error results.

c) $z = \dfrac{.48 - .44}{.0165} = 2.42$

Since $z > 2$, H_0 is rejected and (the true hypothesis) H_1 is accepted.
No error results.

d) $z = \dfrac{.45 - .44}{.0165} = .61$

Since $z < 2$, H_0 is not rejected and a type II error results (since H_1
is true).

5) a) For this two tail test, H_0 is rejected if $z > 1.65$ or $z < -1.65$.
The probability of rejecting H_0 when H_0 is true is thus

$$P(Z > 1.65 \text{ or } Z < -1.65) = P(Z > 1.65) + P(Z < -1.65)$$
$$= .0495 + .0495 = .0990 \approx .10$$

b) $P(Z > 2.0 \text{ or } Z < -2.0) = .0228 + .0228 = .0456 \approx .05$

c) $P(Z > 2.33 \text{ or } Z < -2.33) = .0099 + .0099 = .0198 \approx .02$

7) $H_0:\ \mu = 9240$ $\text{S.E.} = \dfrac{820}{\sqrt{30}} = 149.71$

     $H_1:\ \mu > 9240$

a) The critical value of Z is $z_{.05} = 1.64$

     The critical value of $\overline{X}$ is located $1.64(\text{S.E.})$ above $\mu = 9240$ at

$$\overline{x} = 9240 + (1.64)(149.71) = 9485.52.$$

b) The critical value of Z is $z_{.01} = 2.33$.

     The critical value of $\overline{X}$ is

$$\overline{x} = 9240 + (2.33)(149.71) = 9588.82.$$

c) The critical value of Z is $z_{.10} = 1.28$.

     The critical value of $\overline{X}$ is

$$\overline{x} = 9240 + (1.28)(149.71) = 9431.63.$$

9) Let p be the proportion of cars having only 1 occupant.

a) $H_0:\ p = .67$

     $H_1:\ p \neq .67$ (no direction of a possible change is indicated)

b) For this two tail test the critical values are $\pm z_{.005} = \pm 2.58$.

     The critical values of $\hat{P}$ are located $2.58(\text{S.E.})$ above and below

     $p = .67$ at

$$.67 \pm 2.58\sqrt{\frac{(.67)(.33)}{1200}} \ .$$

These values are $.7050$ and $.6350$.

11) $z = \dfrac{123 - 118}{\dfrac{12}{\sqrt{36}}} = 2.5$

p value $= P(\overline{X} \geq 123) = P(Z \geq 2.5)$

$$= .5000 - .4938 = .0062$$

13) S.E. $= \dfrac{18}{\sqrt{36}} = 3$

   a)

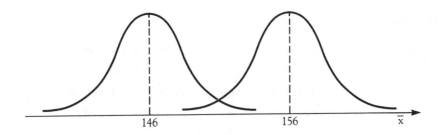

   b) A type II error results if H_1: $\mu = 156$ is the true state of affairs
       and yet H_0: $\mu = 146$ is not rejected. This will occur if $\overline{x} \leq 153$.
       Thus

$$\beta = P(\overline{X} \leq 153 | \mu = 156)$$
$$z = \frac{153 - 156}{3} = -1$$
$$\beta = P(Z \leq -1) = .5000 - .3413 = .1587$$

   c) H_0: $\mu = 156$
       A type I error results when $\mu = 156$ is true and yet $\overline{x} \leq 153$ (so
       that H_0 is rejected).

$$\alpha = P(\overline{X} \leq 153 | \mu = 156)$$
$$z = \frac{153 - 156}{3} = -1$$
$$\alpha = P(Z \leq -1) = .1587$$

       A type II error results when $\mu = 146$ is the true state of affairs
       and yet H_0: $\mu = 156$ is not rejected. This occurs if $\overline{x} > 153$.

$$\beta = P(\overline{X} > 153 | \mu = 146)$$
$$z = \frac{153 - 146}{3} = 2.33$$
$$\beta = P(Z > 2.33) = .5000 - .4901 = .0099.$$

15) Note that in a two tail test, H_0 will be rejected for any α larger than
     .01. Thus H_0 would be rejected for $\alpha = .02$, $\alpha = .03$, $\alpha = .05, \ldots$

   a) Yes

b) No

Note that in a one tail test, H_0 will be rejected for any α larger

than $\dfrac{.01}{2}$ = .005 .

c) Yes

d) Yes

17) a) α = .035. For a single tail test, the p value is the smallest signi-
      ficance level at which H_0 can be rejected.

     b) α = .07. For a two tail test, the p value is one half the smallest
      significance level at which H_0 can be rejected.

Exercise Set IX.3

1) a) This is a two tailed test due to the lack of direction in H_1. The
critical values are $\pm z_{\alpha/2} = \pm z_{.025} = \pm 1.96$.

$$S.E. = \frac{14}{\sqrt{64}} = 1.75$$

$$z = \frac{150 - 146}{1.75} = 2.29$$

Since $z > z_{.025}$, reject H_0 and accept H_1: $\mu \neq 146$

$$p \text{ value} = P(\overline{X} \geq 150 | \mu = 146)$$
$$= P(Z > 2.29) = .5000 - .4890 = .0110$$

b) This is a one sided test to the left as a result of the direction in H_1.
The critical value is $-z_{.05} = -1.64$.

$$S.E. = \frac{15}{\sqrt{30}} = 2.74$$

$$z = \frac{210 - 218}{2.74} = -2.92$$

Since $z < -z_{.05}$, reject H_0 and accept H_1

$$p \text{ value} = P(Z < -2.92) = .5000 - .4982 = .0018.$$

c) This is a one sided test to the right as a result of the direction in H_1.
The critical value is $z_{.01} = 2.33$.

$$S.E. = \frac{15}{\sqrt{36}} = 2.5$$

$$z = \frac{29.5 - 24}{2.5} = 2.2$$

Since $z < z_{.01}$, do not reject H_0

$$p \text{ value} = P(Z \geq 2.2) = .5000 - .4861 = .0139.$$

3) H_0: $\mu = 16$
H_1: $\mu \neq 16$
This is a two tailed test due to the lack of direction in H_1.
The critical values are $\pm z_{.01} = \pm 2.33$

$$\text{S.E.} = \frac{.8}{\sqrt{100}} = .08$$

$$z = \frac{15.6 - 16.0}{.08} = -5.0$$

Since $z < -2.33$ reject H_0 and conclude that $\mu < 16$.

$$p \text{ value} = P(Z < -5.0) < .0000003$$

5) H_0: $\mu = 44.5$
    H_1: $\mu > 44.5$
    Critical value: $z_{.05} = 1.64$

$$\text{S.E.} = \frac{7.4}{\sqrt{30}} = 1.35$$

$$z = \frac{45.75 - 44.50}{1.35} = .93$$

Do not reject H_0.

7) H_0: $\mu = 45$
    H_1: $\mu > 45$

$$\text{S.E.} = \frac{3.6}{\sqrt{39}} = .576$$

Critical value: $z_{.025} = 1.96$

$$z = \frac{46.4 - 45}{.576} = 2.43$$

Reject H_0 and conclude that $\mu > 45$.

9) Convert the updating times to minutes.
    H_0: $\mu = 134$
    H_1: $\mu < 134$
    This is a one tail test.
    The critical value is $-z_{.01} = -2.33$.

$$\text{S.E.} = \frac{36}{45} = 5.37$$

$$z = \frac{123 - 134}{5.37} = -2.05$$

Do not reject H_0.

11) For this one tail test the critical value is $z_{.05} = 1.64$.

Solve for n:

$$1.64 = \frac{46.6 - 45}{\frac{6.0}{\sqrt{n}}}$$

$$1.64 = \frac{(1.1)\sqrt{n}}{6.0}$$

$$\sqrt{n} = \frac{(6.0)(1.64)}{1.} = 6.15$$

$$n = 37.82$$

The sample size should be at least 38 if H_0 is to be rejected.

13) H_0: $\mu = 45$

H_1: $\mu \neq 45$

Using $\alpha = .05$, the critical values are $\pm z_{.025} = \pm 1.96$.

$$S.E. = \frac{20}{\sqrt{600}} = .816$$

$$z = \frac{46.3 - 45}{.816} = 1.59$$

Do not reject H_0.

$$p \text{ value} = P(Z \geq 1.59) = .0559.$$

15) This is a one tail test. The smallest significance level for which H_0 can be rejected is equal to the p value.

$$S.E. = \frac{38.5}{\sqrt{225}} = 2.57$$

$$z = \frac{144 - 150}{2.57} = -2.33$$

$$p \text{ value} = P(Z < -2.33) = .0099.$$

17) H_0: $\mu = 300$

H_1: $\mu < 300$

The critical value is $-z_{.001} = -3.08$

$$S.E. = \frac{100}{\sqrt{50}} = 14.14$$

$$z = \frac{278 - 300}{14.14} = -1.56$$

Do not reject H_0. We cannot conclude that learning is impaired by stress.

p value = $P(z \leq -1.56)$ = .0594. The results are not significant even at α = .05.

Exercise Set IX.4

1) This is a one tail test to right.

$\nu = 12 - 1 = 11$ d.f.

Critical value: $t_{.01} = 2.718$

S.E. $= \dfrac{2.3}{\sqrt{12}} = .664$

$t = \dfrac{11.2 - 10}{.664} = 1.81$

Do not reject H_0.

$t_{.05} = 1.796$ and $t_{.025} = 2.201$
Thus p value $< .05$.

3) This is a two tailed test.

$\nu = 20 - 1 = 19$ d.f.

Critical values: $\pm t_{.005} = \pm 2.861$

S.E. $= \dfrac{12.5}{\sqrt{20}} = 2.80$

$t = \dfrac{91.4 - 100}{2.80} = -3.07$

Reject H_0.

5) Let μ be the mean self image score of science students.

H_0: $\mu = 80$
H_1: $\mu > 80$
This is a one tail test.

$\nu = 20 - 1 = 19$

Critical value: $t_{.05} = 1.729$

S.E. $= \dfrac{24}{\sqrt{20}} = 5.37$

$t = \dfrac{88 - 80}{5.37} = 1.49$

Do not reject H_0. We cannot conclude that the science students have a better self image.

7) Let μ be the mean hydrocarbon emission level for the new engines.

H_0: $\mu = 29.2$
H_1: $\mu > 29.2$
This is a one tail test to the right.

$\nu = 12 - 1 = 11$ d.f.

Critical value: $t_{.05} = 1.796$

S.E. $= \dfrac{8.1}{\sqrt{12}} = 2.34$

$t = \dfrac{40.6 - 29.2}{2.34} = 4.88$

Reject H_0 and conclude that the new engines have a higher emissions level than 29.2 ppm.

9) First the mean $\bar{x}$ and standard deviation s of the sample are needed. These can be found using the minor statistical package of your calculator (if you have it) or proceed as in Chapter III:

x	x^2
38.2	1459.24
40.5	1640.25
49.1	2410.81
41.2	1697.44
30.6	936.36
199.6	8144.10

$\bar{x} = \dfrac{199.6}{5} = 39.92$

$s^2 = 8144.10 - \dfrac{(199.6)^2}{5} = 44.017$

$s = \sqrt{44.017} = 6.63$

This is a one tail test to the left.

$\nu = 5 - 1 = 4$ d.f.

Critical value: $-t_{.01} = -3.747$

S.E. $= \dfrac{6.63}{\sqrt{5}} = 2.97$

$t = \dfrac{39.92 - 45}{2.97} = -1.71$

Do not reject H_0.

11) H_0: $\mu = 28$ (days)

H_1: $\mu \neq 28$

This is a two tail test.

$\nu = 9 - 1 = 8$ d.f.

The critical values are $\pm t_{.025} = \pm 2.306$.

S.E. $= \dfrac{4.0}{\sqrt{9}} = 1.33$

$t = \dfrac{31.4 - 28}{1.33} = 2.56$

Reject H_0.

13) Note that $t_\alpha > z$ for $\alpha = .1, .05, .01, .001,$ etc.

Let $t = z = \dfrac{\overline{x} - \mu}{S.E.}$.

If $t > t_\alpha$ then $z > t_\alpha > z_\alpha$ and H_0 is also rejected using the normal distribution.

Generally speaking less evidence is required for rejecting H_0 with the normal than with the t.

Exercise Set IX.5

1) a) This is a two tailed test.

The critical values are $\pm z_{.025} = \pm 1.96$.

$$S.E. = \sqrt{\frac{(.4)(.6)}{144}} = .0408$$

$$z = \frac{.43 - .40}{.0408} = .73$$

Since $z < z_{.025}$, do not reject H_0.

$$p \text{ value} = P(Z > .73) = .2327$$

b) This is a one tail test to the right.

The critical value is $z_{.01} = 2.33$.

$$S.E. = \sqrt{\frac{(.7)(.3)}{900}} = .0153$$

$$z = \frac{.74 - .70}{.0153} = 2.62$$

Reject H_0.

$$p \text{ value} = P(Z > 2.62) = .0044$$

3) Let p be the proportion of students who favor the long weekend concept.

$$\hat{p} = \frac{296}{625} = .4736$$

H_0: $p = \frac{4}{9} = .4444$

H_1: $p > .4444$

The critical value is $z_{.01} = 2.33$.

$$S.E. = \sqrt{\frac{(.4444)(.5556)}{625}} = .0199$$

$$z = \frac{.4736 - .4444}{.0199} = 1.47$$

Do not reject H_0. We cannot conclude that the ratio is higher.

5) Let p be the proportion of native born people in the state.

$$\hat{p} = \frac{1145}{1940} = .5902$$

H_0: p = .64

H_1: p < .64

This is a one tail test to the left.

The critical value is $-z_{.01}$ = -2.33.

$$S.E. = \sqrt{\frac{(.64)(.36)}{1940}} = .0109$$

$$z = \frac{.5902 - .64}{.0109} = -4.57$$

Reject H_0 and conclude that less than 64% of the population of this state is native born.

7) $\hat{p} = \frac{82}{400}$ = .205

H_0: p = .16

H_1: p > .16

This is a one tail test to the right.

The critical value is $z_{.10}$ = 1.28.

$$S.E. = \sqrt{\frac{(.16)(.84)}{400}} = .0183$$

$$z = \frac{.205 - .16}{.0183} = 2.46 \qquad \text{Reject } H_0.$$

p value = P(Z $\geq$ 2.46) = .5000 - .4931 = .0069.

9) Let p = the probability of a 7 on the dice.

H_0: $p = \frac{1}{6} \approx .1667$ (The dice are fair.)

H_1: $p < \frac{1}{6}$ (The dice are biased against a 7.)

This is a one tail test to the left.

The critical value is $-z_{.01}$ = -2.33.

$$S.E. = \sqrt{\frac{(\frac{1}{6})(\frac{5}{6})}{300}} = .0215$$

$$\hat{p} = \frac{42}{300} = .1400$$

$$z = \frac{.1400 - .1667}{.0215} = -1.24$$

Do not reject H_0. We cannot conclude a bias against a 7 on this pair of dice.

11) Let p be the probability of a head. If the coin is fair, $p = \frac{1}{2} = .5$.
     If it is biased in favor of "tails," then $p < .5$. Accordingly, we test
     H_0:   p = .5
     H_1:   p < .5
     This is a one tail test to the left.
     Choosing $\alpha = .05$ the critical value is $-z_{.05} = -1.64$.

$$\hat{p} = \frac{100}{250} = .40$$

$$S.E. = \sqrt{\frac{(.5)(.5)}{250}} = .0316$$

$$z = \frac{.40 - .50}{.0316} = -3.16$$

     Reject H_0 and conclude a bias toward "tails."

13) Let p be the proportion of women with a radical mastectomy which have a
     5 year relapse free period.
     H_0:   p = .45
     H_1:   p > .45
     This is a single tail test to the right.
     The critical value is $z_{.05} = 1.64$.

a) $z = \dfrac{.77 - .45}{\sqrt{\dfrac{(.45)(.55)}{78}}} = 5.68$

     Reject H_0 and conclude that the relapse free proportion exceeds .45
     (45%).

b) $z = \dfrac{.50 - .45}{\sqrt{\dfrac{(.45)(.55)}{222}}} = 1.50$

     Do not reject H_0.

Exercise Set IX.6

1) power $= 1 - \beta = 1 - .65 = .35$

3) Yes.

$\beta = P(H_0$ is not rejected$|H_1$ is true$)$

$\quad = P(H_0$ is not rejected$|H_0$ is false$)$

$P(H_0$ is not rejected$|H_0$ is false$) + P(H_0$ is rejected$|H_0$ is false$) = 1$

Thus:

$P(H_0$ is rejected$|H_0$ is false$) = 1 - P(H_0$ is not rejected$|H_0$ is false$)$

$\qquad\qquad\qquad\qquad\qquad\quad = 1 - P(H_0$ is not rejected$|H_1$ is true$)$

$\qquad\qquad\qquad\qquad\qquad\quad =$ power.

5) $H_0:\ p = .5$

$H_1:\ p = .44$

A single tail test to the left is required.

H_0 will be rejected if $\hat{p}$ is more than $z_{.05}$ standard errors below $p = .5$. Thus the critical value is located at

$$\hat{p} = .5 - (1.64)\sqrt{\frac{(.5)(.5)}{900}} = .4723.$$

A type II error results if $H_1:\ p = .44$ is true and $\hat{p} \geq .4723$.

$$\beta = P(\hat{p} \geq .4723 | p = .44)$$

$$z = \frac{.4723 - .44}{\sqrt{\dfrac{(.44)(.56)}{900}}} = 1.95 \qquad \text{(Note the standard error is now computed with } p = .44 \text{ since } H_1 \text{ is assumed true.)}$$

$$\beta = P(Z \geq 1.95) = .0256$$

$$\text{power} = 1 - \beta = 1 - .0256 = .9744$$

An alternate approach is to use problem #3.

$$\text{power} = P(\hat{p} < .4723 | p = .44) = P(Z < 1.95) = .5000 + .4744$$

$$= .9744$$

7) a) Decreases

As α decreases, β increases. Thus the power, power $= 1 - \beta$, decreases.

b) Increases

As α increases, β decreases, and $1 - \beta$ increases.

9) Since $\alpha = .05$, the critical value of $\hat{P}$ is located at

$$\hat{p} = .55 - (1.64)\sqrt{\frac{(.5)(.5)}{n}} \ .$$

Since power $= .95 = 1 - \beta$, $\beta = .05$.

$$\beta = .05 = P(Z \geq 1.64) = P(\hat{P} \geq \hat{p}|p = .40)$$

Thus the critical value is also given by

$$\hat{p} = .40 + 1.64\sqrt{\frac{(.40)(.60)}{n}}$$

Equating these two expressions for the critical value yields

$$.40 + 1.64\sqrt{\frac{(.40)(.60)}{n}} = .55 - (1.64)\sqrt{\frac{(.5)(.5)}{n}}$$

Solving:

$$.40 + \frac{.8034}{\sqrt{n}} = .55 - \frac{.82}{\sqrt{n}}$$

$$\frac{1.6234}{\sqrt{n}} = .15$$

$$\sqrt{n} = \frac{1.6234}{.15} = 10.823$$

$$n = 117.13..$$

Take $n \geq 118$.

11) The garage owner is testing

H_0: $\mu = 39$

H_1: $\mu > 39$

The critical value of $\overline{X}$ is $\overline{x} = 39 + (1.64)\dfrac{10}{\sqrt{36}} = 41.73$.

The required probability is $p(H_0$ is rejected$|\mu = 45)$ which is simply
the power of the test:

$$H_0: \ \mu = 39$$
$$H_1: \ \mu = 45.$$

$$\text{power} = P(\overline{X} > 41.73 \mid \mu = 45)$$
$$= P(Z > -1.96) = .5000 + .4750 = .9750$$

Here

$$z = \frac{41.73 - 45}{\dfrac{10}{\sqrt{36}}} = -1.96.$$

13) $\alpha = .10$ and $\beta = 1 - .83 = .17$

a) .10. The probability of rejecting H_0 when H_0 is true is just the significance level α.

b) .90.

$$P(H_0 \text{ is not rejected} \mid H_0 \text{ is true}) = 1 - P(H_0 \text{ is rejected} \mid H_0 \text{ is true})$$
$$= 1 - \alpha = 1 - .10 = .90$$

c) .17. The probability of not rejecting H_0 when H_1 is true is the probability β of a type II error.

d) .83. The probability of rejecting H_0 when H_1 is true is the power of the test.

e) .90. If H_0 is rejected, then H_1 is accepted. Thus this is the probability of a type I error.

f) .83 (see part d).

Chapter Test

1) a) A type I error results if the null hypothesis is rejected when in fact
   it is true. A type II error results if the null hypothesis is not rejected
   when in fact it is false, or equivalently, the alternate hypothesis is
   not accepted when, in fact, it is the alternate hypothesis which is true.

   b) The significance level α of a test is the probability of a type I
   error, i.e.,

   $$\alpha = P(H_0 \text{ is rejected}|H_0 \text{ is true}).$$

   c) The p value of a test is the probability of obtaining a value of the
   test statistic as extreme as actually resulted when in fact H_0 is true.

   d) Let β be the probability of a type II error. Then power $= 1 - \beta$.
   This is simply the probability of accepting H_1 when H_1 is true or
   equivalently, of rejecting H_0 when H_1 is true.

2) a) $\alpha = .04$ is the probability of a type I error.

   b) $1 - \alpha = 1 - .04 = .96$

   Note that this event is the complement of the event of part (a).

   c) $\beta = .15$ is the probability of a type II error.

   $\quad = P(H_1 \text{ is not accepted}|H_1 \text{ is true}).$

   We need

   $P(H_0 \text{ is rejected}|H_1 \text{ is true}) = 1 - \beta = 1 - .15 = .85.$

3) a) For any $\alpha \geq .02$, H_0 will certainly be rejected.
   Thus for $\alpha = .10$, H_0 will be rejected.

   b) We cannot be sure. We need to know the power of the test.

   c) We know the sample proportion $\hat{p}$ produced a $z \geq z_{.02}$. For a two tailed
   test with $\alpha = .15$, the critical values are $\pm z_{.025}$. Since
   $z \geq z_{.02} > z_{.025}$, the null hypothesis would also be rejected for the
   two tailed test.

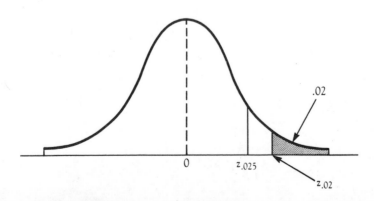

4) a) $\alpha = .08 = 2$ (p value)

    b) $\alpha = .04 = $ p value

5) $\hat{p} = \dfrac{130}{185} = .7027$

    H_0: p = .75
    H_1: p < .75
    Critical value = $-z_{.05} = -1.64$

    S.E. = $\sqrt{\dfrac{(.75)(.25)}{185}} = .0318$

    $z = \dfrac{.7027 - .75}{.0318} = -1.49$

    Since z > -1.64, do not reject H_0.
    p value = $P(\hat{p} \le .7027) = P(Z \le -1.49) = .5000 - .4319 = .0681$

6) H_0: p = .60
    H_1: p ≠ .60
    $\hat{p} = .622$ and n = 1200
    Critical values: $\pm z_{.025} = \pm 1.96$

    $z = \dfrac{.622 - .60}{\sqrt{\dfrac{(.60)(.40)}{1200}}} = 1.56$

    Do not reject H_0.
    p value = $p(\hat{P} \ge .622) = P(Z \ge 1.56) = .5000 - .4406$

    $\qquad\qquad\qquad\qquad\qquad\qquad\quad = .0594$

7) Small sample test requires the t distribution!
    H_0: μ = 55
    H_1: μ ≠ 55
    For $\alpha = .05$ and $\nu = 16 - 1 = 15$, the critical values are $\pm t_{.025} = \pm 2.131$

    $t = \dfrac{52.8 - 55}{\dfrac{8}{\sqrt{16}}} = -1.1$

    Do not reject H_0. Do not reject the claim.

8) H_0: μ = 16
    H_1: μ < 16
    Critical value: $-z_{.05} = -1.96$

    $z = \dfrac{14.3 - 16}{\dfrac{3.4}{\sqrt{40}}} = 3.16$

Reject H_0 and conclude that the dissolved oxygen content is below 16 ppm.

p value = $P(Z \geq 3.16)$ = .5000 - .4992 = .0008

9)   When $\bar{x} = 85$, $z = \dfrac{85 - 80}{\dfrac{10}{\sqrt{36}}} = 3.00 = z_{.0013}$

Thus the critical value $\bar{x} = 85$ corresponds to $z_{.0013}$. For a single tail test of significance level α, the critical value is z_α. Thus $z_\alpha = z_{.0013}$ and $\alpha = .0013 = .13\%$.

10) A type II error occurs if $\mu = 450$ and $\bar{x} < 425$ so that H_0 is not rejected.

$\beta = P(\bar{X} < 425 | \mu = 450)$ is needed

$z = \dfrac{425 - 450}{\dfrac{100}{\sqrt{50}}} = -1.77$

$\beta = P(X < 425 | \mu = 450) = P(Z < -1.77) = .5000 - .4616 = .0384$

Power = $1 - \beta = 1 - .0384 = .9616$

Chapter X
The Comparison of Two Populations

Important techniques for testing hypotheses concerning means or proportions of two populations are developed in this chapter. Many of these techniques are extensions of those developed for simple hypothesis testing in Chapter 9. Tests based on both independent samples and matched pairs or dependent samples are considered.

Concepts/Techniques

1) Two Population Notation

When working with 2 populations, subscripts such as 1 and 2 are used to identify corresponding parameters, sample statistics, observations, etc. The following table illustrates this use.

	Population 1	Population 2
Mean	μ_1	μ_2
Proportion	p_1	p_2
Standard Deviation	σ_1	σ_2
Sample Size	n_1	n_2
Sample Mean	$\bar{x}_1$	$\bar{x}_2$
Sample Standard Deviation	s_1	s_2
Standard Error	$S.E._1$	$S.E._2$
Observations of the Sample	$x_{11}, x_{12}, \ldots, x_{1,n_1}$	$x_{21}, x_{22}, \ldots, x_{2,n_2}$

2) Independent vs. Dependent Samples

In testing for the equality of two population means, we may choose to select two independent random samples and compare their means $\bar{x}_1$ and $\bar{x}_2$. If these are quite different, we conclude that the null hypothesis $H_0: \mu_1 = \mu_2$ should be rejected. Another approach is to try and match or pair up subjects from the 2 populations according to variables which would be expected to have an influence on the variable under study. The two samples are no longer independent. Inferences are then based on the differences of the observations from the matched pairs. If these differences $d_1, d_2, \ldots, d_n$ have a mean that is significantly different from 0 then this is taken as evidence that the null hypothesis $H_0: \mu_1 = \mu_2$ should be rejected.

Example: Two samples of test grades are given below.

Test 1	Test 2	Difference
$x_{11} = 70$	$x_{21} = 73$	$d_1 = x_{11} - x_{21} = -3$
$x_{12} = 80$	$x_{22} = 85$	$d_2 = x_{12} - x_{22} = -5$
$x_{13} = 50$	$x_{23} = 65$	$d_3 = x_{13} - x_{23} = -15$
$x_{14} = 92$	$x_{24} = 72$	$d_4 = x_{14} - x_{24} = 20$
$x_{15} = 68$	$x_{25} = 50$	$d_5 = x_{15} - x_{25} = 18$
$\bar{x}_1 = 72$	$\bar{x}_2 = 69$	$\bar{d} = 3$

If these are independent samples, then inferences about the difference $\mu_1 - \mu_2$ will be based on $\bar{x}_1 - \bar{x}_2 = 72 - 69 = 3$. If the test subjects are matched pairs, then inferences are based on the mean $\bar{d} = 3$ of the paired differences.

3) The Distribution of $\bar{X}_1 - \bar{X}_2$ (Large Sample Case)

When the sample sizes n_1 and n_2 are both at least 30, the difference $\bar{X}_1 - \bar{X}_2$ is approximately normal with mean $\mu_1 - \mu_2$ and standard error

$$S.E. = \sqrt{\frac{\sigma_1^2}{n_1} + \frac{\sigma_2^2}{n_2}} \; .$$

The z score corresponding to $\bar{x}_1 - \bar{x}_2$ is then computed by

$$z = \frac{(\bar{x}_1 - \bar{x}_2) - (\mu_1 - \mu_2)}{\sqrt{\dfrac{\sigma_1^2}{n_1} + \dfrac{\sigma_2^2}{n_2}}}$$

<u>Example:</u> Gasoline mileages of two makes of light trucks, call them
1 and 2, are reported to have means and standard deviations of
$\mu_1 = 28$, $\sigma_1 = 6$, $\mu_2 = 24$, and $\sigma_2 = 9$. If random samples of size
$n_1 = 35$ and $n_2 = 40$ are tested, what is the probability that the
difference $\bar{x}_1 - \bar{x}_2$ of their means will be 2 miles per gallon or
less? (Note we would expect it to be around $\mu_1 - \mu_2 = 4$.)
The standard error is

$$S.E. = \sqrt{\frac{6^2}{35} + \frac{9^2}{40}} = 1.75.$$

Then

$$z = \frac{(\bar{x}_1 - \bar{x}_2) - (\mu_1 - \mu_2)}{S.E.} = \frac{2 - 4}{1.75} = \frac{-2}{1.75} = -1.14$$

Thus $P(\bar{X}_1 - \bar{X}_2 \leq 2) = P(Z \leq -1.14) = .1271.$

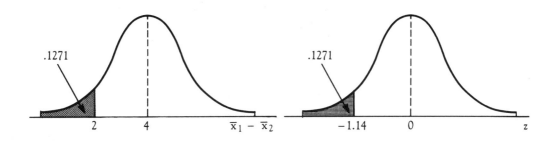

In inferential applications the population variances σ_1^2 and σ_2^2 are
generally not known and must be approximated by s_1^2 and s_2^2. The
standard error is then estimated by

$$S.E. = \sqrt{\frac{s_1^2}{n_1} + \frac{s_2^2}{n_2}}.$$

When hypothesis testing with the null hypothesis H_0: $\mu_1 = \mu_2$, the
difference $\mu_1 - \mu_2$ is 0 and the z score is computed by

$$z = \frac{\overline{x}_1 - \overline{x}_2}{\sqrt{\dfrac{s_1^2}{n_1} + \dfrac{s_2^2}{n_2}}} \, .$$

4) <u>The Distribution of</u> $\overline{X}_1 - \overline{X}_2$ <u>(Small Sample Case)</u>

When the sample sizes are below 30, the normal distribution can no longer be used with $\overline{X}_1 - \overline{X}_2$. If both populations are normal and have the same variance then the t distribution with $\nu = n_1 + n_2 - 2$ degrees of freedom may be used. A pooled estimate of the common variance $\sigma^2 = \sigma_1^2 = \sigma_2^2$ is furnished by

$$s_p^2 = \frac{(n_1 - 1)s_1^2 + (n_2 - 1)s_2^2}{n_1 + n_2 - 2} \, .$$

The standard error is then estimated by

$$\text{S.E.} = \sqrt{\frac{\sigma_1^2}{n_1} + \frac{\sigma_2^2}{n_2}} = \sqrt{s_p^2 \left(\frac{1}{n_1} + \frac{1}{n_2} \right)} = s_p \sqrt{\frac{1}{n_1} + \frac{1}{n_2}}$$

Thus

$$t = \frac{(\overline{x}_1 - \overline{x}_2) - (\mu_1 - \mu_2)}{\sqrt{s_p^2 \left(\frac{1}{n_1} + \frac{1}{n_2} \right)}}$$

and when working under the null hypothesis $H_0: \mu_1 = \mu_2$ this reduces to

$$t = \frac{\overline{x}_1 - \overline{x}_2}{\sqrt{s_p^2 \left(\frac{1}{n_1} + \frac{1}{n_2} \right)}} \, .$$

<u>Example</u>: Using the following data, compute the pooled estimate to the common population variance, compute the standard error, and interpret the observed difference $\overline{x}_1 - \overline{x}_2$ assuming $\mu_1 = \mu_2$.

Sample 1	Sample 2
$n_1 = 22$	$n_2 = 24$
$\overline{x}_1 = 100$	$\overline{x}_2 = 94$
$s_1 = 18$	$s_2 = 20$

$$s_p^2 = \frac{(22 - 1) \cdot 18^2 + (24 - 1) \cdot 20^2}{22 + 24 - 2} = 363.72 \quad \text{and} \quad s_p = 19.07$$

$$\text{S.E.} = \sqrt{363.72 \left(\frac{1}{22} + \frac{1}{24} \right)} = 5.63.$$

The difference of the sample means should be near 0 if the population means are equal. The difference $\bar{x}_1 - \bar{x}_2 = 100 - 94 = 6$ is a little more than 1 standard error from 0. More precisely

$$z = \frac{\bar{x}_1 - \bar{x}_2}{\text{S.E.}} = \frac{6}{5.63} = 1.07.$$

5) Test Procedures Concerning the Difference of 2 Means (Independent Samples)

Hypotheses concerning μ_1 and μ_2 can be rewritten in terms of the difference $\mu_1 - \mu_2$:

H_0: $\mu_1 = \mu_2$ is equivalent to H_0: $\mu_1 - \mu_2 = 0$

H_1: $\mu_1 \neq \mu_2$ is equivalent to H_1: $\mu_1 - \mu_2 \neq 0$

H_1: $\mu_1 > \mu_2$ is equivalent to H_1: $\mu_1 - \mu_2 > 0$

H_1: $\mu_1 < \mu_2$ is equivalent to H_1: $\mu_1 - \mu_2 < 0$ or $(\mu_2 - \mu_1 > 0)$

Once this is done the value of the test statistic is $\bar{x}_1 - \bar{x}_2$ or the equivalent z or t score which are computed by

$$z = \frac{\bar{x}_1 - \bar{x}_2}{\sqrt{\dfrac{s_1^2}{n_1} + \dfrac{s_2^2}{n_2}}} \quad \text{and} \quad t = \frac{\bar{x}_1 - \bar{x}_2}{\sqrt{s_p^2 \left(\dfrac{1}{n_1} + \dfrac{1}{n_2} \right)}}$$

(where s_p^2 is the pooles estimate of the common variance.

From this point on, the procedure is exactly as with the testing done using one sample.

Example: Samples of two brands of "lean" ground beef are tested for their fat content. The results (% of fat) are summarized below. Can we conclude that there is a difference in the fat contents of these 2 brands of lean beef? Use $\alpha = .05$.

Brand A	Brand B
$n = 50$	$n = 46$
$\bar{x} = 26.0$	$\bar{x} = 29.3$
$s = 9.0$	$s = 8.0$

We let μ_1 be the mean fat content of brand B, μ_2 that of brand A. In this way we work with a positive difference $\bar{x}_1 - \bar{x}_2$. We are to test

$$H_0: \mu_1 = \mu_2 \quad \text{or equivalently} \quad H_0: \mu_1 - \mu_2 = 0$$
$$H_1: \mu_1 \neq \mu_2 \qquad\qquad\qquad\quad H_1: \mu_1 - \mu_2 \neq 0$$

The critical values of this two tailed test are $\pm z_{.025} = \pm 1.96$. The value of the test statistic is

$$z = \frac{\bar{x}_1 - \bar{x}_2}{\sqrt{\dfrac{s_1^2}{n_1} + \dfrac{s_2^2}{n_2}}} = \frac{29.3 - 26.0}{\sqrt{\dfrac{8.0^2}{46} + \dfrac{9.0^2}{50}}} = \frac{3.3}{1.74} = 1.90.$$

The null hypothesis is not rejected. We cannot conlcude that there is a difference in the fat contents of these two brands.

Note: p value $= P(z > 1.90) = .0287 = 2.87\%$

Thus the smallest significance level for which H_0 could be rejected with a two tailed test is $2(.0287) = (.0574) = 5.74\%$.

Example: Random samples of male and female state employees are given a job satisfaction survey. The results are summarized below (high scores indicating a greater degree of satisfaction). Can we conclude that women are more satisfied with their jobs than men? Use $\alpha = .01$. What assumptions are necessary?

Men	Women
$n = 20$	$n = 25$
$\bar{x} = 81.4$	$\bar{x} = 95.8$
$s = 18.0$	$s = 15.6$

Let μ_1 be the mean job satisfaction score for the female employees, μ_2 that of the males. We wish to test:

$$H_0: \mu_1 - \mu_2 = 0 \quad (i.e. \quad \mu_1 = \mu_2)$$
$$H_1: \mu_1 - \mu_2 > 0 \quad (i.e. \quad \mu_1 > \mu_2)$$

Since small samples are used we need to assume

1) The job satisfaction scores of both male and female employees are normally distributed.

2) The variances of both the male and female scores are equal.

Using $\nu = 25 + 20 - 2 = 43$ d.f., the critical value for this one tail test is $t_{.01} = 2.423$. (Note, since $\nu = 43$ is not in the table, use the next smallest ν entry, namely $\nu = 40$.)

The pooled estimate of the variance is

$$s_p^2 = \frac{(25 - 1)(15.6)^2 + (20 - 1)(18.0)^2}{25 + 20 - 2} = 278.99$$

$$s_p = \sqrt{278.99} = 16.70$$

Thus

$$t = \frac{\bar{x}_1 - \bar{x}_2}{\sqrt{s_p^2 \left(\frac{1}{n_1} + \frac{1}{n_2}\right)}} = \frac{95.8 - 81.4}{\sqrt{278.99 \left(\frac{1}{25} + \frac{1}{20}\right)}} = 2.87.$$

Clearly H_0 is rejected and we conclude that the female employees have a greater degree of job satisfaction than males.

Note also that p value < .005.

6) <u>Test Procedure Concerning the Difference of 2 Means (Dependent Samples)</u>

Given the two paired samples $x_{11}, x_{12}, \ldots, x_{1n}$ and $x_{21}, x_{22}, \ldots, x_{2n}$, we form the differences

$$d_1 = x_{11} - x_{21}$$
$$d_2 = x_{12} - x_{22}$$
$$\vdots$$
$$d_n = x_{1n} - x_{2n}$$

We regard $d_1, d_2, \ldots, d_n$ as a single sample. Let $\bar{d}$ and s_d be its mean and standard deviation. This mean $\bar{d}$ is the value of a sample statistic $\bar{D}$. When the original populations are normal with equal means $(\mu_1 = \mu_2)$ and equal variances, then $\bar{D}$ has a mean of 0 and a standard error which can be estimated by

$$S.E. = \frac{s_d}{\sqrt{n}} \ .$$

As a result

$$t = \frac{\overline{d}}{\frac{s_d}{\sqrt{n}}}$$

is the value of a variable having the t distribution with $\nu = n - 1$ degrees of freedom. Hypothesis tests concerning μ_1 and μ_2 are now based on the sample mean $\overline{d}$ using the single sample t test. For example, to test

$$\begin{cases} H_0: \ \mu_1 = \mu_2 \\ H_1: \ \mu_1 \neq \mu_2 \end{cases} \text{ we test } \begin{cases} H_0: \ \overline{D} = 0 \\ H_1: \ \overline{D} \neq 0. \end{cases}$$

Similarly to test

$$H_0: \ \mu_1 - \mu_2$$
$$H_1: \ \mu_1 > \mu_2$$

we use

$$H_0: \ \overline{D} = 0$$
$$H_1: \ D > 0.$$

Example: A weight reduction plan was tried on 12 individuals. The differences between their initial and final weights had a mean $\overline{d} = 8.4$ pounds and standard deviation $s_d = 5.6$ pounds. Can we conclude that this plan is effective in producing weight loss. Use $\alpha = .025$.

This is a dependent sample test where each subject is paired with himself.

We let μ_1 be the mean initial weight of subjects starting this plan and μ_2 their mean weight at completion. We will test

$$\begin{cases} H_0: \ \mu_1 = \mu_2 \\ H_1: \ \mu_1 > \mu_2 \end{cases} \text{ or equivalently } \begin{cases} H_0: \ \overline{D} = 0 \\ H_1: \ \overline{D} > 0 \end{cases}$$

Using $\nu = 12 - 1 = 11$ d.f. the critical value of this one tail test is $t_{.025} = 2.201$.
The value of the test statistic is

$$t = \frac{\overline{d}}{\frac{s_d}{\sqrt{n}}} = \frac{8.4}{\frac{5.6}{\sqrt{12}}} = 5.19$$

We reject the hypothesis H_0: $\mu_1 = \mu_2$ and conclude that $\mu_1 > \mu_2$, that is, the plan is effective in achieving a weight loss.

Example: When five autos were tested for wind resistance with 2 types of grills, their drag coefficients were determined as follows:

Auto #	1	2	3	4	5
Grill A	.47	.44	.46	.40	.43
Grill B	.50	.44	.45	.47	.48
$x_B - x_A$	+.03	.00	−.01	+.07	+.05

Using $\alpha = .05$, test for a difference in the drag coefficient due to the type of grill.

Let μ_1 be the mean drag coefficient of the autos with grill B, μ_2 the mean of those with grill A.

We test:

$$\begin{cases} H_0: & \mu_1 = \mu_2 \\ H_1: & \mu_1 \neq \mu_2 \end{cases} \text{ or } \begin{cases} H_0: & \overline{D} = 0 \\ H_1: & \overline{D} \neq 0 \end{cases}$$

For this two tailed test the critical values are $\pm t_{.025} = \pm 2.776$ (use $\nu = 5 - 1 = 4$ d.f.).

We next compute $\overline{d}$ and s_d as in Chapter 3.

d	d^2
.03	.0009
.00	.0000
−.01	.0001
.07	.0049
.05	.0025
Σ d = .14	Σ d^2 = .0084

$$\overline{d} = \frac{\Sigma\, d}{n} = \frac{.14}{5} = .028$$

$$s_d = \sqrt{\frac{\Sigma\, d^2 - \frac{(\Sigma\, d)^2}{n}}{n - 1}}$$

$$= \sqrt{\frac{.0084 - \frac{(.14)^2}{5}}{4}}$$

$$= .0335$$

The value of the test statistic is

$$t = \frac{.028}{\frac{.0335}{\sqrt{5}}} = 1.87.$$

Do not reject H_0. The differences are not sufficient to allow us to conclude that the grills have different effects.

7) <u>The Distribution of $\hat{P}_1 - \hat{P}_2$ (Large Samples)</u>
When the sample sizes are large, the difference $\hat{P}_1 - \hat{P}_2$ of the sample proportions is approximately normal with mean $P_1 - P_2$ and standard error

$$S.E. = \sqrt{\frac{P_1 q_1}{n_1} + \frac{P_2 q_2}{n_2}}.$$

(Here P_1 and P_2 are the corresponding population proportions.)
Thus

$$z = \frac{(\hat{P}_1 - \hat{P}_2) - (P_1 - P_2)}{\sqrt{\frac{P_1 q_1}{n_1} + \frac{P_2 q_2}{n_2}}}$$

is a value of the standard normal variable.

Example: Suppose that 45% of all men and 32% of all women smoke. If random samples of 400 men and 350 women are selected, what is the probability that the proportion of men who smoke will exceed that of the women by at most .04 = 4%?
Let Population 1 correspond to the men and Population 2 to the women. Then $P_1 = .45$, $P_2 = .32$, $n_1 = 400$, $n_2 = 350$ and we need $P(\hat{P}_1 - \hat{P}_2 < .04)$.
The standard error is

$$S.E. = \sqrt{\frac{(.45)(.55)}{400} + \frac{(.32)(.68)}{350}} = .0352 = 3.52\%.$$

Then

$$z = \frac{(\hat{P}_1 - \hat{P}_2) - (P_1 - P_2)}{S.E.} = \frac{.04 - .13}{.0352} = -2.56.$$

Then $P(\hat{P}_1 - \hat{P}_2 < .04) = P(Z < -2.56) = .0052.$

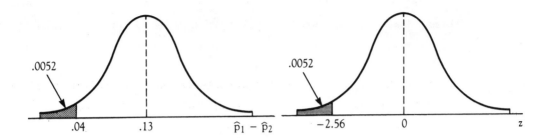

In the event $p = p_1 = p_2$ (as in hypothesis testing) the standard error is given by

$$S.E. = \sqrt{\frac{pq}{n_1} + \frac{pq}{n_2}} = \sqrt{pq\left(\frac{1}{n_1} + \frac{1}{n_2}\right)}$$

and the z score is computed by

$$z = \frac{\hat{p}_1 - \hat{p}_2}{\sqrt{pq\left(\frac{1}{n_1} + \frac{1}{n_2}\right)}}.$$

In hypothesis testing where the null hypothesis is $p_1 = p_2$, it will be necessary to use a pooled estimate p_p of $p = p_1 = p_2$. This is furnished by either of the following:

$$\hat{p} = \frac{n_1\hat{p}_1 + n_2\hat{p}_2}{n_1 + n_2} = \frac{x_1 + x_2}{n_1 + n_2}.$$

Example: Suppose the same proportion of men as women have had chicken pox. Estimate this proportion p if 320 men of a sample of 500 and 430 women of a sample of 700 were found to have chicken pox.

Using $x_1 = 320$, $n_1 = 500$, $x_2 = 430$, and $n_2 = 700$ the pooled estimate is

$$\hat{p} = \frac{x_1 + x_2}{n_1 + n_2} = \frac{320 + 430}{500 + 700} = \frac{750}{1200} = .625 = 62.5\%.$$

Or if we had been given $\hat{p}_1 = \frac{320}{500} = .64$ and $\hat{p}_2 = \frac{430}{700} = .6143$, then we would use

$$p_p = \frac{n_1\hat{p}_1 + n_2\hat{p}_2}{n_1 + n_2} = \frac{(500)(.64) + (700)(.6143)}{500 + 700} = .625 = 62.5\%.$$

8) <u>Test Procedures Concerning the Difference of 2 Proportions</u>

Hypotheses concerning the two proportions p_1 and p_2 can be restated in terms of the difference $\hat{p}_1 - \hat{p}_2$. Once this is done the difference $\hat{p}_1 - \hat{p}_2$ or the corresponding z score is the basis for the decision rule. Here

$$z = \frac{\hat{p}_1 - \hat{p}_2}{\sqrt{\hat{p}\hat{q}\left(\frac{1}{n_1} + \frac{1}{n_2}\right)}}$$

(Note the pooled estimate of p.)

As usual the critical values are $\pm z_{\alpha/2}$ for a two tailed test and z_α or $-z_\alpha$ for a single tailed test.

> Example: A study is made of business support of the immigration enforcement practices of the INS. Suppose 73% of a sample of 300 retailers and 64% of a sample of 400 light manufacturers said they fully supported the policies of the INS. Is there sufficient evidence here to conclude that the INS policies are not equally supported by these 2 business sectors. Use $\alpha = .02$.
>
> Let p_1 and p_2 be the proportions of retailers and light manufacturers respectively who support the INS policies. We test
>
> $$\begin{cases} H_0: & p_1 = p_2 \\ H_1: & p_1 \neq p_2 \end{cases} \text{ or equivalently } \begin{cases} H_0: & p_1 - p_2 = 0 \\ H_1: & p_1 - p_2 \neq 0. \end{cases}$$
>
> The critical values of this two tailed test are $\pm z_{.01} = \pm 2.33$. The pooled estimate of $p = p_1 = p_2$ is
>
> $$\hat{p} = \frac{(300)(.73) + (400)(.64)}{300 + 400} = \frac{219 + 256}{700} = .6786.$$
>
> The standard error is
>
> $$\text{S.E.} = \sqrt{(.6786)(.3214)\left(\frac{1}{300} + \frac{1}{400}\right)} = .03567.$$
>
> Thus
>
> $$z = \frac{\hat{p}_1 - \hat{p}_2}{\text{S.E.}} = \frac{.73 - .64}{.03567} = 2.52.$$
>
> Since $z = 2.52 > z_{.01}$ we reject H_0.
> Further p value $= P(Z > 2.52) = .0059.$

9) <u>Confidence Intervals for</u> $\mu_1 - \mu_2$

Both large and small sample $1 - \alpha$ confidence intervals for the differ-ence $\mu_1 - \mu_2$ of two population means are of the form

$$(\bar{x}_1 - \bar{x}_2) - E < (\mu_1 - \mu_2) < (\bar{x}_1 - \bar{x}_2) + E.$$

When large samples $(n_1, n_2 \geq 30)$ are used the maximum error of the estimate is

$$E = z_{\alpha/2} S.E. \quad \text{where} \quad S.E. = \sqrt{\frac{s_1^2}{n_1} + \frac{s_2^2}{n_2}} \; .$$

For small samples from normal populations with a common variance,

$$E = t_{\alpha/2}(S.E.) \quad \text{where} \quad \nu = n_1 + n_2 - 2,$$

$$S.E. = \sqrt{s_p^2\left(\frac{1}{n_1} + \frac{1}{n_2}\right)} \quad \text{and} \quad s_p = \sqrt{\frac{(n_1 - 1)s_1^2 + (n_2 - 1)s_2^2}{n_1 + n_2 - 1}} \; .$$

<u>Example</u>: Find a 90% confidence interval for $\mu_1 - \mu_2$ using the following information

Sample 1	Sample 2
$n_1 = 60$	$n_2 = 80$
$\bar{x}_1 = 94.3$	$\bar{x}_2 = 82.50$
$s_1 = 8.5$	$s_2 = 9.8$

The computations are as follows:

$$S.E. = \sqrt{\frac{s_1^2}{n_1} + \frac{s_2^2}{n_2}} = \sqrt{\frac{(8.5)^2}{60} + \frac{(9.8)^2}{80}} = 1.55$$

$$E = z_{.05} S.E. = (1.64)(1.55) = 2.54$$

$$(\bar{x}_1 - \bar{x}_2) \pm E = (94.3 - 82.5) \pm 2.54 = 14.34 \quad \text{or} \quad 9.26$$

$$9.26 < \mu_1 - \mu_2 < 14.34$$

Thus we are 90% confident that μ_1 exceeds μ_2 by at least 9.26 and by at most 14.34.

<u>Example</u>: Oxygen content readings (parts per million) are made at random locations of a lake both prior to and following a 3 year

clean up campaign. Using the results below, find a 95% confidence
interval for the improvement in the mean oxygen content of the lake.

Sample 1 (Following Clean Up)	Sample 2 (Prior to Clean Up)
$n_1 = 24$	$n_2 = 19$
$\bar{x}_1 = 12.8$	$\bar{x}_2 = 7.4$
$s_2 = 7.0$	$s_2 = 6.2$

The pooled estimate of the common variance is

$$s_p^2 = \frac{(n_1 - 1)s_1^2 + (n_2 - 1)s_2^2}{n_1 + n_2 - 2} = \frac{(24 - 1)(7.0)^2 + (19 - 1)(6.2)^2}{24 + 19 - 2} = 44.36$$

and $s_p = \sqrt{44.36} = 6.66$.
The standard error is

$$S.E. = \sqrt{s_p^2 \left(\frac{1}{n_1} + \frac{1}{n_2} \right)} = \sqrt{44.36 \left(\frac{1}{24} + \frac{1}{19} \right)} = 2.05.$$

Since $\nu = 24 + 19 - 2 = 41$ is not a table entry, we use the next
smallest entry, $\nu = 40$ and find $t_{.025} = 2.021$.
Then $E = (t_{.025})S.E. = (2.021)(2.05) = 4.14$.
The upper and lower limits are $(\bar{x}_1 - \bar{x}_2) \pm E = (12.8 - 7.4) \pm 4.14$.
These are 9.54 and 1.26.
The confidence interval is

$$1.26 < \mu_1 - \mu_2 < 9.54.$$

Thus the improvement in the oxygen content of the lake is somewhere
between 1.26 and 9.54 ppm (with a 90% certainty).

10) <u>Confidence Intervals for $P_1 - P_2$ Based on Large Samples</u>
A $1 - \alpha$ confidence interval for the difference $P_1 - P_2$ of two
population proportions is of the form

$$(\hat{p}_1 - \hat{p}_2) - E < (p_1 - p_2) < (\hat{p}_1 + \hat{p}_2) + E.$$

The maximum error of the estimate is

$$E = z_{\alpha/2}(S.E.) \quad \text{where} \quad S.E. = \sqrt{\frac{\hat{p}_1 \hat{q}_1}{n_1} + \frac{\hat{p}_2 \hat{q}_2}{n_2}}$$

Note that a pooled estimate is not used since we are not assuming that $P_1 = P_2$.

Example: Using the results below, find a 99% confidence interval for $P_1 - P_2$.

	Sample 1	Sample 2
	$n_1 = 800$	$n_2 = 1000$
	$\hat{p}_1 = .55$	$\hat{p}_2 = .63$

The computations are as follows:

$$S.E. = \sqrt{\frac{(.55)(.45)}{800} + \frac{(.63)(.37)}{1000}} = .0233$$

$$E = (z_{.005})(S.E.) = (2.58)(.0233) = .0601 = 6.01\%.$$

Thus the upper and lower limits are

$$(\hat{p}_1 - \hat{p}_2) + E = (.55 - .63) + .0601 = -.08 + .0601 = -.0199$$

$$\text{and} \quad (\hat{p}_1 - \hat{p}_2) - E = -.08 - .0601 = -.1401.$$

Thus

$$-.1401 < P_1 - P_2 < -.0199.$$

This may also be written as

$$.0199 < P_2 - P_1 < .1401.$$

Example: Prior to an intense advertising campaign a survey of 900 adults produced 360 who were familiar with the advertisers' product. Following the campaign, 600 adults of a sample of 1250 demonstrated a familiarity with the product. Find a 95% confidence interval for the improvement in adult awareness of this product as a result of the advertising campaign.

We arrange the results as follows:

$$\underline{\text{Sample 1 (After)}} \qquad \underline{\text{Sample 2 (Before)}}$$

$$n_1 = 1250 \qquad\qquad n_2 = 900$$

$$x_1 = 600 \qquad\qquad x_2 = 360$$

$$\hat{p}_1 = \frac{600}{1250} = .48 \qquad \hat{p}_2 = \frac{360}{900} = .40$$

$$S.E. = \sqrt{\frac{(.48)(.52)}{1250} + \frac{(.40)(.60)}{900}} = .0216$$

$$E = z_{.025}(S.E.) = (1.96)(.0216) = .0423$$

$$\hat{p}_1 - \hat{p}_2 = .48 - .40 = .08$$

$$.08 - .0423 < p_1 - p_2 < .08 + .0423$$

$$.0377 < p_1 - p_2 < .1223$$

Thus the improvement in product awareness is somewhere between 3.77% and 12.23% of the adult population (with a 95% certainty).

Calculator Usage Illustration

Compute the large sample estimate to the standard error of $\bar{X}_1 - \bar{X}_2$.

Find S.E. $= \sqrt{\dfrac{(25.3)^2}{60} + \dfrac{(31.4)^2}{46}}$.

Method: Direct evaluation.

25.3 $\boxed{x^2}$ $\boxed{\div}$ 60 $\boxed{+}$ 31.4 $\boxed{x^2}$ $\boxed{\div}$ 46 $\boxed{=}$ $\boxed{\sqrt{}}$ 5.6658697

Calculator Usage Illustration

Compute the pooled estimate to variance of $\bar{X}_1 - \bar{X}_2$.

Find $s_p^2 = \dfrac{(12 - 1)(18.5)^2 + (10 - 1)(15.6)^2}{12 + 10 - 2} = \dfrac{(11)(18.5)^2 + (9)(15.6)^2}{20}$.

Method: Direct evaluation.

11 $\boxed{\times}$ 18.5 $\boxed{x^2}$ $\boxed{+}$ 9 $\boxed{\times}$ 15.6 $\boxed{x^2}$ $\boxed{=}$ $\boxed{\div}$ 20 $\boxed{=}$ 297.7495

Calculator Usage Illustration

Compute the estimate of the standard error of $\bar{X}_1 - \bar{X}_2$ using the pooled variance estimate.

Find S.E. $= \sqrt{109.512 \left(\dfrac{1}{12} + \dfrac{1}{10} \right)}$.

Method 1) Arrange the computation under the radical to

$$\left(\frac{1}{12} + \frac{1}{10} \right) \cdot 109.512$$

1 $\boxed{\div}$ 12 $\boxed{+}$ 1 $\boxed{\div}$ 10 $\boxed{=}$ $\boxed{\times}$ 109.512 $\boxed{=}$ $\boxed{\sqrt{}}$ 4.4807589

Method 2) Use the reciprocal key in place of division in the first method.

12 $\boxed{\frac{1}{x}}$ $\boxed{+}$ 10 $\boxed{\frac{1}{x}}$ $\boxed{=}$ $\boxed{\times}$ 109.512 $\boxed{=}$ $\boxed{\sqrt{}}$ 4.4807589

Calculator Usage Illustration

Compute the standard error estimate of $\hat{P}_1 - \hat{P}_2$.

Find $\sqrt{\dfrac{(.24)(.76)}{400} + \dfrac{(.31)(.69)}{500}}$.

Method: Direct evaluation.

.24 $\boxed{\times}$ .76 $\boxed{\div}$ 400 $\boxed{+}$ .31 $\boxed{\times}$ .69 $\boxed{\div}$ 500 $\boxed{=}$ $\boxed{\sqrt{}}$ .02972877

Calculator Usage Illustration

Compute the estimate to the standard error of $\hat{P}_1 - \hat{P}_2$ using the pooled estimate of p.

Find $\sqrt{(.62)(.38)\left(\dfrac{1}{500} + \dfrac{1}{700}\right)}$.

Method 1) Arrange the computation to $\left(\dfrac{1}{500} + \dfrac{1}{700}\right) \cdot (.62) \cdot (.38)$ and do the division using the reciprocal key.

500 $\boxed{\frac{1}{x}}$ $\boxed{+}$ 700 $\boxed{\frac{1}{x}}$ $\boxed{=}$ $\boxed{\times}$ .62 $\boxed{\times}$ .38 $\boxed{=}$ $\boxed{\sqrt{}}$ .0284213

Method 2) Direct evaluation using parentheses.

.62 $\boxed{\times}$ .38 $\boxed{\times}$ $\boxed{(}$ 1 $\boxed{\div}$ 500 $\boxed{+}$ 1 $\boxed{\div}$ 700 $\boxed{)}$ $\boxed{=}$ $\boxed{\sqrt{}}$ .0284213

Exercise Set X.1

1) a) $\mu_{\overline{X}_1 - \overline{X}_2} = \mu_1 - \mu_2 = 200 - 200 = 0$

$\sigma^2_{\overline{X}_1 - \overline{X}_2} = \sigma^2_{\overline{X}_1} + \sigma^2_{\overline{X}_2} = \dfrac{\sigma^2_1}{n_1} + \dfrac{\sigma^2_2}{n_2} = \dfrac{30^2}{100} + \dfrac{45^2}{150} = 9 + 13.5 = 22.5$

S.E. = 4.74

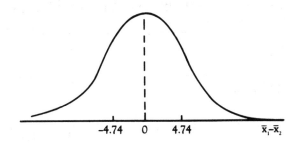

b) $\mu_{\overline{X}_1 - \overline{X}_2} = \mu_1 - \mu_2 = 36.2 - 32.0 = 4.2$

$\sigma^2_{\overline{X}_1 - \overline{X}_2} = \dfrac{\sigma^2_1}{n_1} + \dfrac{\sigma^2_2}{n_2} = \dfrac{(8.95)^2}{50} + \dfrac{(6.53)^2}{64} = 2.27$

S.E. = 1.51

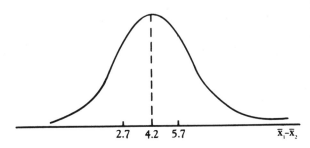

3) a) $\overline{X}_1 - \overline{X}_2$ is approximately normal.

$\mu_{\overline{X}} = 0$

S.E. $= \sqrt{\dfrac{\sigma^2_1}{n_1} + \dfrac{\sigma^2_2}{n_2}} = \sqrt{\dfrac{(100)^2}{64} + \dfrac{(200)^2}{80}} = 25.6$

b) $z = \dfrac{\overline{x}_1 - \overline{x}_2}{S.E.} = \dfrac{40}{25.6} = 1.56$

$$P((\bar{X}_1 - \bar{X}_2) \geq 40) = P(z \geq 1.56) = .5000 = .4406 = .0594$$

5) $\bar{x}_1 = 13.5, \; s_1^2 = 21.67$

$\bar{x}_2 = 9, \; s_2^2 = 9$

$$s_p^2 = \frac{(4 - 1)21.67 + (3 - 1)9}{4 + 3 - 2} = 16.60$$

$$S.E. = \sqrt{16.60\left(\frac{1}{4} + \frac{1}{3}\right)} = 3.11$$

7) a) $\sum_{i=1}^{3} (x_{1i} - \bar{x}_1)^2 = (x_{11} - \bar{x}_1)^2 + (x_{12} - \bar{x}_1)^2 + (x_{13} - \bar{x}_1)^2$

If the sample has only 3 observations, then $\sum_{i=1}^{3} (x_{1i} - \bar{x}_1)^2 = (3 - 1)s_1^2 =$

sum of the squares of the deviations of the observations of the first

sample from their mean.

b) $\sum_{i=1}^{4} (x_{2i} - \bar{x}_2)^2 = (x_{21} - \bar{x}_2)^2 + (x_{22} - \bar{x}_2)^2 + (x_{23} - \bar{x}_2)^2 + (x_{24} - \bar{x}_2)^2 =$

$(4 - 1)s_2^2$ if sample 2 has only 4 observations.

c) $$\frac{\sum\limits_{i=1}^{3} (x_{1i} - \bar{x}_1)^2 + \sum\limits_{i=1}^{4} (x_{2i} - \bar{x}_2)^2}{3 + 4 - 2}$$

$$= \frac{(x_{11}-\bar{x}_1)^2+(x_{12}-\bar{x}_1)^2+(x_{13}-\bar{x}_1)^2+(x_{21}-\bar{x}_2)^2+(x_{22}-\bar{x}_2)^2+(x_{23}-\bar{x}_2)^2+(x_{24}-\bar{x}_2)^2}{3 + 4 - 2}$$

$$= \frac{(3 - 1)s_1^2 + (4 - 1)s_2^2}{3 + 4 - 2} = s_p^2$$

(provided sample 1 has 3 elements and sample 2 four).

9) a) Equipment, vehicles, road course, conditions,... possibly use the same
autos twice with the same drivers over the same road course. Simply
change the fuel system on each auto and repeat the test.

b) job classifications, duration of employment, geographic region of employment, sex of employees ...

c) medium on which the test patches are applied, methods of application,
type of exposure...

11) $$s_p^2 = \frac{(12 - 1)(8.2)^2 + (10 - 1)(7.2)^2}{12 + 10 - 2} = 60.31$$

$$S.E. = \sqrt{60.31\left(\frac{1}{12} + \frac{1}{10}\right)} = 3.33$$

$$z = \frac{3.3}{3.33} = .99$$

$P(Z > .99) = .5000 - .3389 - .1611$

$P(Z < -.99) = .1611$

$P(\overline{X}_1 - \overline{X}_2 > 3.3 \text{ or } \overline{X}_1 - \overline{X}_2 < -3.3) = .1611 + .1611 = .3222$

Exercise Set X.2

1) a) critical values: $\pm z_{.025} = \pm 1.96$

$$S.E. = \sqrt{\frac{(14)^2}{30} + \frac{9^2}{40}} = 2.93$$

$$z = \frac{118 - 113}{2.93} = 1.71$$

Do not reject H_0.

b) $\nu = (10 - 1) + (12 - 1) = 20$, critical value: $\pm t_{.025} = \pm 2.086$

$$s_p^2 = \frac{(10 - 1)(25)^2 + (12 - 1)(23)^2}{10 + 12 - 2} = 572.22$$

$$S.E. = \sqrt{s_p^2 \left(\frac{1}{n_1} + \frac{1}{n_2} \right)} = \sqrt{572.22 \left(\frac{1}{10} + \frac{1}{12} \right)} = 10.24$$

$$t = \frac{164 - 150}{10.24} = 1.37$$

Do not reject H_0.

3) $$S.E. = \sqrt{\frac{(40.5)^2}{50} + \frac{(34.2)^2}{30}} = 8.47$$

$$E = z_{.005} S.E. = (2.58)(8.47) = 21.85$$
$$(\bar{x}_1 - \bar{x}_2) \pm E = (145.2 - 132.4) \pm 21.85$$
$$= 34.65 \text{ or } -9.05$$

Thus $-9.05 < \mu_1 - \mu_2 < 34.65$.

5) $\bar{x}_1 = 101.54$, $s_1^2 = 39.418$, and $n_1 = 5$

$\bar{x}_2 = 111.96$, $s_2^2 = 147.478$, and $n_2 = 5$

For $\nu = 8$, the critical values are $\pm t_{.025} = 2.306$.

$$s_p^2 = \frac{(5 - 1)39.418 + (5 - 1)(147.478)}{5 + 5 - 2} = 93.448$$

$$S.E. = \sqrt{93.448 \left(\frac{1}{5} + \frac{1}{5} \right)} = 6.11$$

$$t = \frac{101.54 - 111.96}{6.11} = -1.71$$

Do not reject H_0.

p value < .1

7) Systolic Analysis

$\nu = 25 + 25 - 2 = 48$

critical values: $\pm t_{.025} = \pm 2.021$

$H_0: \mu_1 = \mu_2$

$H_1: \mu_1 \neq \mu_2$

$$s_p^2 = \frac{(25 - 1)(6.2)^2 + (25 - 1)(6.0)^2}{25 + 25 - 2} = 37.22$$

$$\text{S.E.} = \sqrt{37.22\left(\frac{1}{25} + \frac{1}{25}\right)} = 1.73$$

$$t = \frac{8.9 - 5.0}{1.73} = 2.26$$

Reject H_0 and conclude that the reduction in blood pressure due to meditation is different for men than for women.

9) Let sample 1 be the control group and sample 2 the experimental. We test

$$H_0: \mu_1 = \mu_2$$
$$H_1: \mu_1 > \mu_2$$

For $\alpha = .025$, the critical value is $z_{.025} = 1.96$

$$\text{S.E.} = \sqrt{\frac{(13.4)^2}{36} + \frac{(15.6)^2}{36}} = 3.43$$

$$z = \frac{106.31 - 95.48}{3.43} = 3.16$$

Reject H_0 and conclude an effect due to the distractions.

11) $$\text{S.E.} = \sqrt{\frac{(105.50)^2}{40} + \frac{(84.60)^2}{32}} = 22.40$$

$$\overline{x}_1 - \overline{x}_2 = 630.00 - 548.40 = 81.60$$

$$z_{.05} = 1.64$$

$$81.60 - (1.64)(22.40) < \mu_1 - \mu_2 < 81.60 + (1.64)(22.40)$$
$$44.86 < \mu_1 - \mu_2 < 118.34$$

13) a) Depression Scores

$$\text{S.E.} = \sqrt{\frac{(5.9)^2}{49} + \frac{(5.4)^2}{56}} = 1.11$$

$$\overline{x}_1 - \overline{x}_2 = 3 \quad \text{and} \quad z = \frac{3}{1.11} = 2.70$$

$$\text{p value} = P(\overline{X}_1 - \overline{X}_2 \geq 3) = P(Z \geq 2.70) = .0035$$

Obsession Scores

$$S.E. = \sqrt{\frac{(3.3)^2}{49} + \frac{(3.9)^2}{56}} = .70$$

$$\bar{x}_1 - \bar{x}_2 = .8$$

$$z = \frac{.8}{.70} = 1.14$$

p value $= P((\bar{X}_1 - \bar{X}_2) > .8) = P(Z \geq 1.14) = .1271$

Somatization Scores

$$S.E. = \sqrt{\frac{(5.2)^2}{49} + \frac{(3.6)^2}{56}} = .89$$

$$\bar{x}_1 - \bar{x}_2 = 1.7$$

$$z = \frac{1.7}{.89} = 1.92$$

p value $= P(\bar{X}_1 - \bar{X}_2 > 1.7) = P(Z > 1.92) = .0274$

Anxiety Scores

$$S.E. = \sqrt{\frac{(3.4)^2}{49} + \frac{(3.4)^2}{56}} = .67$$

$$\bar{x}_1 - \bar{x}_2 = 1.1$$

$$z = \frac{1.1}{.67} = 1.64$$

p value $= P((\bar{X}_1 - \bar{X}_2) \geq 1.1) = P(Z > 1.64) = .0495$

b) Confidence interval for the mean of the Depression Scores

S.E. $= 1.11$

$E = (1.96)(1.11) = 2.18$

$3 - 2.18 < \mu_1 - \mu_2 < 3 + 2.18$ or

$.82 < \mu_1 - \mu_2 < 5.18$

Somatization Scores

S.E. $= .89$

$E = (1.96)(.89) = 1.74$

$1.7 - 1.74 < \mu_1 - \mu_2 < 1.7 + 1.74$

$-.04 < \mu_1 - \mu_2 < 3.44$

Anxiety Scores

$-.20 < \mu_1 - \mu_2 < 2.40$

15) a) H_0: $\mu_1 = \mu_2$

H_1: $\mu_1 > \mu_2$

$n_1 = n_2 = 12$

$\nu = 12 + 12 - 2 = 22$ d.f.

For $\alpha = .05$, the critical value is $t_{.05} = 1.717$

$$S.E._{(1)} = .80 = \frac{s_1}{\sqrt{12}} \qquad\qquad S.E._{(2)} = .79 = \frac{s_2}{\sqrt{12}}$$

$$s_1^2 = 12(.80)^2 = 7.68 \qquad\qquad s_2^2 = 12(.79)^2 = 7.49$$

$$s_p^2 = \frac{(12 - 1)7.68 + (12 - 1)7.49}{12 + 12 - 2} = 7.59$$

$$S.E. = \sqrt{7.59\left(\frac{1}{12} + \frac{1}{12}\right)} = 1.12$$

$$t = \frac{23.7 - 20.9}{1.12} = 2.31$$

Reject H_0.

b) H_0: $\mu_1 = \mu_2$

H_1: $\mu_1 \neq \mu_2$ (mean heights are not equal)

$$S.E._{(1)} = 2.4 = \frac{s_1}{\sqrt{12}} \; , \; s_1^2 = 12(2.4)^2 = 69.12$$

$$S.E._{(2)} = 1.7 = \frac{s_2}{\sqrt{12}} \; , \; s_2^2 = 12(1.7)^2 = 34.68$$

$$s_p^2 = \frac{(12 - 1)69.12 + (12 - 1)(34.68)}{12 + 12 - 2} = 51.90$$

$$S.E. = \sqrt{51.90\left(\frac{1}{12} + \frac{1}{12}\right)} = 2.94$$

$$t = \frac{176 - 174}{2.94} = .68$$

Do not reject the null hypothesis of equal mean heights.

Exercise Set X.3

1) a) $\bar{d} = 5$, $s_d = 9.19$, $n = 6$

H_0: $\mu_1 = \mu_2$

H_1: $\mu_1 \neq \mu_2$

$\alpha = .05$, $\nu = 6 - 1 = 5$, $\pm t_{.025} = \pm 2.571$

$t = \dfrac{\bar{d}}{\dfrac{s}{\sqrt{n}}} = \dfrac{5}{\dfrac{9.19}{\sqrt{6}}} = 1.33$

Do not reject H_0.

b) $E = 2.571 \left(\dfrac{9.19}{\sqrt{6}} \right) = 9.65$

$5 - 9.65 < \mu_1 - \mu_2 < 5 + 9.65$ or

$-4.65 < \mu_1 - \mu_2 < 14.65$

3) H_0: $\mu_1 = \mu_2$

H_1: $\mu_1 \neq \mu_2$

A x	B x	d	d^2
1.4	1.7	−.3	.09
4.3	3.9	.4	.16
14.2	14.6	−.4	.16
5.0	4.2	.8	.64
2.2	2.0	.2	.04
		$\Sigma\, d = .7$	1.09

$\bar{d} = \dfrac{.7}{5} = .14$

$s^2 = \dfrac{1.09 - \dfrac{(.7)^2}{5}}{4} = .248$, $s = .50$

For $\nu = 5 - 1 = 4$, the critical values are $\pm t_{.025} = \pm 2.776$

$t = \dfrac{.14}{\dfrac{.50}{\sqrt{5}}} = .62$

Do not reject H_0.

5)

x_1	x_2	d	d^2
1650	1340	310	96100
410	430	-20	400
820	830	-10	100
1010	970	40	1600
300	300	0	0
120	110	10	100
		330	98300

$$\bar{d} = \frac{330}{6} = 55$$

$$s^2 = \frac{98300 - \frac{(330)^2}{6}}{5} = 16030, \ s = 126.61$$

$H_0: \ \mu_1 = \mu_2$

$H_1: \ \mu_1 \neq \mu_2$

$\alpha = .05, \ \nu = 6 - 1 = 5,$ and the critical values are $\pm t_{.025} = \pm 2.571$

$$t = \frac{55}{\frac{126.61}{\sqrt{6}}} = 1.06$$

Do not reject H_0. We cannot conclude a difference in the mean appraisals of these two services.

7) $H_0: \ \mu_A = \mu_B$

$H_1: \ \mu_B < \mu_A$

a) $\alpha = .05, \ \nu = 4,$ and the critical value is $t_{.05} = 2.132$

$$S.E. = \frac{450}{\sqrt{5}} = 201.25$$

$$t = \frac{840}{201.25} = 4.17$$

Reject H_0 and conclude that system B has a lower mean operating cost than does system A.

b) $E = t_{.025}S.E. = (2.776)(201.25) = 558.67$

$840 - 558.67 < \mu_1 - \mu_2 < 840 + 558.67$ or

$281.33 < \mu_1 - \mu_2 < 1398.67$

9) d(worksheet - form)

4, 5, 0, -5, 3, 1, 8, 1, 4, 2

0, 3, 4, 1, -1, 0, 2, 3, -3, 2

$\bar{d} = 1.7$, s = 2.87

S.E. $= \dfrac{2.87}{\sqrt{20}} = .64$

$H_0: \mu_1 = \mu_2$
$H_1: \mu_1 > \mu_2$

For $\alpha = .05$ and $\nu = 19$, the critical value is $t_{.05} = 1.729$

$t = \dfrac{1.7}{.64} = 2.65$

Reject H_0 and conclude a lower score due to the test form.

11) d(before - after)

40	$\bar{d} = 56$
140	s = 104.58
-30	S.E. $= \dfrac{104.58}{\sqrt{10}} = 33.07$
-70	
290	$t = \dfrac{56}{33.07} = 1.69$
100	
-20	
10	
10	
90	

No. The null hypothesis of equal mean usages before and after the installation of the device cannot be rejected even at $\alpha = .10$.

Exercise Set X.4

1) a) critical values: $\pm z_{.025} = \pm 1.96$

$$\hat{p} = \frac{450(.4) + 600(.35)}{450 + 600} = .3714$$

$$S.E. = \sqrt{(.3714)(.6286)\left(\frac{1}{450} + \frac{1}{600}\right)} = .030$$

$$z = \frac{.40 - .35}{.030} = 1.67$$

Do not reject H_0.

b) critical values: $\pm z_{.005} = \pm 2.57$

$$\hat{p} = \frac{64 + 80}{300 + 450} = .1920$$

$$\hat{p}_1 = \frac{64}{300} = .2133, \ \hat{p}_2 = \frac{80}{450} = .1778$$

$$S.E. = \sqrt{(.1920)(.8080)\left(\frac{1}{300} + \frac{1}{450}\right)} = .0294$$

$$z = \frac{.2133 - .1778}{.0294} = 1.21$$

Do not reject H_0.

3) $H_0: \ P_1 = P_2$
$H_1: \ P_1 > P_2$
where p is the proportion who return a positive report.

$$\hat{p} = \frac{320(.420) + 280(.355)}{320 + 280} = .3897$$

$$S.E. = \sqrt{(.3897)(.6103)\left(\frac{1}{320} + \frac{1}{280}\right)} = .0399$$

For $\alpha = .05$, the critical values $z_{.01} = 2.33$

$$z = \frac{.420 - .355}{.0399} = 1.63$$

Do not reject H_0.

5) $H_0: \ P_1 = P_2$ (P_1 is the proportion of model A drivers who survive while P_2
$H_1: \ P_1 > P_2$ is the proportion of drivers of model B who survive.)
For $\alpha = .05$, the critical value is $z_{.05} = 1.64$.

$$\hat{p} = \frac{40(.645) + 40(.51)}{40 + 40} = .5775$$

$$S.E. = \sqrt{(.5775)(.4225)\left(\frac{1}{40} + \frac{1}{40}\right)} = .1105$$

$$z = \frac{.645 - .51}{.1105} = 1.22$$

Do not reject H_0.

7) H_0: $P_1 = P_2$ (Population 1 corresponds to the presweetened cereal users.)
    H_1: $P_1 \neq P_2$
    For $\alpha = .05$, the critical values are $\pm z_{.025} = \pm 1.96$.

$$\hat{P}_1 = \frac{184}{300} = .6133, \quad \hat{P}_2 = \frac{205}{400} = .5125$$

$$\hat{P} = \frac{184 + 205}{300 + 400} = .5557$$

$$\text{S.E.} = \sqrt{(.5557)(.4443)\left(\frac{1}{300} + \frac{1}{400}\right)} = .0380$$

$$z = \frac{.6133 - .5125}{.0380} = 2.65$$

Reject H_0.

No.

9) H_0: $P_1 = P_2$ (P_1 is the proportion of those which respond to flier 1.)
    H_1: $P_1 > P_2$
    For $\alpha = .01$, the critical values are $\pm z_{.005} = \pm 2.57$.

$$\hat{P} = \frac{(1000)(.18) + (1000)(.145)}{1000 + 1000} = .1625$$

$$\text{S.E.} = \sqrt{(.1625)(.8375)\left(\frac{1}{1000} + \frac{1}{1000}\right)} = .0165$$

$$z = \frac{.18 - .145}{.0165} = 2.12$$

Do not reject at $\alpha = .01$ significance level.

p value $= P(Z > 2.12) = .5000 - .4830 = .0170$

11) $\hat{P}_1 = \frac{245}{400} = .6125, \quad \hat{P}_2 = \frac{176}{300} = .5867$

$\hat{P}_1 - \hat{P}_2 = .0258$

$$\text{S.E.} = \sqrt{\frac{(.6125)(.3875)}{400} + \frac{(.5867)(.4133)}{300}} = .0374$$

E $= (1.96)(.0374) = .0733$

$.0258 - .0733 < P_1 - P_2 < .0258 + .0733$

$\qquad -.0475 < P_1 - P_2 < .0991$

Since this interval includes 0 it is by no means clear that P_1 and P_2 are different.

13) Let $\hat{p}_0 = \hat{p}_1 = \hat{p}_2$ be the common value.

Then $\hat{p} = \dfrac{n_1\hat{p}_1 + n_2\hat{p}_2}{n_1 + n_2} = \dfrac{n_1\hat{p}_0 + n_2\hat{p}_0}{n_1 + n_2} = \dfrac{(n_1 + n_2)\hat{p}_0}{(n_1 + n_2)} = \hat{p}_0.$

15) $\hat{p} = \dfrac{n(.40) + n(.40)}{n + n} = .40$

S.E. $= \sqrt{(.40)(.60)\left(\dfrac{1}{n} + \dfrac{1}{n}\right)} = \sqrt{\dfrac{(.40)(.60)(2)}{n}} = \dfrac{.6928}{\sqrt{n}}$

To be significant, we would need $z > 1.96$.

$z = \dfrac{p_1 - p_2}{\dfrac{.6928}{\sqrt{n}}} = \dfrac{.03}{\dfrac{.6928}{\sqrt{n}}} > 1.96$

Solve $\dfrac{.03}{\dfrac{.6928}{\sqrt{n}}} = 1.96$

$\dfrac{.03\sqrt{n}}{.6928} = 1.96$

$\sqrt{n} = \dfrac{(1.96)(.6928)}{.03} = 45.26$

$n = 2048.48$

Take $n \geq 2049$.

Chapter Test

1) critical values: $\pm z_{.025} = \pm 1.96$

   S.E. $= \sqrt{\dfrac{(500)^2}{80} + \dfrac{(550)^2}{75}} = 84.61$

   $z = \dfrac{2150 - 1930}{84.61} = 2.60$

   Reject H_0 and conclude that $\mu_1 \neq \mu_2$.

2) critical value: $z_{.05} = 1.64$

   $\hat{p} = \dfrac{(400)(.600) + (500)(.660)}{400 + 500} = .6333$

   S.E. $= \sqrt{(.6333)(.3667)\left(\dfrac{1}{400} + \dfrac{1}{500}\right)} = .0323$

   $z = \dfrac{.660 - .600}{.0323} = 1.86$

   Since $z > 1.86$, reject H_0 and conclude that $p_1 < p_2$.

3) Small samples require the t distribution.
   For $\alpha = .01$ and $\nu = 10 + 8 - 2 = 16$, the critical value is $t_{.01} = 2.583$
   (next largest table entry used).

   $s_p^2 = \dfrac{(10 - 1) \cdot 7^2 + (8 - 1) \cdot 8^2}{10 + 8 - 2} = 55.5625$

   S.E. $= \sqrt{55.5625\left(\dfrac{1}{10} + \dfrac{1}{8}\right)} = 3.54$

   $t = \dfrac{21 - 16}{3.54} = 1.41$

   Do not reject H_0.

4) $c = .95$, $\alpha = .05$, $\dfrac{\alpha}{2} = .025$, $z_{.025} = 1.96$

   S.E. $= 84.61$ (from problem 1)

   $E = (1.96)(84.61) = 165.84$

   $(2150 - 1930) - 165.84 < \mu_2 - \mu_1 < (2150 - 1930) + 165.84$ or
   $$54.16 < \mu_2 - \mu_1 < 385.84$$

5) Since we now assume $p_1 \neq p_2$, the standard error must be recomputed:

   S.E. $= \sqrt{\dfrac{(.600)(.400)}{400} + \dfrac{(.660)(.340)}{500}} = .0324$

$c = .90$, $\alpha = .10$, $\frac{\alpha}{2} = .05$, $z_{.05} = 1.64$

$E = (1.64)(.0324) = .0531$

$(.660 - .600) - .0531 < p_2 - p_1 < (.660 - .600) + .0531$ or

$.0069 < p_2 - p_1 < .1131$

6) A paired difference test is needed. We assume the test scores (both pre and post test) are normally distributed (or at least their differences are such). Let μ_1 be the mean of student pretest scores and μ_2 the mean of the post test scores. We test

$$H_0: \quad \mu_1 = \mu_2$$
$$H_1: \quad \mu_2 > \mu_1$$

For $\alpha = .01$ and $\nu = 5 - 1 = 4$ d.f., the critical value is $t_{.01} = 2.132$. Let $d = $ (past test score - pretest score)

Student	d	d^2
1	15	225
2	2	4
3	14	196
4	-2	4
5	7	49
	36	478

$\overline{d} = \dfrac{36}{5} = 7.2$

$s = \sqrt{\dfrac{478 - \dfrac{(36)^2}{5}}{4}} = 7.40$

S.E. $= \dfrac{7.40}{\sqrt{5}} = 3.31$

$t = \dfrac{d}{S.E.} = \dfrac{7.2}{3.31} = 2.18$

Reject H_0. We conclude an improvement in the algebraic skills.

7) $H_0: \quad \mu_1 = \mu_2$
$H_1: \quad \mu_1 \neq \mu_2$
For $\alpha = .01$ and $\nu = 10 - 1 = 9$, the critical values are $\pm t_{.005} = \pm 3.250$.

$t = \dfrac{\overline{d}}{S.E.} = \dfrac{400}{\dfrac{420}{\sqrt{10}}} = 3.01$

Do not reject H_0.

8) $H_0: \mu_1 = \mu_2$
$H_1: \mu_1 < \mu_2$
critical value: $z_{.05} = 1.64$

$$\text{S.E.} = \sqrt{\frac{(730)^2}{60} + \frac{(670)^2}{70}} = 123.67$$

$$z = \frac{24960 - 24690}{123.67} = 2.18$$

Reject H_0 and conclude that the salaries of area I are less than those of area II.

p value $= P(Z > 2.18) = .5000 - .4854 = .0146$

9) $H_0: P_1 = P_2$
$H_1: P_1 \neq P_2$
critical values: $z_{.025} = 1.96$

$$p = \frac{(.42)(525) + (.46)(400)}{525 + 400} = .4372$$

$$\text{S.E.} = \sqrt{(.4372)(.5628)\left(\frac{1}{525} + \frac{1}{400}\right)} = .0329$$

$$z = \frac{.46 - .42}{.0329} = 1.22$$

Do not reject H_0. We cannot conclude that there has been a change of opinion.

10) Assuming the restart times (or at least their differences) are normally distributed for both procedures and have the same variance, the t distribution may be used.

$H_0: \mu_1 = \mu_2$
$H_1: \mu_1 \neq \mu_2$
For $\alpha = .05$ and $\nu = 16$ d.f., the critical values are $\pm t_{.025} = \pm 2.120$.

$$s_p^2 = \frac{(10 - 1)(12.0)^2 + (8 - 1)(10.7)^2}{10 + 8 - 2} = 131.09$$

$$\text{S.E.} = \sqrt{131.09\left(\frac{1}{10} + \frac{1}{8}\right)} = 5.43$$

$$t = \frac{26.8 - 18.3}{5.43} = 1.57$$

Do not reject H_0. We cannot conclude a difference in the restart times due to the installation procedure.

Chapter XI
Correlation and Regression

Correlation and regression analysis are the subjects of this chapter. In the context of the development here, correlation is concerned with the strength of the linear relationship that exists between two or more variables. Often we know that one variable is related in a dependent fashion to the remaining variables. Linear regression analysis is concerned with finding the best linear equation with which values of the dependent variable can be predicted from the values of these other, so called, predictor variables.

Concepts/Techniques

1) Linear Equations

A linear equation relating y to x is an equation of the form

$y = b_0 + b_1 x.$

The graph of a linear equation is a straight line. The number b_0 is the y intercept and the number b_1 is the slope of the line. The slope is the change in y, denoted by Δy, for a unit change in x, that is, for $\Delta x = 1$.

Example: Suppose an individuals salary y is $400 per week plus $15 for each hour of overtime. Then the salary y is related to the number of hours of overtime x by the equation

$$y = 400 + 15x.$$

The y intercept is b_0 = 400 and the slope is b_1 = 15. Clearly
for a 1 hour change in the overtime x, the salary y changes by
15 dollars.

To graph this equation we need 2 points. One is the y intercept
(0,400). Choosing, say, x = 5, we find y = 400 + 15(5) = 475.
Thus (5,475) is a second point. The graph is given below.

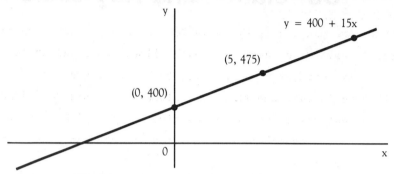

If the slope of a line is positive, then y increases as x increases
and decreases as x decreases. Such lines point up to the right.
The graph of y = 400 + 15x above has slope b_1 = 15 and has this
feature.

Example: Graph y = 60 - 3x.

The y intercept is b_0 = 60 and the slope is b_1 = -3. Thus
when x increases by 1 ($\Delta x = 1$), the change in y is $\Delta y = -3$.
One point of the line is the intercept (0,60).

When x = 10, y = 60 - 3 · 10 = 30. Thus (10,30) is a second
point. Plot these 2 points and draw the line to obtain the graph.

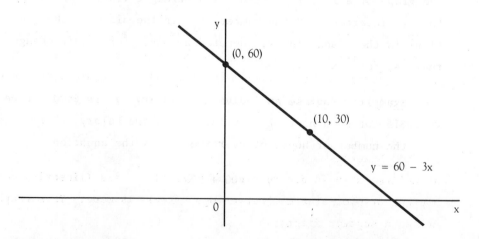

Remember: If a line has positive slope then x and y have the same
behavior: if one increases so does the other; if one decreases, so does
the other. On the other hand, if the slope is negative, x and y
have opposite types of behavior; if one increases the other decreases –
if one decreases the other increases.

2) Linear Correlation

There are many pairs of variables X and Y whose corresponding values
(x,y) fall on or near a straight line. Such variables are said to be
linearly correlated, or simply, to be correlated.

If there is evidence that the values of Y are dependent on X in the
sense that the value x gives rise to the value y, then we speak of Y
as being the dependent or criterion variable and X as the independent
or predictor variable.

If the variables are linearly correlated and a positive slope is
suggested by the observations (x,y) then, we speak of a positive
correlation. If a negative slope is suggested, then we speak of a
negative correlation. Some examples are shown below.

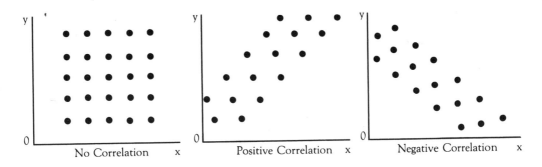

An important point: When two variables have a positive correlation the
tendency is for y to increase as x increases and for y to decrease
as x decreases. When they are negatively correlated, the tendency is
for y to decrease as x increases and for y to increase as x
decreases.

3) Scatter Diagrams

As a first step in determining if X and Y are (linearly) correlated,
a graph or plot of the observations (x,y) is made. This graph is
called a scatter diagram.

Example: Give the scatter diagram for the following data.

x	0	1	1	2	2	3	3	4	5	6
y	2	3	6	6	5	7	5	8	12	13

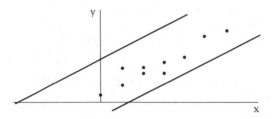

A positive correlation is suggested.

In correlation studies a scatter diagram should always be prepared to
determine if, indeed, there is any possibility of a linear correlation.
The development of this chapter depends on the existence of a linear
correlation between X and Y.

4) The Covariance of X and Y

The covariance of X and Y, denoted by σ_{XY}, is defined as the
expected value of the product $(X - \mu_X)(Y - \mu_Y)$, that is

$$\sigma_{XY} = E[(X - \mu_X)(Y - \mu_Y)].$$

The covariance is a measure of the correlation between X and Y in the
population. It is rarely available and must be estimated by the covariance
s_{XY} of a sample.
The sample covariance is defined by

$$s_{XY} = \frac{\sum\limits_{i=1}^{n} (x_i - \bar{x})(y_i - \bar{y})}{n - 1}.$$

Positive and negative covariances correspond to positive and negative
correlations respectively.

Example: Compute the sample covariance for the data at the
right on the top of the next page.

First find the means:

x	y
1	2
2	5
3	3
4	6
2	4

$$\bar{x} = \frac{1 + 2 + 3 + 4 + 2}{5} = 2.4$$

$$\bar{y} = \frac{2 + 5 + 3 + 6 + 4}{5} = 4$$

Next compute the differences $(x - \bar{x})$ and $(y - \bar{y})$ and their products as we have done in the table below.

$x - \bar{x}$	$y - \bar{y}$	$(x - \bar{x})(y - \bar{y})$
$1 - 2.4 = -1.4$	$2 - 4 = -2$	2.8
$2 - 2.4 = -.4$	$5 - 4 = 1$	$-.4$
$3 - 2.4 = .6$	$3 - 4 = -1$	$-.6$
$4 - 2.4 = 1.6$	$6 - 4 = 2$	3.2
$2 - 2.4 = -.4$	$4 - 4 = 0$	0
		5

Thus

$$s_{XY} = \frac{\Sigma(x - \bar{x})(y - \bar{y})}{n - 1} = \frac{5}{5 - 1} = 1.25$$

Thus the covariance is positive as is the correlation. See the scatter diagram below.

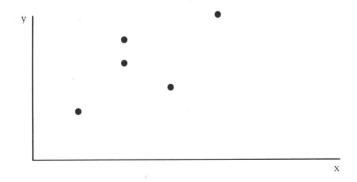

The covariance can be quite large or small depending on the data. Thus it does not prove to be too useful as a measure of the degree of correlation that exists.

5) The Pearson Product Correlation Coefficient

Using the means $\bar{x}$ and $\bar{y}$ and standard deviations s_X and s_Y of the sample $(x_1,y_1),(x_2,y_2),\ldots,(x_n,y_n)$, standardized z scores can be computed:

$$z_x = \frac{x - \bar{x}}{s_X} \quad \text{and} \quad z_y = \frac{y - \bar{y}}{s_Y} .$$

The covariance of these pairs (z_x, z_y) of standardized scores is called the (Pearson) correlation coefficient and is denoted by r. Thus

$$r = \frac{\Sigma \, z_x z_y}{s_X s_Y} .$$

On substituting for z_x and z_y we find the more convenient formula:

$$r = \frac{n(\Sigma \, xy) - (\Sigma \, x)(\Sigma \, y)}{\sqrt{n(\Sigma \, x^2) - (\Sigma \, x)^2} \, \sqrt{n(\Sigma \, y^2) - (\Sigma \, y)^2}}$$

The correlation coefficient r, unlike s_{XY}, varies between -1 and $+1$. A positive correlation corresponds to a positive value of r, a negative correlation to a negative value of r.

If $r = 1$ or $r = -1$ then all the pairs of points (x_i, y_i) fall on a straight line. The closer r is to 1 or -1, the more nearly this is the case. If $r = 0$, there is no linear correlation.

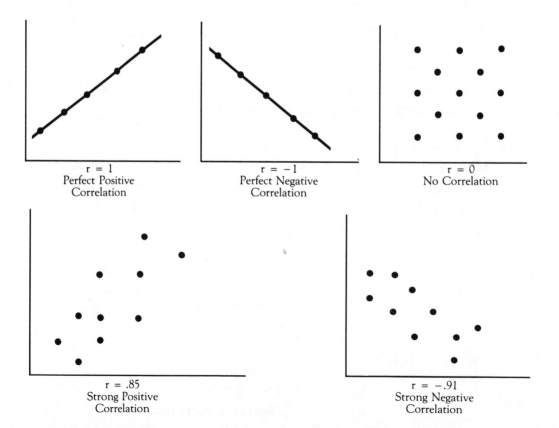

r = 1
Perfect Positive
Correlation

r = −1
Perfect Negative
Correlation

r = 0
No Correlation

r = .85
Strong Positive
Correlation

r = −.91
Strong Negative
Correlation

Example: Compute the correlation coefficient r for the following
data

x	1	2	3	4	2
y	2	5	3	6	4

Since Σx, Σy, Σx^2, Σy^2 and Σxy are needed, the table below
is first prepared

x	y	x^2	y^2	xy
1	2	1	4	2
2	5	4	25	10
3	3	9	9	9
4	6	16	36	24
2	4	4	16	8
Σ 12	20	34	90	53

The column sums are then substituted in to the formula for r.

$$r = \frac{n(\Sigma\ xy) - (\Sigma\ x)(\Sigma\ y)}{\sqrt{n(\Sigma\ x^2) - (\Sigma\ x)^2}\ \sqrt{n(\Sigma\ y^2) - (\Sigma\ y)^2}} = \frac{5(53) - (12)(20)}{\sqrt{5(34) - (12)^2}\ \sqrt{5(90) - (20)^2}} = .693$$

The correlation coefficient between X and Y in the population
is denoted by the Greek letter (rho) ρ. In most instances it must
be approximated by the coefficient r of a sample.

6) Tests for a Correlation Based on r

When r is close to 0, we can conclude that there is no (linear)
correlation between the variables X and Y in the population.   To
test whether r is of such magnitude that a correlation may be inferred,
we compare r with the appropriate critical value (Table VIII).

H_0: $\rho = 0$ (no correlation)

H_1: $\rho \neq 0$ (correlation)

The critical values are $\pm r_{\alpha/2}$.

For the single tail test:

H_0: $\rho = 0$ (no correlation)

H_1: $\rho > 0$ (positive correlation)

the critical value is r_{α} .
For the single tailed test:

$$H_0: \quad \rho = 0 \quad \text{(no correlation)}$$
$$H_1: \quad \rho < 0 \quad \text{(negative correlation)}$$

The critical value is $-r_{\alpha}$.
(In all cases, the significance level is α .)

> Example: A study involving the grade point average X at graduation
> and the starting salary Y of 50 college graduates produced the
> coefficient r = .58. Is this evidence to allow us to conclude a
> positive correlation between these variables? Test using $\alpha = .05$.
> We test:
>
> $$H_0: \quad \rho = 0$$
> $$H_1: \quad \rho > 0$$
>
> where ρ is the correlation coefficient between X and Y.

Using $\nu = n - 2 = 50 - 2 = 48$ d.f., Table VIII yields $r_{.06} = .264$.
(Use next smallest ν entry, namely $\nu = 40$.) Since r = .58 > .264 we
reject H_0 and conclude that a positive correlation exists between
starting salaries and grade point averages.

7) Correlation and Causality
A correlation between 2 variables X and Y does not imply that a
"cause and effect" relationship exists between the variables.

> Example: Since textbook costs and doctor's office visit charges
> have both increased gradually over the last 3 decades, we might be
> able to establish a positive correlation between these variables.
> But this would not imply that, say, increased textbook costs are
> responsible for increased office charges by doctors.

8) The Regression Line of y on x
A set of points exhibiting a linear correlation is plotted on the diagram
below. If a line is drawn as shown, the vertical distances from the
point to the line, $d_1, d_2, d_3, \ldots$ can be computed. If the line is close
to all the data points, then the quantity

$$R = d_1^2 + d_2^2 + d_3^2 + \cdots$$

will be small.

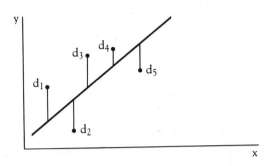

The line which best fits the data in the <u>least squares sense</u> has a smaller value of R than any other line which can be drawn. This unique line is called the least squares line or the regression line of y on x. The equation of the regression line may be written as

$$y = b_0 + b_1 x.$$

The slope b_1 and the y intercept of this line are found from

$$b_1 = \frac{n(\Sigma\ xy) - (\Sigma\ x)(\Sigma\ y)}{n(\Sigma\ x^2) - (\Sigma\ x)^2}$$

$$b_0 = \overline{y} - b_1\overline{x}$$

<u>Example</u>: Find the equation for the regression line of y on x using the data below. Then find the value of y that is predicted by $x = 5$.

x	1	2	2	3	5	5
y	4	6	5	8	9	10

First a table is prepared to obtain $\Sigma\ x$, $\Sigma\ y$, $\Sigma\ xy$, and $\Sigma\ x^2$

x	y	x^2	xy
1	4	1	4
2	6	4	12
2	5	4	10
3	8	9	24
5	9	25	45
5	10	25	50
Σ 18	42	68	145

All computations are now easily made

$$\overline{x} = \frac{18}{6} = 3, \ \overline{y} = \frac{42}{6} = 7$$

$$b_1 = \frac{n(\Sigma \ xy) - (\Sigma \ x)(\Sigma \ y)}{n(\Sigma \ x^2) - (\Sigma \ x)^2} = \frac{6(145) - (18)(42)}{6(68) - (18)^2} = 1.35714$$

$$b_0 = \overline{y} - b_1\overline{x} = 7 - (1.35714)(3) = 2.92857$$

These formulas are sensitive to rounding so do not do so until both
coefficients have been found.

Now we use $b_0 = 2.93$ and $b_1 = 1.36$ and write the regression
line as

$$y = 2.93 + 1.36x.$$

The predicted value of y when x = 5 is denoted by y' and is

$$y' = 2.93 + (1.36)5 = 9.73.$$

The scatter diagram and the regression line are shown below.

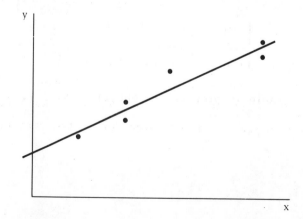

The following are important facts about the regression line

a) The regression line passes through the point $(\bar{x},\bar{y})$.

b) An alternate formula for the regression line is

$$y = \bar{y} + b_1 (x - \bar{x}).$$

For example, the line just found above can be written as

$$y = 7 + 1.36 (x - 3).$$

c) The correlation coefficient r for a set of data is related to the slope of the corresponding regression line by

$$b_1 = r \left(\frac{s_Y}{s_X} \right).$$

d) If each data pair (x,y) is replaced by the standardized pair (z_x, z_y), then the regression equation of z_y on z_x is

$$z_y = r z_x$$

i.e., the slope is the correlation coefficient and the intercept is 0.

e) The square r^2 of the correlation coefficient, called the coefficient of determination. It can be shown that

$$r^2 = \frac{s_{Y'}^2}{s_Y^2}$$

where $s_{Y'}^2$ and s_Y^2 are the variances in the predicted and observed Y values. Thus r^2 is the proportion of the variance in Y which is accounted for or which is explained by the regression with the variable X (which produced the y' values).

Example: Suppose a study of interest rates Y and the M1 money supply X produced a correlation coefficient of $r = -.70$. Then $r^2 = .49$ and this states that 49% of the variation (observed) in interest rates is accounted for by the variation in M1.

9) The Theoretical Regression Line

When two variables X and Y vary jointly, the values of Y corresponding to a particular value x of X have a probability distribution.
Generally speaking, we obtain a different distribution for each x. We
use the notation $\mu_{Y|x}$ for the mean of the Y distribution corresponding
to x.

> Example: Consider a study involving the weight Y and height X
> of 25 year old men. If we fix the height at x = 66 inches, then
> the weights of these (66 inch tall) men has a probability distribution.
> Its mean is denoted by $\mu_{Y|66}$.
> Similarly $\mu_{Y|68}$ would indicate the mean weight of men of height
> x = 68 inches.

Linear regression analysis makes the following assumptions about these Y
distributions:

a)   Each Y distribution is approximately normal.

b)   These distributions have a common variance σ^2.

c)   The means $\mu_{Y|x}$ of these distributions lie on a straight line whose
     equation is

$$y = \beta_0 + \beta_1 x$$

i.e., for a specific x

$$\mu_{Y|x} = \beta_0 + \beta_1 x.$$

This line $y = \beta_0 + \beta_1 x$ is called the theoretical regression line of y
on x.

> Example: Suppose the values of the weight Y and height X of
> 25 year old men are related (theoretically) by
>
> $$y = -115 + 4x.$$
>
> The distribution of the weight Y of men of hieght x = 70 has
> mean
>
> $$\mu_{Y|70} = -115 + 4(70) = 165 \text{ (pounds)}.$$
>
> Similarly, for men of height x = 63, we find

$$\mu_{Y|63} = 137 \text{ (pounds)}.$$

These two distributions are shown below separately and then super-
imposed on a single graph containing the regression line.

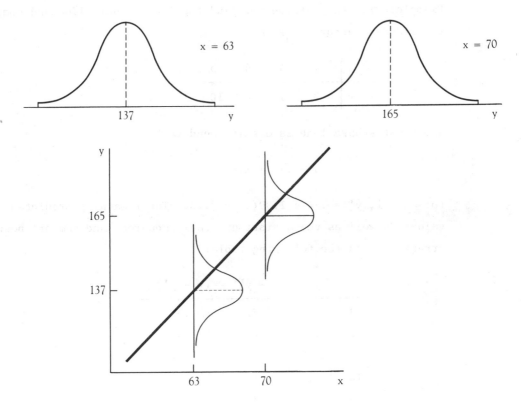

When the regression analysis assumptions hold, the sample least
square line $y = b_0 + b_1x$ is an approximation to the theoretical
line $y = \beta_0 + \beta_1x$.

10) The Standard Error of Estimate

Let $y_1', y_2', \ldots, y_n'$ be the values of y predicted by the least square
line $y = b_0 + b_1x$ for $x = x_1, x_2, \ldots, x_n$. The standard error of estimate
is defined as

$$S_e = \sqrt{\frac{(y_1 - y_1')^2 + (y_2 - y_2')^2 + \cdots + (y_n - y_n')^2}{n - 2}}$$

The standard error of estimate has two important uses:

a) Since S_e is sort of an average of the deviations between observed
and predicted y values, the size of S_e provides an indication of
how close the observed points $(x_1, y_1), (x_2, y_2), \ldots, (x_n, y_n)$ are to

the least square line.

b) The number S_e^2 is an estimate of the common variance σ^2 of the Y distributions which arise by fixing different values x of X.

Example: For the data below, find the least square line and compute the standard error of estimate.

x	1	4	4	5	7
y	3	7	8	10	15

The least square line is easily found to be

$$y = .24 + 1.99x.$$

For x = 1, y' = .24 + 1.99(1) = 2.23. The remaining predicted values as well as the deviations, their squares, and sum are best arranged as in the following table:

x	y	y'	y - y'	$(y - y')^2$
1	3	2.23	0.77	.59
4	7	8.20	-1.20	1.44
4	8	8.20	-0.20	0.04
5	10	10.19	-0.19	0.04
7	15	14.17	0.83	0.69
				Σ 2.80

Thus

$$S_e^2 = \frac{\Sigma (y - y')^2}{n - 2} = \frac{2.80}{3} = .933 \approx .93$$

and

$$S_e = \sqrt{.933} = .97.$$

Thus the common variance of the Y distribution is estimated as

$$\sigma^2 = .93$$

An alternate formula for the standard error of estimate which does not require the computation of the predicted y values is as follows:

$$S_e = \sqrt{\frac{\Sigma y^2 - b_0 (\Sigma y) - b_1 (\Sigma xy)}{n - 2}} \; .$$

Example: Using the alternate formula, compute S_e for the data of the last example.

The following table contains all needed sums.

x	y	xy	y^2
1	3	3	9
4	7	28	49
4	8	32	64
5	10	50	100
7	15	105	225
Σ		43	218	447

Using $b_0 = .24$, $b_1 = 1.99$, we have

$$S_e^2 = \frac{\Sigma y^2 - b_0(\Sigma y) - b_1(\Sigma xy)}{n - 2} = \frac{447 - (.24)(43) - (1.99)(218)}{3}$$

$$= \frac{2.86}{3} = .953.$$

Thus

$$S_e = \sqrt{.953} = .98.$$

The small discrepency between this and the previous value of .97 is due to rounding.

11) <u>Confidence Intervals for</u> $\mu_{Y|x}$

A $(1 - \alpha)$ confidence interval for the mean $\mu_{Y|x}$ is of the form

$$y' - E < \mu_{Y|x} < y' + E$$

where

$$y' = b_0 + b_1 x,$$

$$E = t_{\alpha/2} S_e \sqrt{\frac{1}{n} + \frac{n(x - \bar{x})^2}{n(\Sigma x_i^2) - (\Sigma x_i)^2}} \; ,$$

$\nu = n - 2$, and n is the number of pairs of observations.

Example: Using the data below find a 95% confidence interval for
$\mu_{Y|4}$

x	y	x^2
1	3	1
4	7	16
4	8	16
5	10	25
7	15	49

Σ 21   107

Previous calculations have shown the regression line and standard
error of estimate to be $y = .24 + 1.99x$ and $S_e = .97$. Also we
see that $\bar{x} = \frac{21}{5} = 4.2$, $\Sigma x_i^2 = 107$ and $\Sigma x_i = 21$.
Using $\nu = 5 - 2 = 3$, we find $t_{.025} = 3.182$.
Thus

$$E = (3.182)(.97)\sqrt{\frac{1}{5} + \frac{5(4 - 4.2)^2}{5(107) - (21)^2}} = 1.39.$$

For $x = 4$, the predicted y value is

$$y' = .24 + 1.99(4) = 8.20.$$

The 95% confidence interval is

$$8.20 - 1.39 < \mu_{Y|4} < 8.20 + 1.39 \quad \text{or}$$

$$6.81 < \mu_{Y|4} < 9.59$$

12) <u>Confidence Intervals for the Slope</u> β_1 <u>of the Theoretical Regression</u>
<u>Line</u>

A $(1 - \alpha)$ confidence interval for β_1 is of the form

$$b_1 - E < \beta_1 < b_1 + E$$

where b_1 is the slope of the (sample) regression line,

$$E = t_{\alpha/2} \frac{S_e}{\sqrt{\frac{n(\Sigma x^2) - (\Sigma x)^2}{n}}} \quad \text{and} \quad \nu = n - 2.$$

Example: Find a 90% confidence interval for the slope β_1 of the theoretical regression line using the data below. Interpret the results.

x	y	x^2
1	3	1
4	7	16
4	8	16
5	10	25
7	15	49
Σ 21		107

As already shown, the least squares line is $y = .24 + 1.99x$ and the standard error of estimate is $S_e = .97$.

Using $\nu = 5 - 2 = 3$, $t_{.05} = 2.353$ and

$$E = 2.343 \; \frac{.97}{\sqrt{\dfrac{5(107) - (21)^2}{5}}} = .52 .$$

Using $b_1 = 1.99$, the 90% confidence interval is

$$1.99 - .52 < \beta_1 < 1.99 + .52 \quad \text{or simply}$$
$$1.47 < \beta_1 < 2.51$$

Thus for a unit change in x, the change in y is predicted to be between 1.47 and 2.51.

13) **Multivariate Correlation**

Multivariate correlation is concerned with a correlation between 3 or more variables $Y, X_1, X_2, \ldots$. We restrict the discussion to the case of two predictor variables, X_1 and X_2.

14) **The Correlation Matrix**

A correlation coefficient can be computed from the sample data for each pair of variables. For X_1 and X_2 this coefficient is denoted by r_{12}, for X_1 and Y by r_{1Y}, and for X_2 and Y by r_{2Y}. We note that $r_{1Y} = r_{Y1}$, $r_{2Y} = r_{Y2}$, and $r_{12} = r_{21}$ since the correlation between 2 variables does not depend on which variable is selected first. Additionally, since a variable correlates perfectly with itself, we have

$$r_{11} = r_{22} = r_{YY} = 1.$$

These correlation coefficients are displayed in an array, called a correlation matrix, as shown below

$$
\begin{array}{c}
 & \begin{array}{ccc} X_1 & X_2 & Y \end{array} \\
\begin{array}{c} X_1 \\ X_2 \\ Y \end{array} &
\left[\begin{array}{ccc}
1 & r_{12} & r_{1Y} \\
r_{21} & 1 & r_{2Y} \\
r_{Y1} & r_{Y2} & 1
\end{array} \right]
\end{array}
$$

Example: A marketing study produced the correlation matrix below. The criterion variable Y   is the market share (%) of a soft drink company. The predictor variables X_1 and X_2 are the size of the sales force and the advertising budget of the company. Ignoring the effect of the budget, we see that the correlation between market share and sales force is $r_{1Y} = r_{Y1} = .48$. The correlation between market share and advertising budget is much higher, namely $r_{2Y} = r_{Y2} = .59$.   Finally, between sales force and budget, $r_{12} = r_{21} = -.08$ indicates no correlation of any significance.

$$
\begin{array}{c}
 & \begin{array}{ccc} X_1 & X_2 & Y \end{array} \\
\begin{array}{c} X_1 \\ X_1 \\ Y \end{array} &
\left[\begin{array}{ccc}
1 & -.08 & .48 \\
-.08 & 1 & .59 \\
.48 & .59 & 1
\end{array} \right]
\end{array}
$$

15)   Partial Correlation Coefficients

Partial correlation coefficients allow us to determine the correlation between two variables when the effect due to a third variable has been removed. If A, B, and C are 3 variables, we use $r_{AB \cdot C}$ for the partial correlation coefficient between A and B when the effect due to C has been removed or equivalently, when C is held constant. The partial correlation coefficient $r_{AB \cdot C}$ is computed by

$$r_{AB \cdot C} = \frac{r_{AB} - r_{AC} r_{BC}}{\sqrt{1 - r_{AC}^2} \sqrt{1 - r_{BC}^2}}.$$

Example: The sample correlation matrix of a marketing study is given below. Compute and interpret the partial correlation coefficients.

	X_1 (Sales Force)	X_2 (Adv. Budget)	Y (Market Share)
X_1	1	-.08	.48
X_2	-.08	1	.59
Y	.48	.59	1

$$r_{1Y\cdot 2} = \frac{r_{1Y} - r_{12}r_{Y2}}{\sqrt{1 - r_{12}^2}\sqrt{1 - r_{Y2}^2}} = \frac{.48 - (-.08)(.59)}{\sqrt{1 - (-.08)^2}\sqrt{1 - (.59)^2}} = .66$$

The correlation between sales force and market share is .66 when the budget is held constant

$$r_{2Y\cdot 1} = \frac{r_{2y} - r_{21}r_{Y1}}{\sqrt{1 - r_{21}^2}\sqrt{1 - r_{Y1}^2}} = \frac{.59 - (-.08)(.48)}{\sqrt{1 - (-.08)^2}\sqrt{1 - (.48)^2}} = .72.$$

When the sales force is held constant, the correlation between market share and advertising budget is .72. Finally, for completeness

$$r_{12\cdot Y} = \frac{r_{12} - r_{1Y}r_{2Y}}{\sqrt{1 - r_{1Y}^2}\sqrt{1 - r_{2Y}^2}} = \frac{(-.08) - (.48)(.59)}{\sqrt{1 - (.48)^2}\sqrt{1 - (.59)^2}} = -.51$$

16) The Multilinear Least Squares Equation

The extension of the least squares line to the case of two predictor variables results in a plane in 3 dimensions whose equation is

$$y = b_0 + b_1x_1 + b_2x_2.$$

The coefficients b_0, b_1, and b_2 are found by solving the system of "normal" equations

$$\begin{cases} nb_0 + (\Sigma x_1)b_1 + (\Sigma x_2)b_2 = (\Sigma y) \\ (\Sigma x_1)b_0 + (\Sigma x_1^2)b_1 + (\Sigma x_1 x_2)b_2 = (\Sigma x_1 y) \\ (\Sigma x_2)b_0 + (\Sigma x_2 x_1)b_1 + (\Sigma x_2^2)b_2 = (\Sigma x_2 y) \end{cases}$$

Here n is the number of subjects or observations, Σx_1 is the sum of the observed values of the variable X_1, and so forth.

Example: The data of a study of the weight Y (pounds), the height X_1 (inches) and the age X_2 (months) of 7 teenage girls is given below. Find the coefficients of the least squares equation $y = b_0 + b_1 x_1 + b_2 x_2$.

X_1 (Height)	X_2 (Age)	Y (Weight)	x_1^2	x_2^2	$x_1 x_2$	$x_1 y$	$x_2 y$
60	180	104	3600	32400	10800	6240	18720
65	175	109	4225	30625	11375	7085	19075
67	194	116	4489	37636	12998	7772	22504
62	200	110	3844	40000	12400	6820	22000
70	204	124	4900	41616	14280	8680	25296
56	160	93	3136	25600	8960	5208	14880
55	158	95	3025	24964	8690	5225	15010
Σ 435	1271	751	27219	232841	79503	47030	137485

The system of normal equations is

$$\left. \begin{array}{l} 7b_0 + 435b_1 + 1271b_2 = 751 \\ 435b_0 + 27219b_1 + 79503b_2 = 47030 \\ 1271b_0 + 79503b_1 + 232841b_2 = 137485 \end{array} \right\}$$

Divide the first equation through by 7, the coefficient of b_0 so that the new coefficient of b_0 is 1. We now have

$$\left. \begin{array}{l} b_0 + 62.1429b_1 + 181.5714b_2 = 107.2857 \\ 435b_0 + 27219b_1 + 79503b_2 = 47030 \\ 1271b_0 + 79503b_1 + 232841b_2 = 137485 \end{array} \right\}$$

Now use the first equation to eliminate b_0 from the second and third equations. The new second equation is obtained by multiplying the first equation by 435 and subtracting from the second. Similarly the new third equation is obtained by multiplying the first equation by 1271 and subtracting from the third.

$$b_0 + 62.1429b_1 + 181.5714b_2 = 107.2857$$

$$186.8390b_1 + 519.4410b_2 = 360.7210$$

$$519.3750b_1 + 2063.7600b_2 = 1124.8800$$

Now divide the second equation by 186.8390, the coefficient of b_1, so that its new coefficient is 1.

$$\left.\begin{array}{l} b_0 + 62.1429b_1 + 181.5714b_2 = 107.2857 \\ b_1 + 2.7802b_2 = 1.9307 \\ 519.3750b_1 + 2063.7600b_2 = 1124.880 \end{array}\right\}$$

Now use the second equation to eliminate b_1 from the third equation: multiply the second equation by 519.3750 and subtract from the third.

$$\left.\begin{array}{l} b_0 + 62.1429b_1 + 181.5714b_2 = 107.2857 \\ b_1 + 2.7802b_2 = 1.9307 \\ 619.7937b_2 = 122.1227 \end{array}\right\}$$

Dividing the last equation through by 619.7937, the coefficient of b_2, we obtain

$$\left.\begin{array}{l} b_0 + 62.1429b_1 + 181.5714b_2 = 107.2857 \\ b_1 + 2.7802b_2 = 1.9307 \\ b_2 = 1.9704 \end{array}\right\}$$

Substituting $b_2^2 = 1.9704$ into the second equation yields

$$b_1 + 5.4781 = 1.9307 \quad \text{or}$$
$$b_1 \qquad = -3.5474$$

Finally we substitute $b_2 = 1.9704$ and $b_1 = -3.5474$ into the first equation to obtain

$$b_0 - 220.4457 + 357.7683 = 107.2857$$

or

$$b_0 = -30.0368.$$

Thus the least squares equation is

$$y = -30.0368 - 3.5474x_1 + 1.9704x_2.$$

Normal equations are quite sensitive to rounding. If, for example, six decimal places are used, in the above example, we find

$$b_0 = -30.10374, \quad b_1 = -3.6191368 \quad and \quad b_2 = 1.9953193.$$

These differ in either the first or second decimal place from the results obtained with 4 decimal places. In practice, multilinear least square problems are solved with the aid of special purpose computer programs.

Calculator Usage Illustration

Compute the covariance of a sample.

Find $\dfrac{(1.4 - 3.5)(8.3 - 6.8) + (4.1 - 3.5)(5.1 - 6.8) + (5.0 - 3.5)(7 - 6.8)}{2}$.

Method 1)   Direct summation using parentheses when computing the products.

(1.4 − 3.5) × (8.3 − 6.8) + (4.1 − 3.5) ×

(5.1 − 6.8) + (5.0 − 3.5) × (7 − 6.8)

= ÷ 2 = <u>−1.935</u>

Method 2)   Accumulate the sum of the products in the memory.

(1.4 − 3.5) × (8.3 − 6.8) = [STO]

(4.1 − 3.5) × (5.1 − 6.8) = [SUM]

(5.0 − 3.5) × (7 − 6.8) = [SUM] [RCL] ÷ 2 = <u>−1.935</u>

Calculator Usage Illustration

Compute the correlation coefficient r.

$$\text{Find } r = \frac{(12)(330) - (55)(58)}{\sqrt{(12)(355) - (55)^2} \ \sqrt{(12)(414) - (58)^2}}$$

Method 1)   Compute the denominator first using the fact that $\sqrt{A} \ \sqrt{B} = \sqrt{AB}$
            and store the result. Then compute the numerator and do the
            division by recalling the denominator from memory.

$(\!($ 12 $\times$ 355 $-$ 55 $\boxed{x^2}$ $)$ $\times$ $(\!($ 12 $\times$ 414 $-$ 58 $\boxed{x^2}$ $)$

$=$ $\boxed{\sqrt{}}$ $\boxed{\text{STO}}$ 12 $\times$ 330 $-$ 55 $\times$ 58 $=$ $\div$ $\boxed{\text{RCL}}$ $=$ .5470853

Method 2)   Proceed as in method (1). In place of storing the denominator,
            however, obtain its reciprocal and multiply the numerator by it.
            Parentheses must be used in computing the numerator.

$(\!($ 12 $\times$ 355 $-$ 55 $\boxed{x^2}$ $)$ $\times$ $(\!($ 12 $\times$ 414 $-$ 518 $\boxed{x^2}$ $)$

$=$ $\boxed{\sqrt{}}$ $\boxed{\frac{1}{x}}$ $\times$ $(\!($ 12 $\times$ 330 $-$ 55 $\times$ 58 $)$ $=$ .5470853

Calculator Usage Illustration

Compute the standard error of estimate.

$$\text{Find } S_e = \sqrt{\frac{215 - (3.4)(18) - (2.7)(38)}{12}} \ .$$

Method: Direct evaluation

215 $-$ 3.4 $\times$ 18 $-$ 2.7 $\times$ 38 $=$ $\div$ 12 $=$ $\boxed{\sqrt{}}$ 2.0655911

Calculator Usage Illustration

Compute a partial correlation coefficient.

Find $r_{1Y \cdot 2} = \dfrac{.97 - (.98)(.97)}{\sqrt{(1 - (.98)^2)(1 - (.97)^2)}}$.

Method: Direct evaluation replacing division with multiplication by the reciprocal.

$\boxed{(}\ 1\ \boxed{-}\ .98\ \boxed{x^2}\ \boxed{)}\ \boxed{\times}\ \boxed{(}\ 1\ \boxed{-}\ .97\ \boxed{x^2}\ \boxed{)}\ \boxed{=}\ \boxed{\sqrt{\ }}\ \boxed{\dfrac{1}{x}}$

$\boxed{\times}\ \boxed{(}\ .97\ \boxed{-}\ .98\ \boxed{\times}\ .97\ \boxed{)}\ \boxed{=}\ \underline{.4010148}$

Exercise Set XI.1

1) a) Negative. As x increases, the tendency is for y to decrease.

   b) Positive. As x increases, the tendency is for y to increase.

   c) Positive. As x increases, y increases.

3) a)

x	y	$x - \bar{x}$	$y - \bar{y}$	$(x - \bar{x})(y - \bar{y})$
0	2	-2.5	-1.5	3.75
1	2	-1.5	-1.5	2.25
2	3	-.5	-.5	.25
3	3	.5	-.5	-.25
4	6	1.5	2.5	3.75
5	5	2.5	1.5	3.75
				13.5

$$\bar{x} = \frac{15}{6} = 2.5$$

$$\bar{y} = \frac{21}{6} = 3.5$$

$$s_{xy} = \frac{\Sigma(x - \bar{x})(y - \bar{y})}{n - 1} = \frac{13.5}{6 - 1} = 2.7$$

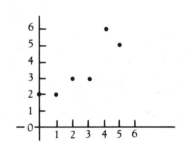

b)

x	y	$x - \bar{x}$	$y - \bar{y}$	$(x - \bar{x})(y - \bar{y})$
0	4	-1.5	0	0
1	4	-.5	0	0
2	4	.5	0	0
3	4	1.5	0	0
				0

$$\bar{x} = \frac{6}{4} = 1.5$$

$$\bar{y} = 4$$

$$s_{xy} = \frac{0}{4 - 1} = 0$$

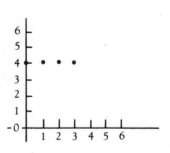

ċ)

x	y	x − $\bar{x}$	y − $\bar{y}$	(x − $\bar{x}$)(y − $\bar{y}$)
0	0	−2	−2	4
0	4	−2	+2	−4
4	0	+2	−2	−4
4	4	+2	+2	4
				0

$\bar{x} = \dfrac{8}{4} = 2$

$\bar{y} = \dfrac{8}{4} = 2$

$s_{xy} = \dfrac{0}{4 - 1} = 0$

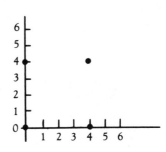

5) a)

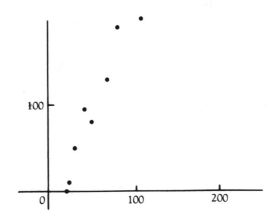

b)

x	y	x^2	y^2	xy
44	95	1936	9025	4180
22	10	484	100	220
30	50	900	2500	1500
110	200	12100	40000	22000
70	130	4900	16900	9100
50	80	2500	6400	4000
20	0	400	0	0
80	190	6400	36100	15200
426	755	29620	111025	56200

$$r = \frac{8(56200) - (426)(755)}{\sqrt{8(29620) - (426)^2} \sqrt{8(111025) - (755)^2}} = .9631$$

c) H_0: $\rho = 0$

H_1: $\rho > 0$

For $\alpha = .05$ and $n = 8$, the critical value is $r_{.05} = .621$.
Since $r = .9631 > .621$, we reject H_0 and conclude that a positive correlation exists.

7) a)

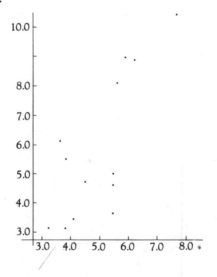

b) H_0: $\rho = 0$

H_1: $\rho \neq 0$

For $n = 13$, the critical values are $\pm r_{.025} = \pm .553$.

x	y	x^2	y^2	xy
3.8	3.2	14.44	10.24	12.16
3.2	3.2	10.24	10.24	10.24
5.5	3.7	30.25	13.69	20.35
4.1	3.5	16.81	12.25	14.35
5.5	4.7	30.25	22.09	25.85
5.5	5.1	30.25	26.01	28.05
4.5	4.8	20.25	23.04	21.60
3.8	5.6	14.44	31.36	21.28
3.6	6.2	12.96	38.44	22.32
5.9	9.1	34.81	82.81	53.69
5.6	8.2	31.36	67.24	45.92
7.7	10.6	59.29	112.36	81.62
6.0	9.0	36.00	81.00	54.00
64.7	76.9	341.35	530.77	411.43

$$r = \frac{(13)(411.43) - (64.7)(76.9)}{\sqrt{(13)(341.35) - (64.7)^2}\sqrt{(13)(530.77) - (76.9)^2}} = .7493$$

Since $r = .7493 > .533$, reject H_0 and conclude that a correlation exists between unemployment and interest rates.

9) The points would all fall on the line $y = x$.

$r = 1$

11)

Quadrant	$x - \bar{x}$	$y - \bar{y}$	$(x - \bar{x})(y - \bar{y})$	$\dfrac{(x - \bar{x})(y - \bar{y})}{s_x s_y}$
I	+	+	+	+
II	−	+	−	−
III	−	−	+	+
IV	+	−	−	−

13) H_0: $\rho = 0$

H_1: $\rho > 0$

critical value: $r_{.025} = .361$ (using $n = 30$)

a) $r = .82 > r_{.025}$. Reject H_0 and conclude a positive correlation exists between the WISCR (verbal) and the SIT scores.

b) $r = .50 > r_{.025}$. Reject H_0 and conclude a positive correlation exists between the WISCR (performance) and the SIT scores.

15) $r_{.05} = .412$ for $n = 17$ and $r_{.05} = .426$ for $n = 16$.

For $n \leq 16$, $r = .42$ is not significant.

Exercise Set XI.2

1) b)

x	y	x^2	y^2	xy
0	0	0	0	0
1	0	1	0	0
2	2	4	4	4
1	2	1	4	2
4	4	6	8	6

a)c)

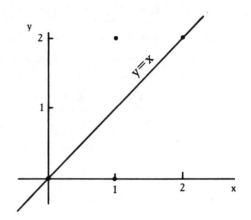

$\bar{x} = \dfrac{4}{4} = 1$

$\bar{y} = \dfrac{4}{4} = 1$

$b_1 = \dfrac{n \Sigma\, xy - (\Sigma\, x)(\Sigma\, y)}{n \Sigma\, x^2 - (\Sigma\, x)^2} = \dfrac{(4)(6) - (4)(4)}{(4)(6) - (4)^2} = 1$

$b_0 = \bar{y} - b_1\bar{x} = 1 - 1 \cdot 1 = 0$

$y' = 0 + 1 \cdot x = x$

d) $S_e = \sqrt{\dfrac{\Sigma\, y^2 - b_0\, \Sigma\, y - b_1\, \Sigma\, xy}{n - 2}}$

$= \sqrt{\dfrac{8 - (0)(4) - (1)(6)}{4 - 2}} = 1$

3) a)

x	y	x^2	y^2	xy
62	253	3844	64009	15686
70	260	4900	67500	18200
80	300	6400	90000	24000
50	225	2500	50625	11250
75	280	5625	78400	21000
65	255	4225	65025	16575
402	1573	27494	415659	106711

$\bar{x} = \dfrac{402}{6} = 67$, $\bar{y} = \dfrac{1573}{6} = 262.1667$ (Keep as many decimal places as possible until all computations are completed.)

$b_1 = \dfrac{(6)(106711) - (402)(1573)}{(6)(27494) - (402)^2} = 2.3571429 \approx 2.36$

$b_0 = 262.1667 - (2.3571429)(67) = 104.23813 \approx 104.24$

y' = 104.24 + 2.36x

b) For x = 50, the predicted F.T.E. is y' = 104.24 + (2.36)(50) = 222.24.

c) For x = 100, the predicted F.T.E. is y' = 104.24 + (2.36)(100) = 340.24.

d) $S_e = \sqrt{\dfrac{415659 - (104.23813)(1573) - (2.3571429)(106711)}{6 - 2}}$ = 6.31

5) a)

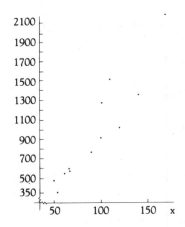

b) The number of employees increases in a "linear" fashion as the city size
increases.

c)

x	y	x^2	y^2	xy
102	1290	10404	1664100	131580
90	780	8100	608400	70200
140	1370	19600	1876900	191800
99	925	9801	855625	91575
51	495	2601	245025	25245
120	1040	14400	1081600	124800
110	1530	12100	2340900	168300
67	610	4489	372100	40870
61	560	3721	313600	34160
68	590	4624	348100	40120
54	360	2916	129600	19440
170	2200	28900	4840000	374000
1132	11750	121656	14675950	1312090

$\bar{x} = \dfrac{1132}{12} = 94.33333$, $\bar{y} = \dfrac{11750}{12} = 979.16667$

$b_1 = \dfrac{(12)(1312090) - (1132)(11750)}{(12)(121656) - (1132)^2} = 13.696315 \approx 13.70$

$b_0 = 979.1667 - (13.696315)(94.33333) - -312.85237 \approx -312.85$

$y' = -312.85 + 13.70x$

d) $S_e = \sqrt{\dfrac{14675950 + (312.85237)(11750) - (13.696315)(1312090)}{12 - 2}}$

$= 195.235$

e) When $x = 110$, $y' = -312.85 + (13.70)(110) = 1194.15$

7)

x	y	x^2	xy
40	140	1600	5600
55	205	3025	11275
20	65	400	1300
28	104	784	2912
45	150	2025	6750
15	30	225	450
70	305	4900	21350
60	265	3600	15900
333	1264	16559	65537

$\bar{x} = \dfrac{333}{8} = 41.625, \ \bar{y} = \dfrac{1264}{8} = 158$

$b_1 = \dfrac{(8)(65537) - (333)(1264)}{(8)(16559) - (333)^2} = 4.7900663 \approx 4.79$

$b_0 = 158 - (4.7900663)(41.625)$

$= -41.386508 \approx -41.39$

$y' = -41.39 + 4.79x$

9) a)

x	y	x^2	y^2	xy
45	12	2025	144	540
40	13	1600	169	520
35	16	1225	256	560
30	18	900	324	540
25	22	625	484	550
175	81	6375	1377	2710

$\bar{x} = \dfrac{175}{5} = 35, \ \bar{y} = \dfrac{81}{5} = 16.2$

$$b_1 = \frac{(5)(2710) - (175)(81)}{6375 - (175)^2} = -.50$$

$$b_0 = 33.70$$

$$y' = 33.70 - .50x$$

b)   When $x = 38$, $y' = 33.70 - (.50)(38) = 14.7$

c) $S_e = \sqrt{\dfrac{1377 - (33.70)(81) + (.50)(2710)}{5 - 2}} = .86$

11) $b'_1 = \dfrac{n \Sigma\, xy - (\Sigma\, y)(\Sigma\, x)}{n \Sigma\, y^2 - (\Sigma\, y)^2}$

$$b'_0 = \bar{x} - b'_1 \bar{y}$$

Using the data of number 9,

$$b'_1 = \frac{(5)(2710) - (175)(81)}{(5)(1377) - (81)^2} = -1.9290123 \approx -1.93$$

$$b'_0 = 35 + (1.9290123)(16.2) = 66.25$$

$$x' = 66.25 - 1.93y$$

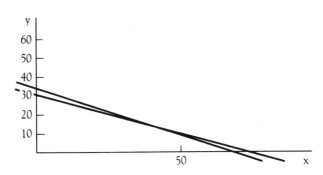

13)   a)   For $\Delta x - 1$, $\Delta y = \text{slope} = 4.25$.

     b)   For $\Delta x = 3$, $\Delta y = 3(\text{slope}) = 3(4.25) = 12.75$.

     c)   For $\Delta x = -4$, $\Delta y = -4(\text{slope}) = -4(4.25) - -17.00$.

Exercise Set XI.3

1) For $\nu = 20 - 2 = 18$ d.f., $t_{.025} = 2.101$

$$\text{S.E.} = \frac{s_e}{\sqrt{\dfrac{n \Sigma x^2 - (\Sigma x)^2}{n}}} = \frac{4.283}{\sqrt{\dfrac{(20)(22500) - (660)^2}{20}}} = .1596$$

$E = (t_{.025})(\text{S.E.}) = (2.101)(.1596) = .3353 \approx .34$

$.83 - .34 < \beta_1 < .83 + .34$ or

$\qquad .49 < \beta_1 < 1.17$

3) When $x = 68$, $y' = -1.60 + (.06)(68) = 2.48$

$\bar{x} = \dfrac{1730}{25} = 69.2$

$$\text{S.E.} = 3.48 \sqrt{\frac{1}{25} + \frac{25(68 - 69.2)^2}{(25)(124,200) - (1750)^2}} = .70331$$

For $\nu = 25 - 2 = 23$ d.f., $t_{.025} = 2.069$

$E = (t_{.025})(\text{S.E.}) = (2.069)(2.069)(.70331) = 1.4552 \approx 1.46$

$2.48 - 1.46 < \mu_{Y|68} < 2.48 + 1.46$ or

$\qquad 1.02 < \mu_{Y|68} < 3.92$

5) a)

x	y	y'	$(y - \bar{y})^2$	$(y' - \bar{y})^2$	$(y - y')^2$
1	0	1.38	17.39	7.78	1.90
2	5	2.50	.69	2.79	6.25
3	4	3.60	.03	.32	.16
4	3	4.72	1.37	.30	2.96
5	5	5.84	.69	2.79	.71
6	8	6.95	14.67	7.73	1.10
			34.84	21.71	13.08

$\bar{y} = \dfrac{25}{6} = 4.17$

Apart from rounding,

$\qquad$ total variation $= 34.84$

$\qquad$ explained variation $= 21.71$

$\qquad$ error variation $= 13.08$

b) total variation = explained variation + error variation

21.71 + 13.08 = 34.79 $\approx$ total variation, the error is due to rounding.

c) We should find that $r^2 = \dfrac{\text{explained variation}}{\text{total variation}}$

$\dfrac{\text{explained variation}}{\text{total variation}} = \dfrac{21.71}{34.84} = .6231 \approx .62$

By direct computation we find $r = .7898017$ and $r^2 = .6237867 \approx .62$.

Again the small difference is due to rounding.

7) a)

x	y	x^2	y^2	xy	y'
.32	28	.1024	784	8.96	29.91
.35	30	.1225	900	10.50	27.56
.38	25	.1444	625	9.50	25.20
.40	24	.1600	576	9.60	23.63
.45	19	.2025	361	8.55	19.70
1.90	126	.7318	3246	47.11	

$b_1 = \dfrac{5(47.11) - (1.90)(126)}{5(.7318) - (1.90)^2} = -78.571429 \approx -78.57$

$b_0 = 25.2 + (78.571429)(.38) = 55.057143 \approx 55.06$

$y = 55.06 - 78.57x$

b) $S_e = \sqrt{\dfrac{3246 - (55.057143)(126) + (78.571429)(47.11)}{5 - 2}} = 1.8529258 \approx 1.85$

c) S.E. $= \dfrac{1.85}{\sqrt{\dfrac{5(.7318) - (1.90)^2}{5}}} = 18.69$

$E = (t_{.025})(\text{S.E.}) = (3.182)(18.69) = 59.46$

$-78.57 - 59.46 < \beta_1 < -78.57 + 49.46$

$-138.03 < \beta_1 < -19.11$

If the metering jet is changed by +1 unit, the mileage will decrease by an amount between 138.03 and 19.11 miles per gallon. More realistically, for $\Delta x = .01$, the predicted change is somewhere between 1.3803 and .1911 mpg.

9) The coefficient of correlation is $r = .9997428.$
   $r^2 = .9994856 \approx 99.95\%$

11) $z_y = rz_x = .9932z_x$

13) total variation = explained variation + error variation
   error variation = total variation − explained variation
$$= \text{total variation} - r^2(\text{total variation})$$
$$= (1 - r^2)\text{total variation}$$
$$\approx 0 \quad \text{if} \quad r^2 \quad \text{is near} \quad 1.$$

15) a) For $\alpha = .05$ and $\nu = 20 - 2 = 18$ d.f., the critical value is
   $t_{.05} = 1.734$
$$t = \cfrac{.83 - .75}{\cfrac{4.283}{\sqrt{\dfrac{20(22500) - (660)^2}{20}}}} = .501$$

   Do not reject $H_0.$

   b) For $\alpha = .01$ and $\nu = 25 - 2 = 23$ d.f., the critical values are
   $\pm t_{.005} = \pm 2.807.$
   From problem 3, $b_1 = .06$ and $s_e = 3.48.$

$$t = \cfrac{.06 - .15}{\cfrac{3.48}{\sqrt{\dfrac{(25)(124200) - (1730)^2}{25}}}} = -1.732$$

   Do not reject $H_0.$

17) The variable Z with the larger coefficient r is the better predictor.
   For Z, $r_Z^2 = (.84)^2 = .7056.$

   For X, $r_X^2 = (.62)^2 = .3844.$

   $\dfrac{r_Z^2}{r_X^2} = \dfrac{.7056}{.3844} = 1.84$

   Z is about 1.84 times as good as X.

19) We find r − =.9564 and r^2 = .9148 ≈ 91.48%.

Thus about 91.48% of the variation in the cancer rate is explained by the vitamin C content of the diet.

Exercise Set XI.4

1) a) The correlation coefficient between high school g.p.a. and S.A.T. scores
        is $r_{1,2} = .51$.

        The correlation coefficient between high school and college g.p.a.'s is
        $r_{1,y} = .45$.

        The correlation between S.A.T. scores and college g.p.a.'s is $r_{2,y} = .97$.

    b) $r_{1y \cdot 2} = \dfrac{.45 - (.51)(.97)}{\sqrt{(1 - (.51)^2)(1 - (.97)^2)}} = -.2138 \approx -.21$

        $r_{2y \cdot 1} = \dfrac{.97 - (.51)(.45)}{\sqrt{(1 - (.51)^2)(1 - (.45)^2)}} = .9640 \approx .96$

    c) The correlation coefficient between high school and college g.p.a.'s
        when the S.A.T. score is held constant is $-.21$.

        The correlation coefficient between S.A.T. scores and college g.p.a.'s
        is $.96$ when the high school g.p.a. is held constant.

3)

x_1	x_2	y	x_1^2	x_2^2	$x_1 x_2$	$x_1 y$	$x_2 y$
0	0	2	0	0	0	0	0
1	1	6	1	1	1	6	6
2	1	8	4	1	2	16	8
1	2	9	1	4	2	9	18
0	4	6	0	16	0	0	24
2	0	8	4	0	0	16	0
6	8	39	10	22	5	47	56

$n = 6$

$6b_0 + 6b_1 + 8b_2 = 39$

$6b_0 + 10b_1 + 5b_2 = 47$

$8b_0 + 5b_1 + 22b_2 = 56$

Eliminating b_0 from equations (2) and (3) using the first equation yields
the new system:

$6b_0 + 6b_1 + 8b_2 = 39$

$\qquad 4b_1 - 3b_2 = 8$

$\qquad 18b_1 - 68b_2 = -104$

Eliminating b_0 from equations (2) and (3) using the first equation yields the new system

$$6b_0 + 6b_1 + 8b_2 = 39$$
$$4b_1 - 3b_2 = 8$$
$$9b_1 - 34b_2 = -12$$

Using the second equation to eliminate b_1 from the third equation yields the new system

$$6b_0 + 6b_1 + 8b_2 = 39$$
$$4b_1 - 3b_2 = 8$$
$$109b_2 = 120$$

The third equation yields $b_2 = \dfrac{120}{109} = 1.1009174$.

Substituting this into the second equation yields

$$b_1 = \frac{8 + 3(1.1009174)}{4} = 2.8256881.$$

Substituting for b_1 and b_2 in the first equation and solving yields

$$b_0 = \frac{39 - 6(2.8256881) - 8(1.1009174)}{6} = 2.206422.$$

Thus $y' = 2.21 + 2.83x_1 + 1.10x_2$.

c)

x_1	x_2	y	y'	$(y - y')^2$
0	0	2	2.21	.0441
1	1	6	6.14	.0196
2	1	8	8.97	.9409
1	2	9	7.24	3.0976
0	4	5	6.61	.3721
2	0	8	7.87	.0169
				4.4912

When $x_1 = 0$ and $x_2 = 0$

$y' = 2.21$

When $x_1 = 1$ and $x_2 = 1$

$y' = 2.21 + (2.83)(1) + 1.10(1) = 6.14$

The remaining predicted values are obtained in a similar fashion.

$$S_e = \sqrt{\frac{4.4912}{6 - (2 + 1)}} = 1.22$$

5) a) $y' = 14.20 + (.02)100 + (.04)(250) + (13)(450) = 84.7$ (hours)

  b) For a one unit change $(\Delta x_3 = 1)$, the predicted change in y is .13
      hours when x_1 and x_2 are constant. Thus for $\Delta x_3 = 50$, $\Delta y = (50)(.13)$
      $= 6.5$ (hours).

7) a) the age of the woman

  b) age, job stress, social stress, job security

9) a) For $x_2 = 2$, $y = .34 + .030x_1 + 1.5 = 1.84 + .030x_1$.
      For $x_2 = 10$, $y = .34 + .030x_1 + 7.5 = 7.84 + .030x_1$.
      For $x_2 = 20$, $y = .34 + .030x_1 + 15 = 15.34 + .030x_1$.

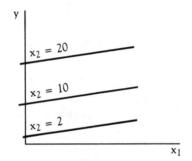

  b) When $x_1 = 100$, $y = .34 + 3 + .75x_2 = 3.34 + .75x_2$.
      When $x_1 = 400$, $y = .34 + 12 + .75x_2 = 12.34 + .75x_2$.
      When $x_1 = 600$, $y = .34 + 18 + .75x_2 = 18.34 + .75x_2$.

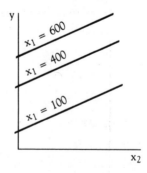

  c) When x_2 is constant, b_1 is the slope of the resulting line.
      When x_1 is constant, b_2 is the slope of the resulting line.

Chapter Test

1) a) positive for (2)

        negative for (1)

    b) positive for (2)

        negative for (1)

    c) positive for (2)

        negative for (1)

Note that all three of these quantities, the slope, correlation coefficient, and covariance have the same sign.

2)

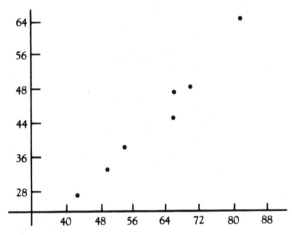

    b) For n = 7, $r_{.025}$ = .754.

        As shown below, r = .9877. Thus reject the hypothesis of no correlation and conclude that a positive correlation exists.

x	y	x^2	y^2	xy
66	48	4356	2304	3168
70	49	4900	2401	3430
66	46	4396	2116	3036
50	34	2500	1156	1700
43	28	1849	784	1204
54	39	2916	1521	2106
82	65	6724	4225	5330
431	309	27601	14507	19974

$$r = \frac{7(19974) - (431)(309)}{\sqrt{7(27601)-(431)^2}\ \sqrt{7(14507)-(309)^2}}$$

$$= .9879$$

3)

x	y	xy	x^2	y'	$(y - y')^2$
68	2.4	163.2	4624	2.48	.0064
75	2.7	202.5	5625	2.84	.0196
40	0.9	36.0	1600	1.04	.0025
80	3.1	248.0	6400	3.09	.0001
86	3.5	301.0	7396	3.39	.0121
55	1.6	88.0	3025	1.81	.0100
90	3.7	333.0	8100	3.60	.0100
68	2.1	142.8	4624	2.48	.1444
70	2.6	182.0	4900	2.58	.0004
60	2.9	174.0	3600	2.07	.6889
96	3.8	364.8	9216	3.91	.0121
788	29.3	2235.3	59110		.9401

$$\bar{x} = \frac{788}{11} = 71.636364, \quad \bar{y} = \frac{29.3}{11} = 2.6636364$$

$$b_1 = \frac{11(2235.3) - (788)(29.3)}{11(59110) - 788} = .0512506$$

$$b_0 = 2.6636364 - (.0512506)(71.636364) = -1.0077702$$

a) $y' = -1.008 + .051x$

b) When $x = 70$, $y' = -1.008 + (.051)(70) = 2.56$.

c) $s_e = \sqrt{\dfrac{\Sigma(y - y')^2}{n - 2}} = \sqrt{\dfrac{9406}{9}} = .3233$ or

$$s_e = \sqrt{\frac{\Sigma y^2 - b_0(\Sigma y) - b_1 \Sigma xy}{n - 2}}$$

$$= \sqrt{\frac{85.99 + (1.0077702)(29.3) - (.0512506)(2235.3)}{9}}$$

$$= .3261$$

The small discrepency is due to the rounding used in the intermediate results, mainly in the y' values.

4) H_0: $\rho = 0$

H_1: $\rho > 0$

For $n = 20$, the critical value is $r_{.005} = .561$.

Since $r = .65 > r_{.005}$, reject H_0 and conclude that a positive correlation exists between educational achievement and family incomes.

5) For $\nu = 10 - 2 = 8$ d.f., $t_{.025} = 2.306$

$$E = (2.306)(4.1) \sqrt{\frac{1}{10} + \frac{10(40 - 14.5)^2}{10(2785) - (145)^2}} = 9.70$$

When $x = 40$, $y' = 6.7 + (1.0)(40) = 46.7$

$46.7 - 9.70 < \mu_{Y|40} < 46.7 + 9.70$ or

$\quad\quad 37.00 < \mu_{Y|40} < 56.40$

6) S.E. $= \dfrac{4.1}{\sqrt{\dfrac{10(2785) - (145)^2}{10}}} = .1569$

$E = t_{.025} S.E. = (2.306)(.1569) = .3618$

$1.0 - .3618 < \beta_1 < 1.0 + .3618$ or

$\quad\quad .6382 < \beta_1 < 1.3618$

7) a) unchanged $r = .9879$

    b) $z_y = .9879 z_x$ (Note: The correlation coefficient is now the slope of the regression level.)

8) By direct computation the correlation coefficient is found to be $r = .9718$. The coefficient of determination is $r^2 = (.9718)^2 = .9443$. Thus about 94.43% of the variation in the pulse rates is explained by the regression with the dosage size.

9) a) $y = -180 + 5(65) - .4(30) = 133$

    b) $b_1 = 5$ predicts an increase of $\Delta y = 5$ pounds for each unit change $\Delta x_1 = 1$ in the height of the individual. When the age x_2 is held constant.

       $b_2 = -4$ predicts a decrease of .4 pounds for each unit change $\Delta x_2 = 1$ of age when the height is held constant.

10) By direct computation we find:

$$r_{1y} = .7314$$
$$r_{2y} = .9091$$
$$r_{12} = .6349$$

The correlation matrix is as follows:

a)

	x_1	x_2	y
x_1	1.0000	.6349	.7314
x_2	.6349	1.0000	.9091
y	.7314	.9091	1.0000

b) $r_{1y \cdot 2} = \dfrac{.7314 - (.6349)(.9091)}{\sqrt{(1 - (.6349)^2)(1 - (.9091)^2)}} = .4789$

When any effect due to the number of promotions is removed (or the number is held constant), the correlation coefficient between job satisfaction and years of employment is .4789.

c) $r_{2y \cdot 1} = \dfrac{(.9091) - (.6349)(.7314)}{\sqrt{(1 - (.6349)^2)(1 - (.7314)^2)}} = .8441$

When the effect due to years of employment is removed (or the years of employment is held constant), the correlation coefficient between job satisfaction and number of promotions is .8441.

11) Since $\beta_2 = .45$ (largest coefficient in magnitude), z_{x_2} is most important in accounting for the total variation in z_y. By the same token, z_{x_1} is of least importance since its coefficient is the smallest in magnitude of the three.

$\dfrac{(\beta_2')^2}{(\beta_1')^2} = \dfrac{(.45)^2}{(.10)^2} = 20.25$ implies that z_{x_2} is about 20.25 times as important as z_{x_1} in accounting for the total variation in z_{y_1}.

Chapter XII
The Analysis of Variance

The analysis of variance or ANOVA is a procedure for testing the equality of
several population means. One way ANOVA is concerned with the case where the
populations correspond to different treatment levels of a single factor. A
difference in means then indicates an effect due to the different treatment levels.
Two way ANOVA extends the analysis to two factors each having several treatment
levels. The procedure then provides a means of testing for an effect due to either
factor as well as the possibility of an interaction effect.

Concepts/Techniques

1) Treatment Levels

The analysis of variance (ANOVA) is concerned with comparing the means
of several populations. These populations are generally identified with
the values of qualitative or categorical variables such as (type of)
procedure, job (classification), (type of) degree, etc. The values of
such variables are called treatment levels.

2) The Null and Alternate Hypothesis for the One Way ANOVA

One way ANOVA is concerned with the case of a single variable with several
treatment levels. The null hypothesis is the hypothesis of no effect
due to the various treatment levels. If the means of the treatment
populations are $\mu_1, \mu_2, \ldots, \mu_k$, then we test

$$H_0: \quad \mu_1 = \mu_2 = \cdots = \mu_k \quad \text{(No treatment effect)}$$
$$H_1: \quad \text{Not all means are equal (A treatment effect)}$$

3) <u>Within Group and Between Group Variances</u>

The ANOVA test proceeds by comparing estimates of the population variance (which indirectly involve the mean). If there is no treatment effect (null hypothesis) we obtain an estimate of the (common) population variance by pooling the variances of the different treatment samples. Since this estimate depends on the "within" groups variation, it is often called the "within groups" variance estimate. If there is no treatment effect, the "within groups" variance is related to random or experimental error and as a result is often called the error variation and is denoted by MSSE. It is also called the mean error sum of squares.

A second measure of the population variance is obtained from the variation of the sample means about the grand mean of all the observations. This produces a "between groups" estimate MSSTr of the variance. If this differs appreciably from the within groups estimate, there must be a treatment effect. As a result, MSSTr is called the treatment variation. It is also called the mean treatment sum of squares.

When there is a treatment effect, we expect the ratio

$$\frac{\text{MSSTr}}{\text{MSSE}} = \frac{\text{treatment variation}}{\text{error variation}}$$

to be considerably greater than 1. We test for this using the F distribution.

4) <u>Computing MSSTr and MSSE</u>

Suppose we have k samples of size $n_1, n_2, \ldots, n_k$ with means $\bar{x}_1, \bar{x}_2, \ldots, \bar{x}_k$ and a grand mean $\bar{x}$ as shown below:

Sample 1	Sample 2 ...	Sample k
x_{11}	x_{21}	x_{k1}
x_{12}	x_{22}	x_{k2}
x_{13}	x_{23}	x_{k3}
$\vdots$	$\vdots$	$\vdots$
x_{1n_1}	x_{2n_2}	x_{kn_k}

The first step is to compute the treatment sum of squares

$$SSTr = n_1(\bar{x}_1 - \bar{x})^2 + n_2(\bar{x}_2 - \bar{x})^2 + \cdots + n_k(\bar{x}_k - \bar{x})^2.$$

The mean treatment sum of squares is then

$$MSSTr = \frac{SSTr}{k - 1}.$$

Next the error treatment sum of squares is computed:

$$SSE = (x_{11} - \bar{x}_1)^2 + (x_{12} - \bar{x}_1)^2 + \cdots + (x_{1n_1} - \bar{x}_1)^2 + \cdots + (x_{k1} - \bar{x}_k)^2$$

$$+ (x_{k2} - \bar{x}_k)^2 + \cdots + (x_{k_1 n_k} - \bar{x}_k)^2$$

Finally the mean error sum of squares is found:

$$MSSE = \frac{SSE}{n - k}.$$

Here

$$n - k = (n_1 - 1) + (n_2 - 1) + \cdots + (n_k - 1)$$

$$= (n_1 + n_2 + \cdots + n_k) - (1 + 1 + \cdots + 1)$$

$$= (n_1 + n_2 + \cdots + n_k) - k$$

where $n = n_1 + n_2 + \cdots + n_k$. Note this is much like two population hypothesis testing where a pooled variance estimate was obtained using $(n_1 - 1) + (n_2 - 1) = (n_1 + n_2) - 2$ degrees of freedom.
In practice these sums of squares are computed as follows:

$$SSTr = \frac{T_1^2}{n_1} + \frac{T_2^2}{n_2} + \cdots + \frac{T_k^2}{n_k} - \frac{T^2}{n} = \Sigma \frac{T_i^2}{n_i} - \frac{T^2}{n}$$

$$SSE = x_{11}^2 + x_{12}^2 + \cdots + x_{k_1 n_k}^2 - \left[\frac{T_1^2}{n_1} + \frac{T_2^2}{n_2} + \cdots + \frac{T_k^2}{n_k} \right]$$

$$= \Sigma_i \Sigma_j x_{ij}^2 - \Sigma \frac{T_i^2}{n_i}.$$

Here T_j is the sum of the observations of the jth treatment sample and T is the sum of all the observations.

Example: Compute the treatment and error variance estimates using the 3 samples below:

	Sample 1	Sample 2	Sample 3
	1	2	3
	2	3	4
	3	4	5
		5	6
			7

We arrange the computations as follows:

Sample 1		Sample 2		Sample 3		Grand Totals
x	x^2	x	x^2	x	x^2	
1	1	2	4	3	9	
2	4	3	9	4	16	
3	9	4	16	5	25	
		5	25	6	36	
				7	49	

	Sample 1	Sample 2	Sample 3	Grand Totals
Sample Size	$n_1 = 3$	$n_2 = 4$	$n_3 = 5$	12
Sum	$T_1 = 6$	$T_2 = 14$	$T_3 = 25$	45
Sum of Squares	$\Sigma\, x_{1i}^2 = 14$	$\Sigma\, x_{2i}^2 = 54$	$\Sigma\, x_{3i}^2 = 135$	203
Quotients	$\dfrac{T_1^2}{n_1} = \dfrac{6^2}{3} = 12$	$\dfrac{T_2^2}{n_2} = \dfrac{14^2}{4} = 49$	$\dfrac{T_3^2}{n_3} = \dfrac{25^2}{5} = 125$	186

$$\text{SSTr} = \frac{T_1^2}{n_1} + \frac{T_2^2}{n_2} + \frac{T_3^2}{n_3} - \frac{T^2}{n} = 186 - \frac{(45)^2}{12} = 17.25$$

Note that the grand totals from rows 4, 2, and 1 were used. The mean treatment sum of squares or "between groups" estimate of the population variance is

$$\text{MSSTr} = \frac{\text{SSTr}}{k - 1} = \frac{17.25}{3 - 1} = 8.625$$

$$\text{SSE} = \Sigma\Sigma\, x_{ij}^2 - \Sigma\left(\frac{T_i^2}{n_i}\right) = 203 - 186 = 17$$

Here the sum of squares and quotients grand totals from rows 3 and 4 were used. The mean error sum of squares or "within groups" estimate of the variance is

$$MSSE = \frac{SSE}{n - k} = \frac{17}{12 - 3} = 1.889.$$

5) Total Sum of Squares   SS

The total sum of squares of the deviations of the observations from the grand mean is

$$SS = \Sigma\Sigma\,(x_{ij} - \bar{x})^2.$$

It can be shown that this is the sum of the treatment and error sums of squares, that is

$$SS = SSTr + SSE.$$

6) The One Way ANOVA Test

The null hypothesis of no treatment effect is performed by comparing the ratio

$$f = \frac{MSSTr}{MSSE}$$

with $f_\alpha(k - 1, n - k)$. If $f > f_\alpha(k - 1, n - k)$, then reject H_0 and conclude a treatment effect exists. The results are often summarized in a ANOVA table whose form is shown below:

Source	Sum of Squares	d.f.	Mean	f
Treatment	SSTr	k − 1	MSSTr	$\frac{MSSTr}{MSSE}$
Error	SSE	n − k	MSSE	
Total	SS	n − 1		

Example: The ANOVA table below summarizes a study of grades issued by different departments of a college.

Source	Sum of Squares	d.f.	Mean	f	p
Treatment	7200	8	900	6	< .001
Error	66150	441	150		
Total	73350	449			

From this we can see that then $n - 1 = 449$ or $n = 450$. Thus there were 450 grade observations in the study.

Since $k - 1 = 8$, $k = 9$ is the number of departments or treatment levels. Thus if the samples from the various departments are of equal size, then each contains $\frac{450}{9} = 50$ grades.

The treatment and error estimates of the variance are MSSTr = 900 and MSSE = 150. The f value is

$$f = \frac{MSSTr}{MSSE} = \frac{900}{150} = 6.00.$$

For say $\alpha = .05$, the critical value is $f_{.05}(8,441) = 1.94$ and since $f > 1.94$ the null hypothesis of no difference in the means of the grades issued by these 9 departments is rejected. We see in fact that $f = 6.00$ is significant even for $\alpha = .001$.

Example: A study of the years of formal education by workers in 3 professions resulted in the following data. Test for a difference in the mean years of formal education by workers in these professions using $\alpha = .01$.

	I		II		III		Grand Total
	x	x^2	x	x^2	x	x^2	
	10	100	14	196	12	144	
	14	196	14	196	12	144	
	10	100	16	256	12	144	
	12	144	15	225	10	100	
	10	100	17	289	14	196	
	13	169			12	144	
Sample Size	6		5		6		17
Sum	69		76		72		217
Sum of Squares	809		1162		872		2843
Quotients	793.5		1155.2		864		2812.7

$$SSTr = 2812.7 - \frac{(217)^2}{17} = 42.76$$

$$MSSTr = \frac{SSTr}{k - 1} = \frac{42.76}{2} = 21.38$$

$$SSE = 2843 - 2812.7 = 30.3$$

$$MSSE = \frac{SSE}{n - k} = \frac{30.3}{14} = 2.16$$

$$f = \frac{MSSTr}{MSSE} = \frac{21.38}{2.16} = 9.879$$

The critical value is $f_\alpha(k - 1, n - k) = f_{.01}(2,14) = 6.51$.
Since $f > 6.51$ the null hypothesis is rejected. We conclude a
difference exists in the mean number of years of formal education
of workers in these 3 professions. Note the sample means are

$$\bar{x}_1 = \frac{69}{6} = 11.5, \quad x_2 = \frac{76}{5} = 15.2 \quad \text{and} \quad x_3 = \frac{72}{6} = 12.0$$

The ANOVA table is given below

Source	Sum of Squares	d.f.	Mean	f
Treatment	42.76	2	21.38	9.879
Error	30.30	14	2.16	
Total	73.06	16		

Example: Complete the following ANOVA table:

Source	Sum of Squares	d.f.	Mean	f
Treatment	60		20	
Error		20		
Total	200			

We see SS = 200. Also SS = SSTr + SSE. Thus SSTr + SSE = 200.
Substituting SSTr = 60 we find SSE = 140.
We next find

$$MSSE = \frac{SSE}{20} = \frac{140}{20} = 7$$

Since $20 = MSSTr = \frac{SSTr}{k - 1} = \frac{60}{k - 1}$, we conclude that $k - 1 = 3$
or $k = 4$.
Finally

$$f = \frac{MSSTr}{MSSE} = \frac{20}{7} = 2.86$$

The complete ANOVA table follows:

Source	Sum of Squares	d.f.	Mean	f
Treatment	60	3	20	2.86
Error	140	20	7	
Total	200	23		

This study involved 4 treatment populations and made use of 24 observations.

7) Two Way Analysis of Variance

Two way ANOVA is concerned with simultaneously testing for an effect due to each of two variables as well as a possible interaction effect. If the two variables or factors have r and c treatment levels respectively and if each level of one variable is studied in combination with each level of the second, then we speak of an $r \times c$ experimental design or simply an $r \times c$ ANOVA. To test for an interaction effect, we must have at least 2 observations for each of these combinations.

Example: Suppose we wish to study whether the age or sex of an individual has an effect on scores of a test of consumer awareness. If we use the age classes: young, middle aged, old, then we have a 2×3 ANOVA (sex by age). If 3 observations are obtained for each of these $2 \times 3 = 6$ sex by age categories, the design of the experiment is shown below.

		Age (c = 3)			
		Young	Middle	Old	
Sex (r=2)	Male	3 scores of young males	3 scores of middle aged males	3 scores of old males	9 scores of males
	Female	3 scores of young females	3 scores of middle aged females	3 scores of old females	9 scores of females
		6 scores of young category	6 scores of middle aged category	6 scores of old category	18 scores in all

The computational layout for an $r \times c$ ANOVA is shown below for a 2×3 design.

		Columns 1		Columns 2		Columns 3		Total
		x	x^2	x	x^2	x	x^2	
Rows	1	x_{111}	x_{111}^2	x_{121}	x_{121}^2	x_{131}	x_{131}^2	$T_1\cdot$ (the sum of the observations in row 1)
		x_{112}	x_{112}^2	x_{122}	x_{122}^2	x_{132}	x_{132}^2	
		x_{113}	x_{113}^2	x_{123}	x_{123}^2	x_{133}	x_{133}^2	
	2	x_{211}	x_{211}^2	x_{221}	x_{221}^2	x_{231}	x_{231}^2	$T_2\cdot$ (the sum of the observations in row 2)
		x_{212}	x_{212}^2	x_{222}	x_{222}^2	x_{232}	x_{232}^2	
		x_{213}	x_{213}^2	x_{223}	x_{223}^2	x_{233}	x_{233}^2	
		$T_{\cdot 1}$ (the sum of the observations in column 1)		$T_{\cdot 2}$ (the sum of the observations in column 2)		$T_{\cdot 3}$ (the sum of the observations in column 3)		T (the sum of all the observations)

In general, the following notation is used.

N = total number of observations

n = number of observations per cell

r = number of rows

c = number of columns

$T_{i\cdot}$ = sum of the observations in row i

$T_{\cdot j}$ = sum of the observations in column j

T = sum of all the observations in the table

When 3 subscripts are used, such as an $x_{i,j,k}$, the first locates the row, the second the column, and the third the observation within the (i,j) cell located at the intersection of the ith row and jth columns. Thus

$x_{1,2,3}$ is the 3rd observation of the $(1,2)$ cell

$\sum_k x_{1,2,k}$ is the sum of the observations of the $(1,2)$ cell

$\sum_i \sum_j \sum_k x_{i,j,k}^2$ is the sum of the squares of all observations in the table.

8) <u>Variance Estimates in an r × c Anova</u>

The sum of squares, SS, of the deviations of the individual observations about the grand mean can be split up as follows:

$$SS = SSR + SSC + SSRC + SSE$$

Here SSE is the error or "within groups" variation and, taken together, SSR, SSC, and SSRC comprise the "between groups" variation. We identify SSR as the variation due to the row variable, SSC the variation due to the column variable and SSRC the variation due to the interaction, if any, of the row and column variables.

When there are no effects nor interaction, we obtain, from these, four variance estimates:

$$MSSR = \frac{SSR}{r - 1}$$

$$MSSC = \frac{SSC}{c - 1}$$

$$MSSRC = \frac{SSRC}{(r - 1)(c - 1)}$$

$$MSSE = \frac{SSE}{rc(n - 1)}$$

The computation of the sums of squares is as follows:

$$SS = \Sigma\Sigma\Sigma \, x_{ijk}^2 - \frac{T^2}{N}$$

$$SSR = \Sigma \, \frac{T_{i\cdot}^2}{cn} - \frac{T^2}{N} = \frac{T_{1\cdot}^2}{cn} + \frac{T_{2\cdot}^2}{cn} + \cdots + \frac{T_{r\cdot}^2}{cn} - \frac{T^2}{N}$$

$$SSC = \Sigma \, \frac{T_{\cdot j}^2}{rn} - \frac{T^2}{N} = \frac{T_{\cdot 1}^2}{rn} + \frac{T_{\cdot 2}^2}{rn} + \cdots + \frac{T_{\cdot c}^2}{rn} - \frac{T^2}{N}$$

$$SSE = \Sigma\Sigma\Sigma \, x_{ijk}^2 - \Sigma_i \, \Sigma_j \, \frac{\left(\Sigma_k x_{ijk}\right)^2}{n}$$

$$SSRC = SS - SSR - SSC - SSE$$

9) <u>The Two Way ANOVA Test</u>

Reject the null hypothesis of no effect due to the row variable if

$$f_R = \frac{MSSR}{MSSE} > f_\alpha(r - 1, rc(n - 1)).$$

Reject the hypothesis of no effect due to the column variable if

$$f_C = \frac{MSSC}{MSSE} > f_\alpha(c - 1, rc(n - 1))$$

Reject the hypothesis of no interaction effect if

$$f_{RC} = \frac{MSSRC}{MSSE} > f_\alpha((r - 1)(c - 1), rc(n - 1))$$

Example: Three surgical procedures I, II, and III are available for a common type of back surgery. Recovery times of patients are studied for these in conjunction with the age of the patient. Using the data below, test for an effect due to age, to procedure, and an interaction. Use $\alpha = .05$.

		Procedure						
		I		II		III		Totals
		x	x^2	x	x^2	x	x^2	
		12	144	11	121	12	144	
Age	under 50	10	100	9	81	15	225	$T_{1.} = 99$
		8	64	10	100	12	144	
		10	100	12	144	14	196	
	over 50	12	144	11	121	15	225	$T_{2.} = 118$
		16	256	13	169	15	225	
		$T_{.1} = 68$		$T_{.2} = 66$		$T_{.3} = 83$		$T = 217$

The sum of the squares of all the elements is $\Sigma\Sigma\Sigma\, x_{ijk}^2 = 2703$

$$SS = \Sigma\Sigma\Sigma\, x_{ijk}^2 - \frac{T^2}{N} = 2703 - \frac{(217)^2}{18} = 86.94$$

$$SSR = \Sigma\, \frac{T_{i.}^2}{nc} - \frac{T^2}{N} = \frac{99^2}{3.3} + \frac{118^2}{3.3} - \frac{(217)^2}{18} = 20.06$$

$$SSC = \Sigma\, \frac{T_{.j}^2}{nr} - \frac{T^2}{N} = \frac{68^2}{3.2} + \frac{66^2}{3.2} + \frac{83^2}{3.2} - \frac{(217)^2}{18} = 28.78$$

In computing SSE, we need the terms $(\Sigma_k x_{ijk})$ which are the cell totals.

$$SSE = \Sigma\Sigma\Sigma \; x_{ijk}^2 \; - \; \Sigma\Sigma \; \frac{(\Sigma_k \; x_{ijk})^2}{n}$$

$$= 2703 - \frac{[30^2 + 30^2 + 39^2 + 38^2 + 36^2 + 44^2]}{3} = 37.33$$

$$SSRC = SS - SSR - SSC - SSE$$

$$= 86.94 - 20.06 - 28.78 - 37.33 = .77$$

Next we compute the variance estimates:

$$MSSR = \frac{SSR}{r - 1} = \frac{20.06}{2 - 1} = 20.06$$

$$MSSC = \frac{SSC}{c - 1} = \frac{28.78}{3 - 1} = 14.39$$

$$MSSRC = \frac{SSRC}{(r - 1)(c - 1)} = \frac{.77}{(2 - 1)(3 - 1)} = .385$$

$$MSSE = \frac{SSE}{rc(n - 1)} = \frac{37.33}{2.3(3 - 1)} = 3.11$$

We test the null hypothesis of no effect on recovery time due to age by comparing.

$$f_R = \frac{MSSR}{MSSE} = \frac{20.06}{3.11} = 6.45 \quad \text{with} \quad f_{.05}(1,12) = 4.75.$$

The null hypothesis is rejected.

We test the null hypothesis of no effect on recovery time due to procedure by comparing

$$f_C = \frac{MSSC}{MSSE} = \frac{14.39}{3.11} = 4.62 \quad \text{with} \quad f_{.05}(2,12) = 3.89.$$

The null hypothesis is rejected.

The null hypothesis of no interaction effect is tested by comparing

$$f_{RC} = \frac{MSSRC}{MSSE} = \frac{.385}{3.11} = .12 \quad \text{with} \quad f_{.05}(2,12) = 3.89.$$

The null hypothesis is not rejected.

Thus recovery times for this type of back surgery are effected both by the age of the patient as well as by the surgical procedure. There is no evidence to support an interaction effect between the age of the patient and the type of procedure.

10) The Two Way ANOVA Test with No Interaction

When there is no interaction effect we may pool the interaction sum   (SSRC)
and error sum (SSE) to obtain a pooled variance estimate:

$$\text{PMSSE} = \frac{\text{SSRC} + \text{SSE}}{(r - 1)(c - 1) + rc(n - 1)} \ .$$

The hypothesis of no row effect is then tested by comparing

$$f = \frac{\text{MSSR}}{\text{PMSSE}} \ \text{with} \ f_\alpha(r - 1, (r - 1)(c - 1) + rc(n - 1)).$$

Similarly the hypothesis of no column effect is tested by comparing

$$f = \frac{\text{MSSC}}{\text{PMSSE}} \ \text{with} \ f_\alpha(c - 1, (r - 1)(c - 1) + rc(n - 1)).$$

Example: In the age   surgical procedure study on recovery times,
we had $r = 2$, $c = 3$, SSRC = .77, SSE = 37.33, MSSR = 20.06 and
MSSC = 14.39. The pooled error or "within groups" variance estimate
is

$$\text{PMSSE} = \frac{\text{SSRC} + \text{SSE}}{(r - 1)(c - 1) + rc(n - 1)} = \frac{.77 + 37.33}{(2 - 1)(3 - 1) + 2.3(3 - 1)} = 2.72.$$

Then

$$f_R = \frac{\text{MSSR}}{\text{PMSSE}} = \frac{20.06}{2.72} = 7.35$$

and

$$f_C = \frac{\text{MSSC}}{\text{PMSSE}} = \frac{14.39}{2.72} = 5.29.$$

These are both larger than before so again the null hypothesis of
no row and no column effects are rejected. Only now the p values
would be much smaller – for two reasons: f_R and f_C are larger
and PMSSE has more degrees of freedom which acts to reduce the
critical value.

A two way ANOVA with only one observation per cell may be carried out
using the pooled variance estimate approach. We now recommend that the
"within groups" sum of squares SSR + SSE be computed by

$$\text{SSRC} + \text{SSE} = \text{SS} - \text{SSR} - \text{SSC}$$

and since the number of entries per cell is $n = 1$, $\nu = (r - 1)(c - 1)$
is the number of degrees of freedom of PMSSE. Thus

$$PMSSE = \frac{SSRC + SSE}{(r - 1)(c - 1)} \; .$$

Example: A 3×3 ANOVA study of container $\times$ label color on sales (1000's of units) of a laundry detergent involving 9 suburban areas contained the data below. Using $\alpha = .05$, test for an effect on sales due to each of these variables.

		Container						
		Cylindrical Box		Rectangular Box		Glass Jar		Total
		x	x^2	x	x^2	x	x^2	
Dominant Label Color	Red	15	225	22	484	23	529	$T_{1.} = 60$
	Blue	10	100	12	144	16	256	$T_{2.} = 38$
	Green	12	144	10	100	18	324	$T_{3.} = 40$
		$T_{.1} = 37$		$T_{.2} = 44$		$T_{.3} = 57$		T = 138

$$\Sigma\Sigma\Sigma \; x_{ijk}^2 = 2306$$

$$SS = \Sigma\Sigma\Sigma \; x_{ijk}^2 - \frac{T^2}{N} = 2306 - \frac{(138)^2}{9} = 190$$

$$SSR = \Sigma \; \frac{T_{i\cdot}^2}{nc} - \frac{T^2}{N} = \frac{60^2}{1\cdot3} + \frac{38^2}{1\cdot3} + \frac{40^2}{1\cdot3} - \frac{(138)^2}{9} = 98.67$$

$$SSC = \Sigma \; \frac{T_{\cdot j}^2}{nr} - \frac{T^2}{N} = \frac{37^2}{1\cdot3} + \frac{44^2}{1\cdot3} + \frac{57^2}{1\cdot3} - \frac{(138)^2}{9} = 68.67$$

$$SSE + SSRC = SS - SSR - SSC$$
$$= 190 - 98.67 - 68.67 = 22.66$$

$$MSSR = \frac{SSR}{r - 1} = \frac{98.67}{3 - 1} = 49.34$$

$$MSSC = \frac{SSC}{c - 1} = \frac{68.67}{3 - 1} = 34.34$$

$$PMSSE = \frac{SSRC + SSE}{(r - 1)(c - 1)} = \frac{22.66}{(3 - 1)(3 - 1)} = 5.67$$

We compare $f_R = \dfrac{MSSR}{PMSSE} = \dfrac{49.34}{5.67} = 8.70$ with $f_\alpha(r - 1,(r - 1)(c - 1))$
$= f_{.05}(2,4) = 6.94$.
We reject the hypothesis of no effect due to the color of the label.
Next we compare

$$f_C = \frac{MSSC}{PMSSE} \doteq \frac{34.34}{5.67} = 6.06 \quad \text{with}$$

$$f_\alpha(c - 1, (r - 1)(c - 1)) = f_{.05}(2, 4) = 6.94.$$

The null hypothesis is not rejected. We cannot conclude an effect due to the container.

Exercise Set XII.1

1) a) Square footage of the home

        Lot size

    b) I.Q.'s

        High school g.p.a.

    c) The size of the community served by the store

        The mean income of residents of the community served by the store

    d) Prevailing temperature

        The number of inches of rain or snow during the week

    e) The level of anxiety of the individual

        The age of the individual

        The weight of the individual

    f) The age of the child

        The I.Q. of the child

        The diet of the child

3) grand mean: $\bar{x} = \dfrac{(16 + 23 + 19) + (14 + 21 + 23) + (12 + 14 + 18) + (11 + 15 + 20)}{12}$

$= 17.25$

total sum of squares:

$$\begin{aligned}
SS = {} & (16 - 17.25)^2 + (23 - 17.25)^2 + (19 - 17.25)^2 \\
& + (15 - 17.25)^2 + (21 - 17.25)^2 + (23 - 17.25)^2 \\
& + (12 - 17.25)^2 + (14 - 17.25)^2 + (18 - 17.25)^2 \\
& + (11 - 17.25)^2 + (15 - 17.25)^2 + (20 - 17.25)^2 = 180.25
\end{aligned}$$

sample means:

$$\bar{x}_1 = \frac{16 + 23 + 19}{3} = 19.33$$

$$\bar{x}_2 = \frac{15 + 21 + 23}{3} = 19.66$$

$$\bar{x}_3 = \frac{12 + 14 + 18}{3} = 14.66$$

$$\bar{x}_4 = \frac{11 + 15 + 20}{3} = 15.33$$

treatment sum of squares:

$$SSTr = 3(19.33 - 17.25)^2 + 3(19.66 - 17.25)^2 + 3(14.66 - 17.25)^2 + 3(15.33 - 17.25)^2$$

$$= 61.58$$

error sum of squares

$$SSE = (16 - 19.33)^2 + (23 - 19.33)^2 + (19 - 19.33)^2$$
$$+ (15 - 19.66)^2 + (21 - 19.66)^2 + (23 - 19.66)^2$$
$$+ (12 - 14.66)^2 + (14 - 14.66)^2 + (18 - 14.66)^2$$
$$+ (11 - 15.33)^2 + (15 - 15.33)^2 + (20 - 15.33)^2 = 118.67$$

check:

$$SSTr + SSE = 61.58 + 118.67 = 180.25 \approx SS$$

The small discrepancy is due to rounding used with the means.

5)

	Sample 1		Sample 2		Sample 3		Grand Total
	x	x^2	x	x^2	x	x^2	
	205	42025	240	57600	260	67600	
	240	57600	285	81225	280	78400	
	255	65025	340	115600	360	129600	
	280	78400	360	129600	410	168100	
	340	115600	370	136900	450	202500	
Sample Size	$n_1 = 5$		$n_2 = 5$		$n_3 = 5$		$n = 15$
Sum	$T_1 = 1320$		$T_2 = 1595$		$T_3 = 1760$		$T = 4675$
Sum of Squares	358650		520925		646200		1525775
Quotient	$\frac{(1320)^2}{5} = 348480$		$\frac{(1595)^2}{5} = 508805$		$\frac{(1760)^2}{5} = 619520$		1476805

The treatment sum of squares is obtained from the totals of rows 2 and 4.

$$SSTr = 1476805 - \frac{(4675)^2}{15} = 19763.33$$

The error sum of squares is obtained from the totals of rows 3 and 4.

$$SSE = 1525775 - 1476805 = 48970$$

The treatment and error mean squares are thus:

$$MSSTr = \frac{19763.33}{3 - 1} = 9881.67$$

$$MSSE = \frac{48970}{15 - 3} = 4080.83$$

Thus

$$f = \frac{MSSTr}{MSSE} = \frac{9881.67}{4080.83} = 2.421$$

The critical value of the test is

$$f_{.05}(3 - 1, 15 - 3) = f_{.05}(2,12) = 3.89$$

Do not reject the null hypothesis of no effect on sales due to the package design.

The ANOVA table follows:

Source of Variation	SS	d.f.	MS	f
Package Design	19763.33	2	9881.67	2.421
Error	48970	12	4080.83	
Total	68733.33	14		

7) a) Since $k - 1 = 5$, $k = 6$.

   b) Since $n - 1 = 53$, $n = 54$.

      or

      Since $n - k = 48$, $n = 48 + k = 48 + 6 = 54$.

   c) There are six samples with a total of 54 subjects. Thus each sample contains $\frac{54}{6} = 9$ subjects if all are of equal size.

   d) $f_{.01}(k - 1, n - k) = f_{.01}(5,48) \approx 3.51$ (largest adjacent entry)

      $f = 2.83 < f_{.01}$

      The null hypothesis should not be rejected.

9)	Sample 1		Sample 2		Sample 3		Grand Total
	x	x^2	x	x^2	x	x^2	
	15.3	234.09	16.0	256.00	14.3	204.49	
	15.8	249.64	16.3	265.69	15.2	231.04	
	16.4	268.96	16.8	282.24	15.9	252.81	
	16.9	285.61	18.2	331.24	16.1	259.21	
	17.3	299.29	19.1	364.81	16.4	268.96	
Sample Size	$n_1 = 5$		$n_2 = 5$		$n_3 = 5$		$n = 15$
Sum	$T_1 = 81.7$		$T_2 = 86.4$		$T_3 = 77.9$		$T = 246$
Sum of Squares	1337.59		1499.98		1216.51		4054.08
Quotient	1334.98		1492.99		1213.68		4041.65

$$\text{SSTr} = 4041.65 - \frac{(246)^2}{15} = 7.25, \quad \text{MSSTR} = \frac{7.25}{2} = 3.625$$

$$\text{SSE} = 4054.08 - 4041.65 - 12.43, \quad \text{MSSE} = \frac{12.43}{12} = 1.036$$

$$f = \frac{\text{MSSTr}}{\text{MSSE}} = \frac{3.625}{1.036} = 3.500$$

$$f_{.025}(2,12) = 5.10.$$

Since $f < f_{.025}$, do not reject the null hypothesis of equal means.

Source	SS	d.f.	MS	f
Makes	7.25	2	3.625	3.50
Error	12.43	12	1.036	
Total	19.68	14		

11)

	Sample 1		Sample 2		Sample 3		Grand Total
	x	x^2	x	x^2	x	x^2	
	205	42025	200	40000	215	46226	
	180	32400	180	32400	195	38025	
	190	36100	170	28900	220	48400	
	225	50625	155	24025	245	60025	
	230	52900	160	25600	200	40000	

Sample Size	$n_1 = 5$	$n_2 = 5$	$n_3 = 5$	$n = 15$
Sum	$T_1 = 1030$	$T_2 = 865$	$T_3 = 1075$	$T = 2970$
Sum of Squares	214050	150925	232675	597650
Quotient	212180	149645	231125	592950

$$SSTr = 592950 - \frac{(2970)^2}{15} = 4890, \quad MSSTr = \frac{4890}{3-1} = 2445$$

$$SSE = 597650 - 592950 = 4700, \quad MSSE = \frac{4700}{15-3} = 391.67$$

$$f = \frac{MSSTr}{MSSE} = \frac{2445}{391.67} = 6.24$$

critical value: $f_{.05}(2,12) = 3.89$

Since $f = 6.24 > 3.89$, reject H_0 and conclude a difference in the mean hospital rates for at least two of the areas

Source	SS	d.f.	MS	f
Area	4890	2	2445	6.24
Error	4700	12	391.67	
Total	9590	14		

13) a) heating cost

b)

	Sample 1		Sample 2		Sample 3		Grand Total
	x	x^2	x	x^2	x	x^2	
	189	35721	210	44100	190	36100	
	187	34969	230	52900	200	40000	
	190	36100	195	38025	180	32400	
	207	42849	215	46225	220	48400	
					205	42025	
Sample Size	$n_1 = 4$		$n_2 = 4$		$n_3 = 5$		$n = 13$
Sum	$T_1 = 773$		$T_2 = 850$		$T_3 = 995$		$T = 2618$
Sum of Squares	149639		181250		198925		529814
Quotient	149382.25		180625		198005		528012.25

$$SSTr = 528012.25 - \frac{(2618)^2}{13} = 787.33, \quad MSSTr = \frac{787.33}{2} = 393.66$$

$$SSE = 529814 - 528012.25 = 1801.75, \quad MSSE = \frac{1801.75}{10} = 180.18$$

$$f = \frac{393.66}{180.18} = 2.18$$

critical value: $f_{.05}(2,10) = 4.10$

Since $f < f_{.05}(2,10)$, do not reject H_0. We cannot conclude an effect on the heating cast due to the type of system.

Source	SS	d.f.	MS	f
System	787.33	2	393.66	2.18
Error	1801.75	10	180.18	

15)

	Sample 1	Sample 2	Sample 3	Grand Total
Sample Size	$n_1 = 10$	$n_2 = 10$	$n_3 = 7$	$n = 27$
Sum	$T_1 = 36.8$	$T_2 = 65.5$	$T_3 = 53.5$	$T = 155.8$
Sum of Squares	180.5	554.25	451.63	1186.38
Quotient	135.424	429.025	408.893	973.342

$$\text{SSTr} = 973.342 - \frac{(155.8)^2}{27} = 74.318, \quad \text{MSSTr} = \frac{74.318}{3-1} = 37.159 \approx 37.16$$

$$\text{SSE} = 1186.38 - 973.342 = 213.038 \approx 213.04, \quad \text{MSSE} = \frac{213.04}{27-3} = 8.88$$

$$f = \frac{37.16}{8.88} = 4.19$$

$$f_{.05}(2,24) = 3.40$$

Since $f > f_{.05}(2,24)$, reject H_0 and conclude that at least two of the engine designs have different mean emission levels.

Source	SS	d.f.	MS	f
Design	74.32	2	37.16	4.19
Error	213.04	24	8.88	
Total	287.36	26		

17)

Source	SS	d.f.	MS	f
Treatment	630	4	$\frac{630}{4} = 157.50$	$\frac{157.50}{34.67} = 4.54$
Error	$1150 - 630 = 520$	15	$\frac{520}{15} = 34.67$	
Total	1150	19		

critical value: $f_{.05}(4,15) = 3.06$

Since $f > f_{.05}(4,15)$, reject the null hypothesis.

Number of treatment levels = 5.

Size of each sample = 4 (if all are of the same size).

19)

	Form 1		Form 2		Form 3		Form 4		Grand Total
	x	x^2	x	x^2	x	x^2	x	x^2	
	78	6084	72	5184	71	5041	73	5329	
	84	7056	55	3025	69	4761	61	3721	
	65	4225	73	5329	77	5929	79	6241	
	89	7921	68	4624	82	6724	68	4624	
Sample Size	$n_1 = 4$		$n_2 = 4$		$n_3 = 4$		$n_4 = 4$		$n = 16$
Sum	$T_1 = 316$		$T_2 = 268$		$T_3 = 299$		$T_4 = 281$		$T = 1164$
Sum of Squares	25286		18162		22455		19915		85818
Quotient	24964		17956		22350.25		19740.25		85010.5

$$\text{SSTr} = 85010.5 - \frac{(1164)^2}{16} = 329.5, \quad \text{MSSTr} = \frac{329.5}{4-1} = 109.83$$

$$\text{SSE} = 85818 - 85010.5 = 807.5, \quad \text{MSSE} = \frac{807.5}{16-4} = 67.29$$

$$f = 1.63$$

$$f_{.025}(3,12) = 4.47$$

Since $f < f_{.025}(3,12)$, do not reject H_0. We cannot conclude an effect on the grades due to the arrangement of the problems.

Source	Sum	d.f.	MS	f
Arrangement	329.5	3	109.83	1.63
Error	807.5	12	67.29	
Total	1137.0	15		

Exercise Set XII.2

1) a) $\bar{x} = \dfrac{(10+15)+(15+30)+(30+40)+(20+30)+(20+50)+(70+30)}{12} = 30$

   b) $\bar{x}_{11} = \dfrac{10+15}{12} = 12.5$

   c) $\bar{x}_{.2} = \dfrac{15+30+20+50}{4} = \dfrac{115}{4} = 28.75$

   d) $T_{.1} = (10+15)+(20+30) = 75$

   e) $T_{3.} = (20+30)+(20+50)+(70+30) = 220$

   f) $T = (10+15)+(15+30)+(30+40)+(20+30)+(20+50)+(70+30) = 360$

3)

	C_1		C_2		C_3		
R_1	10	100	15	225	30	900	$T_{1.} = 140$
	15	225	30	900	40	1600	
R_2	20	400	20	400	70	4900	$T_{2.} = 220$
	30	900	50	2500	30	900	
	$T_{.1} = 75$		$T_{.2} = 115$		$T_{.3} = 170$		$T = 360$

$\Sigma\Sigma\Sigma \, x_{ijk}^2 = (100+225)+(225+900)+(900+1600)$
$$\qquad\qquad + (400+900)+(400+2500)+(4900+900)$$
$$\qquad\qquad = 13950$$

$SS = \Sigma\Sigma\Sigma \, x_{ijk}^2 - \dfrac{T^2}{N} = 13950 - \dfrac{(360)^2}{12} = 3150$

$SSR = \Sigma \dfrac{T_{i.}^2}{n\cdot c} - \dfrac{T^2}{N} = \dfrac{(140)^2}{2\cdot 3} + \dfrac{(220)^2}{2\cdot 3} - \dfrac{(360)^2}{12} = 533.33$

$SSC = \Sigma \dfrac{T_{.j}^2}{nr} - \dfrac{T^2}{N} = \dfrac{(75)^2}{2\cdot 2} + \dfrac{(115)^2}{2\cdot 2} + \dfrac{(170)^2}{2\cdot 2} - \dfrac{(360)^2}{12} = 1137.5$

$SSE = \Sigma\Sigma\Sigma \, x_{ijk}^2 - (\Sigma \, \Sigma \, (\Sigma \, x_{ijk})^2)^2$
$\qquad \qquad \qquad \quad {}_{i\ \ j}$

$\quad = 13950 - \left[\dfrac{25^2}{2} + \dfrac{45^2}{2} + \dfrac{70^2}{2} + \dfrac{50^2}{2} + \dfrac{70^2}{2} + \dfrac{100^2}{2} \right]$

$\quad = 1475$

SSRC = SS − SSR − SSC − SSE

$\quad$ = 3150 − [533.33 + 1137.5 + 1475]

$\quad$ = 4.17

$$\text{MSSR} = \frac{\text{SSR}}{r - 1} = \frac{533.3}{2 - 1} = 533.33$$

$$\text{MSSC} = \frac{\text{SSC}}{c - 1} = \frac{1137.5}{3 - 1} = 568.75$$

$$\text{MSSRC} = \frac{\text{SSRC}}{(r - 1)(c - 1)} = \frac{4.17}{(2 - 1)(3 - 1)} = 2.09$$

$$\text{MSSE} = \frac{\text{SSE}}{rc(n - 1)} = \frac{1475}{2 \cdot 3(2 - 1)} = 245.83$$

Test for a row effect:

critical value: $f_{.05}(2 - 1, 2 \cdot 3(2 - 1)) = f_{.05}(1,6) = 5.99$

$$f_R = \frac{\text{MSSR}}{\text{MSSE}} = \frac{533.33}{245.83} = 2.17$$

Since $f_R < f_{.05}(1,6)$, do not reject the null hypothesis of no row effect.

Test for a column effect:

critical value: $f_{.05}(3 - 1, 2 \cdot 3(3 - 1)) = f_{.05}(2,6) = 5.14$

$$f_C = \frac{\text{MSSC}}{\text{MSSE}} = \frac{568.75}{245.83} = 2.31$$

Since $f_C < f_{.05}(1,6)$, do not reject the null hypothesis of no column effect.

Test for an interaction effect:

critical value: $f_{.05}((2 - 1)(3 - 1), 2 \cdot 3(2 - 1)) = f_{.05}(2,6) = 5.14$

$$f_{RC} = \frac{\text{MSSRC}}{\text{MSSE}} = \frac{2.09}{245.83} = .009$$

Do not reject the null hypothesis of no interaction effect.

Source	SS	d.f.	MS	f
Rows	533.33	1	533.33	2.17
Columns	1137.50	2	568.75	2.31
Interaction	4.17	2	2.09	.009
Error	1475.00	6	245.83	
Total	3150.00	11		

5) a) $r - 1 = 4$ so $r = 5$

   b) $c - 1 = 2$ so $c = 3$

   c) $rcn - 1 = 29$ so $5 \cdot 3 \cdot n = 30$, $15n = 30$, and $n = 2$

   d) Row Test:

   critical value: $f_{.05}(4,15) = 3.06$

   $f = \dfrac{11}{2.8} = 3.93$

   Reject the null hypothesis of no row effect.

   Column Test:

   critical value: $f_{.05}(2,15) = 3.68$

   $f = \dfrac{27}{2.8} = 9.64$

   Reject the null hypothesis of no column effect.

   Interaction Test:

   critical value: $f_{.05}(8,15) = 2.64$

   $f = \dfrac{2}{2.8} < 1$

   Do not reject the null hypothesis of no interaction effect.

7)

	C_1		C_2		C_3		
R_1	60.0	3600.00	68.4	4678.56	84.2	7089.64	$T_1. = 441.6$
	65.8	4329.64	77.2	5959.84	86.0	7396.00	
R_2	78.0	6084.00	80.0	6400.00	82.0	6724.00	$T_2. = 468.0$
	75.0	5625.00	76.0	5776.00	77.0	5929.00	
R_3	74.0	5476.00	78.0	6084.00	84.0	7056.00	$T_3. = 468.0$
	70.0	4900.00	76.0	5776.00	86.0	7396.00	
	$T_{.1} = 422.8$		$T_{.2} = 455.6$		$T_{.3} = 499.2$		$T = 1377.6$

$\Sigma\Sigma\Sigma \; x_{ijk}^2 = 106279.68$

$SS = 106279.68 - \dfrac{(1377.6)^2}{18} = 847.36$

$SSR = \dfrac{(441.6)^2}{2\cdot 3} + \dfrac{(468.0)^2}{2\cdot 3} + \dfrac{(468.0)^2}{2\cdot 3} - \dfrac{(1377.6)^2}{18} = 77.44$

$SSC = \dfrac{(422.8)^2}{2\cdot 3} + \dfrac{(455.6)^2}{2\cdot 3} + \dfrac{(499.2)^2}{2\cdot 3} - \dfrac{(1377.6)^2}{18} = 489.65$

$$SSE = 106279.68 - \left[\frac{(125.8)^2}{2} + \frac{(145.6)^2}{2} + \frac{(170.2)^2}{2} + \frac{(153)^2}{2} + \frac{(156)^2}{2} \right.$$
$$\left. + \frac{(159)^2}{2} + \frac{(144)^2}{2} + \frac{(154)^2}{2} + \frac{(170)^2}{2} \right] = 94.16$$

$$SSRC = 847.36 - [77.44 + 489.65 + 94.16] = 186.11$$

$$MSSR = \frac{77.41}{3 - 1} = 38.71$$

$$MSSC = \frac{489.65}{3 - 1} = 244.83$$

$$MSSRC = \frac{186.11}{(3 - 1)(3 - 1)} = 46.53$$

$$MSSE = \frac{94.16}{3 \cdot 3(2 - 1)} = 10.46$$

Test for an effect due to the company:

critical value: $f_{.05}(2,9) = 4.26$

$$f = \frac{38.71}{10.46} = 3.70$$

Do not reject the null hypothesis of no effect on tube life due to the producer of the tube.

Test for an effect due to price:

critical value: $f_{.05}(2,9) = 4.26$

$$f = \frac{244.83}{10.46} = 23.41$$

Reject the null hypothesis of no effect on tube life due to the cost of the tube, i.e., the cost is a factor in tube life.

Test for an interaction effect:

critical value: $f_{.05}(4,9) = 3.63$

$$f = \frac{46.53}{10.46} = 4.45$$

Reject the null hypothesis of no interaction effect on tube life by the cost and the producer.

Source	SS	d.f.	MS	f
Producer	77.44	2	38.71	3.70
Price	489.65	2	244.83	23.41
Interaction	186.11	4	46.53	4.45
Error	94.16	9	10.46	
Total	847.36	17		

9) From problem 3 we have:

$$SSRC = 4.17, \quad SSE = 1475, \quad MSSR = 533.33 \quad \text{and} \quad MSSC = 568.75$$

$$PMSSE = \frac{4.17 + 1475}{(2 - 1)(3 - 1) + 2 \cdot 3(2 - 1)} = 184.90$$

Test for a row effect:

critical value: $f_{.05}(1,8) = 5.32$

$$f_R = \frac{533.33}{184.90} = 2.88$$

Do not reject the hypothesis of no row effect.

Test for a column effect:

critical value: $f_{.05}(2,8) = 4.46$

$$f_C = \frac{568.75}{184.90} = 3.08$$

Do not reject the null hypothesis of no column effect.

11)

		C_1		C_2		C_3		
R_1	140	19600	380	144400	260	67600		
	170	28900	270	72900	200	40000	$R_1 . = 3020$	
	210	44100	310	96100	240	57600		
	200	40000	340	115600	300	90000		
R_2	320	102400	490	240100	500	250000		
	400	160000	580	336400	490	240100	$R_2 . = 5370$	
	380	144400	390	152100	400	160000		
	440	193600	500	250000	480	230400		
	$R_{.1} = 2260$		$R_{.2} = 3260$		$R_{.3} = 2870$		$T = 8390$	

$$\Sigma \ \Sigma \ \Sigma \ x_{ijk}^2 = 3276300$$

$$SS = 3276300 - \frac{(8390)^2}{24} = 343295.83$$

$$SSR = \frac{(3020)^2}{4 \cdot 3} + \frac{(5370)^2}{4 \cdot 3} - \frac{(8390)^2}{24} = 230104.17$$

$$SSC = \frac{(2260)^2}{4 \cdot 2} + \frac{(3260)^2}{4 \cdot 2} + \frac{(2870)^2}{4 \cdot 2} - \frac{(8390)^2}{24} = 63508.33$$

$$SSE = 3276300 - \left[\frac{(720)^2}{4} + \frac{(1300)^2}{4} + \frac{(1000)^2}{4} + \frac{(1540)^2}{4} + \frac{(1960)^2}{4} + \frac{(1870)^2}{4} \right]$$

$$= 46675.00$$

$$SSRC = 343295.83 - [230104.17 + 63508.33 + 46675.00]$$

$$= 3008.33$$

$$MSSR = \frac{230104.17}{2 - 1} = 230104.17$$

$$MSSC = \frac{63508.33}{3 - 1} = 31754.17$$

$$MSSRC = \frac{3008.33}{(2 - 1)(3 - 1)} = 1504.17$$

$$MSSE = \frac{46675.00}{2 \cdot 3 \cdot (4 - 1)} = 2593.06$$

Test for a display rack effect:

critical value: $f_{.05}(1,18) = 4.41$

$$f_R = \frac{230104.17}{2593.06} = 88.74$$

Reject H_0 and conclude that the type of display rack has an effect on sales.

Test for a package design effect:

critical value: $f_{.05}(2,18) = 3.55$

$$f_C = \frac{31754.17}{2593.06} = 12.25$$

Reject H_0 and conclude the package design has an effect on sales.

Test for an interaction effect:

critical value: $f_{.05}(2,18) = 3.55$

$$f_{RC} = \frac{1504.17}{2593.06} = .58$$

Do not reject the null hypothesis of no interaction effect.

Source	SS	d.f.	MS	f
Rack	230104.17	1	230104.17	88.74
Package	63508.33	2	31754.17	12.25
Interaction	3008.33	2	1504.17	.58
Error	46675.00	18	2593.06	
Total	343295.83	23		

Chapter Test

1) The common population variance when there is no treatment effect.

2) $MSSTr = \dfrac{\sum\limits_{i} n_i(\bar{x}_i - \bar{x})^2}{k - 1}$

The value of $MSSTr$ depends on the variation between or among the various sample means; hence the between group name

$MSSE = \dfrac{\sum\limits_{i}\sum\limits_{j} (x_{ij} - \bar{x}_i)^2}{n - k}$

The value of $MSSE$ depends on the variation of the observations within each sample about their respective means; hence the within group name.

3) $SS = SSTr + SSE$

SS is called the total sum of squares.

$SSTr$ is called the treatment sum of squares.

SSE is called the error sum of squares.

SSE is a measure of the variation due to random errors.

The quotient $\dfrac{SSE}{n - k}$ is an estimate of the population variance.

$SSTr$ is a measure of the variation due to a treatment effect if one exists.

If not, $\dfrac{SSTr}{k - 1}$ is an estimate of the population variance.

4) $SS = SSR + SSC + SSRC + SSE$

SS is called the total sum of squares.

SSR is called the treatment sum of squares for the factor R.

SSC is called the treatment sum of squares for the factor C.

$SSRC$ is called the interaction treatment sum of squares.

SSE is called the error sum of squares.

$SS = \sum\limits_{i}\sum\limits_{j}\sum\limits_{k} (x_{ijk} - \bar{x})^2$

$SSR = \sum\limits_{i} n \cdot c(\bar{x}_{i.} - \bar{x})^2$

$SSC = \sum\limits_{j} n \cdot r(\bar{x}_{.j} - \bar{x})^2$

$SSE = \sum\sum\sum (x_{ijk} - \bar{x}_{ij})^2$

$SSRC = SS - SSR - SSC - SSE$

5) a) $k - 1 = 4$ so $k = 5$

   b) $n - 1 = 19$ so there are $n = 20$ observations in all.

   In each treatment group there are $\frac{20}{5} = 4$ observations.

   c) critical value: $f_{.05}(4,15) = 3.06$

   $f = \frac{18.5}{2.6} = 7.12$

   $f > f_{.05}(4,15)$. Reject H_0 and conclude that at least two of the population (treatment) means are different.

6)

Source	SS	d.f.	MS
Treatment	$(5)(20) = 100$	5	20
Error	180	$\frac{180}{5} = 36$	5
Total	280	41	

7)

	A		B		C		D		Grand Total
	x	x^2	x	x^2	x	x^2	x	x^2	
	76	5776	59	3481	64	4096	73	5329	
	72	5184	63	3969	68	4624	71	5041	
	81	6561	60	3600	72	5184	76	5776	
	79	6241	66	4356	70	4900	74	5476	
Sample Size	$n_1 = 4$		$n_2 = 4$		$n_3 = 4$		$n_4 = 4$		$n = 16$
Sum	$T_1 = 308$		$T_2 = 248$		$T_3 = 274$		$T_4 = 294$		$T = 1124$
Sum of Squares	23762		15406		18804		21622		79594
Quotients	23716		15376		18769		21609		79470

$$\text{SSTr} = 79470 - \frac{(1124)^2}{16} = 509$$

$$\text{SSE} = 79594 - 79470 = 124$$

$$\text{MSSTr} = \frac{509}{4 - 1} = 169.67$$

critical value: $f_{.05}(3,12) = 3.49$

$f = \frac{169.67}{10.33} = 16.42$

Reject H_0 and conclude that a difference exists in the mean noise levels for at least 2 of the airlines.

Source	SS	d.f.	MS	f
Airlines	509	3	169.67	16.42
Error	124	12	10.33	
Total	633	15		

8)

	I		II		III		
10	100	7	49	9	81	$T_{1\cdot} = 47$	
8	64	5	25	8	64		
9	81	11	121	9	81	$T_{2\cdot} = 56$	
9	81	7	49	11	121		
11	121	8	64	10	100	$T_{3\cdot} = 58$	
12	144	5	25	12	144		
$T_{\cdot 1} = 59$		$T_{\cdot 2} = 43$		$T_{\cdot 3} = 59$		$T = 161$	

$$\Sigma\Sigma\Sigma\; x_{ijk}^2 = 1515$$

$$SS = 1515 - \frac{(161)^2}{18} = 74.94$$

$$SSR = \frac{(47)^2}{2\cdot 3} + \frac{(56)^2}{2\cdot 3} + \frac{(58)^2}{2\cdot 3} - \frac{(161)^2}{18} = 11.44$$

$$SSC = \frac{(59)^2}{2\cdot 3} + \frac{(43)^2}{2\cdot 3} + \frac{(59)^2}{2\cdot 3} - \frac{(161)^2}{18} = 28.44$$

$$SSE = 1515 - \left[\frac{(18)^2}{2} + \frac{(12)^2}{2} + \frac{(17)^2}{2} + \frac{(18)^2}{2} + \frac{(18)^2}{2} + \frac{(20)^2}{2} + \frac{(23)^2}{2} \right.$$
$$\left. + \frac{(13)^2}{2} + \frac{(22)^2}{2} \right]$$

$$= 21.50$$

$$SSRC = 74.94 - [11.44 + 28.44 + 21.50] = 13.56$$

$$MSSR = \frac{11.44}{3 - 1} = 5.72$$

$$MSSC = \frac{28.44}{3 - 1} = 14.22$$

$$MSSRC = \frac{13.56}{(3 - 1)(3 - 1)} = 3.39$$

$$MSSE = \frac{21.50}{3 \cdot 3(2 - 1)} = 2.39$$

<u>Test for an effect due to the variety:</u>

$f_{.05}(2,9) = 4.26$

$f_R = \frac{5.72}{2.39} = 2.39$

Do not reject the null hypothesis of no effect due to the varieties of poultry used.

<u>Test for an effect due to the dietary program:</u>

$f_{.05}(2,9) = 4.26$

$f_C = \frac{14.22}{2.39} = 5.95$

Reject H_0 and conclude an effect due to the dietary program.

<u>Test for an interaction effect:</u>

$f_{.05}(4,9) = 3.63$

$f_{RC} = \frac{3.39}{2.39} = 1.42$

Do not reject the null hypothesis of no interaction effect.

Source	SS	d.f.	MS	f
Variety	11.44	2	5.72	2.39
Diet	28.44	2	14.22	5.95
Interaction	13.56	4	3.39	1.42
Error	21.50	9	2.39	
Total	74.94	17		

9) From number 8, SSRC = 21.50 and SSE = 13.56.

Thus PMSSE $= \frac{21.50 + 13.56}{4 + 9} = 2.70.$

<u>Test for variety effect:</u>

$f_{.05}(2,13) = 3.81$

$f_R = \frac{5.72}{2.70} = 2.12$

Do not reject H_0.

<u>Test for a diet effect:</u>

$f_{.05}(2,13) = 3.81$

$$f_C = \frac{14.22}{2.70} = 5.27$$

Reject H_0 and conclude an effect on weight gain due to the diet.

Source	SS	d.f.	MS	f
Variety	11.44	2	5.72	2.12
Diet	28.44	2	14.22	5.27
Error	35.06	13	2.70	
Total	74.94	17		

10) a) $(10 + 8) + (7 + 5) + (9 + 8) + (9 + 9) + (11 + 7) + (9 + 11)$

$\qquad + (11 + 12) + (8 + 5) + (10 + 2) = 161$

b) $\dfrac{x_{231} + x_{232}}{2} = \dfrac{9 + 11}{2} = 10$

c) $\bar{x}_{1\cdot} = \dfrac{T_{1\cdot}}{nc} = \dfrac{47}{2 \cdot 3} = \dfrac{47}{6} = 7.83$

d) $\bar{x}_{\cdot 2} = \dfrac{T_{\cdot 2}}{nr} = \dfrac{43}{2 \cdot 3} = \dfrac{43}{6} = 7.17$

Chapter XIII
Non-Parametric Statistics

Tests which do not involve population parameters such as the mean and variance or which make no assumption regarding the form of the distribution are referred to as non-parametric tests. This chapter contains a selection of such tests which parallel the tests developed in the first twelve chapters of the text.

Concepts/Techniques

1) Non Parametric and Distribution Free Hypothesis Tests

Non parametric and distribution free tests do not depend on specific population parameters such as the mean μ or variance σ^2 or do not require that the distribution of the population be of some specific type such as the normal. Both types of tests are generally referred to as non-parametric tests.

2) Contingency Tables

Many sets of data are such that they may be classified or grouped according to the categories of 2 variables. An $r \times c$ contingency table is a table in which the entries are the number of subjects observed in each of the pairs of joint classifications.

Notation: The observed frequency or number of subjects in the (i,j) cell at the intersection of the ith row and jth column is denoted by O_{ij}.

Example: The 2 × 3 contingency table below is for the sex × age classification of a sample of 150 subjects.

Age

	Young	Middle Aged	Old	Marginal Row Totals
Male	40	35	25	100
Female	25	10	15	50
Marginal Column Totals	65	45	40	150

Sex

The marginal row totals state that 100 of the 150 subjects were males and the remaining 50 were females.

The marginal column totals state that 65 of the subjects were classified as young, 45 as middle aged, and 40 as old.

The entry $O_{11} = 40$ at the intersection of row 1 and column 1 states that 40 of the subjects were young males.

The entry $O_{12} = 35$ at the intersection of the 1st row and 2nd column states that 35 of the subjects were middle aged males.

The entry $O_{23} = 15$ at the intersection of the 2nd row and 3rd column states that 15 of the subjects were old females.

And so forth.

3) Predicted Entries for a Contingency Table

The predicted entry e_{ij} for the (i,j) cell of a contingency table is found by multiplying the ith row marginal total by the jth column marginal total and dividing by the total number of subjects.

The predicted entries are the expected cell frequencies for both Chi Square tests to be discussed shortly.

Example: Find the predicted entries for the contingency table of the previous example. We now draw diagonals through each cell and record the observed frequencies above the diagonal and the predicted below.

	Young	Middle Aged	Old	Marginal Totals
Male	40 / 43.33	35 / 30.00	25 / 26.67	100
Female	25 / 21.67	10 / 15.00	15 / 13.33	50
Marginal Totals	65	45	40	150

Sex

$$e_{11} = \frac{100 \cdot 65}{150} = 43.33$$

$$e_{12} = \frac{100 \cdot 45}{150} = 30.00$$

$$e_{13} = \frac{100 \cdot 40}{150} = 26.67$$

$$e_{21} = \frac{50 \cdot 65}{150} = 21.67$$

$$e_{22} = \frac{50 \cdot 45}{150} = 15.00$$

$$e_{23} = \frac{\cdot 40}{150} = 13.33$$

4)   The Chi-Square Test of Independence for 2 Variables

First a random sample is selected, the contingency table is constructed, and the predicted cell frequencies are computed. These predicted values are the expected cell frequencies if the two variables of classification are independent.

Next compute

$$\chi^2 = \Sigma \frac{(O_{ij} - e_{ij})^2}{e_{ij}} = \Sigma \frac{\left(\begin{array}{c}\text{Observed Cell} \\ \text{Frequency}\end{array} - \begin{array}{c}\text{Expected Cell} \\ \text{Frequency}\end{array}\right)^2}{\text{Expected Cell Frequency}} .$$

This is a measure of how well the observed frequencies agree with the expected frequencies under the assumption of independence.

Finally compare χ^2 with χ^2_α where $\nu = (r - 1)(c - 1)$. If $\chi^2 > \chi^2_\alpha$, reject the null hypothesis

H_0:   The variables are independent

and accept

H_1: The variables are not independent

Remember. The expected frequencies are computed under the assumption that the variables are independent. A large χ^2 value, i.e. $\chi^2 > \chi_\alpha^2$, indicates a poor agreement between these expected values and what has been observed.

Example: A sample of 1020 students is classified according to their class standing and their political affiliation. Using the results below, test the hypothesis that political affiliation of students is independent of their class standing. Use $\alpha = .05$.

	Freshman	Sophomore	Junior	Senior	Marginal Totals
	Class Standing				
Democrat	150 / 109.80	85 / 87.50	60 / 82.35	55 / 70.34	350
Republican	90 / 141.17	100 / 112.50	140 / 105.88	120 / 90.44	450
Other	80 / 69.02	70 / 55.00	40 / 51.76	30 / 44.22	220
Marginal Totals	320	255	240	205	1020

The computation of the expected cell frequencies is omitted. For $\nu = (r - 1)(c - 1) = (3 - 1)(4 - 1) = 6$, $\chi_{.05}^2 = 12.59$. The value of the test statistic is

$$\chi^2 = \frac{(150 - 109.80)^2}{109.80} + \frac{(85 - 87.50)^2}{87.50} + \frac{(60 - 82.35)^2}{82.35} + \frac{(55 - 70.34)^2}{70.34}$$
$$+ \frac{(90 - 141.17)^2}{141.17} + \frac{(110 - 112.50)^2}{112.50} + \frac{(140 - 105.88)^2}{105.88} + \frac{(120 - 90.44)^2}{90.44}$$
$$+ \frac{(80 - 69.02)^2}{69.02} + \frac{(70 - 55.00)^2}{55.00} + \frac{(40 - 51.76)^2}{51.76} + \frac{(30 - 44.22)^2}{44.22}$$

$$= 76.92$$

Since $\chi^2 = 76.92 > \chi_{.05}^2$ we reject the null hypothesis and conclude

that the political preference of students is related to their class
standing, that is, class standing and political preference are not
independent.

Note in the above computation that the first term
$\frac{(150 - 110)^2}{110}$ = 14.55 exceeds the critical value $\chi^2_{.05}$ = 12.59. Thus
the entire sum would not need to be computed. With it, however, we
can see that p < .005.

Note: In order to use the chi-square distribution the expected cell
frequencies should all be at least five. If this is not the case, it
will be necessary to collapse 2 or more of the categories of one of the
variables into a single category.

5) The Chi-square Test for Homogeneity
    Suppose two populations are classified according to the several categories
    of a single variable and
        a) The proportions of the populations in the classes corresponding
            to a category are equal.
        b) This (a) is true for each category.
    Then the two populations are said to be homogeneous with respect to the
    variable of classification.

    Example: Suppose the variable (of classification) is class standing
    and that the two populations are science and non science students.
    These two populations are homogeneous with respect to class standing
    if
        i)   The proportion of freshman science students is equal to the
             proportion of freshman non science students.
        ii) The proportion of sophomore science students is equal to the
             proportion of sophomore non-science students.
        iii) The proportion of junior science students is equal to the
             proportion of junior non science students.
        iv) The proportion of senior science students is equal to the
             proportion of senior non science students.
    Note, it is not required that the proportions for different categories
    be equal.

    In this example, it is doubtful if the two populations are homogeneous.
    We wish to test:

H_0: The two populations are homogeneous with respect to the variable of classification

H_1: The proportions of the populations in the classes corresponding to a category are different for at least one category.

Two independent samples are used. These are classified, the contingency table is completed, and the test is carried out exactly as with the chi-square test for independence.

Example: We wish to determine whether rural and city voters have the same attitude toward a political figure. Random samples of each are obtained and classified according to their view (support, oppose, no opinion). Using $\alpha = .05$, test these two populations for homogeneity with respect to attitude toward the candidate.

| | \multicolumn{3}{c}{Attitude} | |
	Support	oppose	no opinion	Marginal Totals
City	80 \ 88.57	50 \ 45.71	70 \ 65.71	200
Rural	75 \ 66.42	30 \ 34.29	45 \ 49.29	150
Marginal Totals	155	80	115	350

The computation of the predicted entries is as follows:

$$e_{11} = \frac{200 \cdot 155}{350} = 88.57 \qquad e_{21} = \frac{150 \cdot 155}{350} = 66.42$$

$$e_{12} = \frac{200 \cdot 80}{350} = 45.71 \qquad e_{22} = \frac{150 \cdot 80}{350} = 34.29$$

$$e_{13} = \frac{200 \cdot 115}{350} = 65.71 \qquad e_{23} = \frac{150 \cdot 115}{350} = 49.29$$

Using $\nu = (2 - 1)(3 - 1) = 2$ d.f. the critical value is found to be $\chi_{.05} = 5.99$.

The value of the test statistic is

$$\chi^2 = \frac{(80 - 88.57)^2}{88.57} + \frac{(50 - 45.71)^2}{45.71} + \frac{(70 - 65.71)^2}{65.71}$$

$$+ \frac{(75 - 66.42)^2}{66.42} + \frac{(30 - 34.29)^2}{34.29} + \frac{(45 - 49.29)^2}{49.29} = 3.53$$

Thus $\chi^2 < \chi^2_{.05}$. We do not reject the homogeneity hypothesis.

6)   The Mann Whitney U Test

This is a two sample test based on ranks for comparing the locations of two populations. The value u of the test statistic is the smaller of the numbers:

$$u_1 = n_1 n_2 + \frac{n_1(n_1 + 1)}{2} - w_1 \quad \text{and}$$

$$u_2 = n_1 n_2 + \frac{n_2(n_2 + 1)}{2} - w_2 \quad \text{where}$$

n_1 and n_2 are the sample sizes.

w_1 and w_2 are the sums of the ranks of the two samples which are obtained when the two samples are combined and the observations are ranked. We then compare u with the critical value $u_{\alpha/2}$. If $u < u_{\alpha/2}$ (note the less than), then reject

H_0:   The two populations have the same location

and conclude

H_1:   The populations differ with respect to location.

Notes:

a)   In place of location the median or the mean is often used.

b)   The Mann Whitney test is often used in place of the two population t test for the equality of means when the normality assumptions are not met.

c)   The Mann Whitney test is used on occasion to test the equality of two distributions.

Example: One sample of experimental rats was fed a vitamin A free diet. A second sample was fed the same diet and was also given daily doses of vitamin A. The weight gains of these two samples are as follows:

Sample 1	Sample 2
Vitamin A Free Diet	Vitamin A Supplemented Diet
Weight Gains x	Weight Gains y
4,12,6,8,17,7	5,14,10,6,18,12,19

First combine the samples and assign ranks to the ordered observations

Observation	4, 5, 6, 6, 7, 8, 10, 12, 12, 14, 17, 18, 19
Rank	1, 2, 3.5, 3.5, 5, 6, 7, 8.5, 8.5, 10, 11, 12, 13
Variable	x, y, x, y, x, x, y, x, y, y, x, y, y

Note that ties are assigned the average of the ranks that would have been used had there been no ties. Thus since the two 6's occupy positions 3 and 4, they are assigned the rank $\frac{3 + 4}{2} = 3.5$. Similarly the two 12's are each assigned the rank $\frac{8 + 9}{2} = 8.5$. The next step is to find the sum of the ranks for each sample.

Sample 1			Sample 2	
x	rank		y	rank
4	1		5	2
6	3.5		6	3.5
7	5		10	7
8	6		12	8.5
12	8.5		14	10
17	11		18	12
			19	13
$w_1 = 35$			$w_2 = 56$	

Next compute u_1 and u_2.

$$u_1 = n_1 n_2 + \frac{n_1(n_1 + 1)}{2} - w_1 = 6 \cdot 7 + \frac{6(6 + 1)}{2} - 35 = 28$$

$$u_2 = n_1 n_2 + \frac{n_2(n_2 + 1)}{2} - w_2 = 6 \cdot 7 + \frac{7(7 + 1)}{2} - 56 = 14$$

Thus $u = 14$ (the smaller of u_1 and u_2).

Using $\alpha = .05$, $n_1 = 6$, and $n_2 = 7$ we find the critical value $u_{.05} = 6$ in Table IX of the text.

Since $u = 14$ is not less than $u_{.05}$, do not reject the hypothesis that these populations have the same location. We interpret this to mean that we cannot conclude a difference in the median weight gains of rats on these two feeding programs.

When both sample sizes exceed 20, the distribution of u may be approximated by the normal distribution. z scores are computed by

$$z = \frac{u - \frac{n_1 n_2}{2}}{\sqrt{\frac{n_1 n_2 (n_1 + n_2 + 1)}{2}}} .$$

Example: A Mann Whitney U value of u = 5 was obtained with samples of size $n_1 = 21$ and $n_2 = 25$. Find the p value of the test.

We seek $P(U \le 5)$

$$z = \frac{5 - \frac{21 \cdot 25}{2}}{\sqrt{\frac{(21)(25)(21 + 25 + 1)}{2}}} = -2.32$$

p value = $P(U \le 5)$ = $P(z < -2.32)$ = .5000 - .4898 = .0102.

Thus the null hypothesis of equal population locations would be rejected for $\alpha = .05$.

7) <u>The Wilcoxon Matched Pair Signed Rank Test</u>
This is a matched pair test for the equality of the locations (say means or medians) of two populations or for the equality of their distributions. In practice, the differences $(y_1 - x_1), (y_2 - x_2), \ldots, (y_n - x_n)$ of the observations of the matched pairs are computed and any zero differences are discarded. Then ignoring any signs on the differences, rank the differences. Once the ranking is completed, compute the sum T_+ of those ranks associated with positive differences and the sum T_- of those belonging to the negative differences.
The value T of the test statistic is the smaller of T_+ and T_-.
Finally compare T with $T_{\alpha/2}$. If $T < T_{\alpha/2}$, reject

H_0: The population distributions are identical

and conclude

H_1: The population distributions are different.

Note: This is a two tailed test with significance level α.

Example: Prior to taking a short course on the use of a hand held calculator, seven students were given a 10 point algebra quiz. Following the course they were given the same test. Their pre course and post course scores, differences, and ranks are shown below. Using $\alpha = .05$, test the hypothesis that this calculator course has an effect on understanding algebra.

Subject	Pre Course Score x	Post Test Score y	Difference d = (y - x)	Unsigned |d|	Rank	Signed Rank
1	6	9	3	3	6	6
2	4	5	1	1	1	1
3	5	7	2	2	3	3
4	10	8	-2	2	3	-3
5	0	6	6	6	8	8
6	4	4	0	*		
7	9	6	-3	3	6	-6
8	8	5	-3	3	6	-6
9	8	10	-2	2	3	-2

$T_+ = 6 + 1 + 3 + 8 = 18$

$T_- = 3 + 6 + 6 + 2 = 17$

$T = 17$ (The Smaller of T_+ and T_-)

Using $n = 8$ (non-zero differences) the critical T value is $T_{.05} = 4$. (See Table X of the text.)

Since $T = 17$ is not less than $T_{.05}$, the null hypothesis of no difference in distribution of scores is not rejected. This can be interpreted to mean that the short course had no effect on the median or average grade.

8) The Spearman Rank Correlation Coefficient

The Spearman correlation coefficient r_s is simply the Pearson correlation coefficient computed on the ranks of the observations rather than the observations themselves.

Starting with the sample $(x_1,y_1),(x_2,y_2),\ldots,(x_n,y_n)$, the first step is to rank the x's and y's separately. Next compute the difference e_i of the ranks assigned to x_i and y_i.
The Spearman Rank Correlation Coefficient is

$$r_s = 1 - \frac{6 \Sigma d_i^2}{n(n^2 - 1)} \, .$$

Notes:

a) r_s has values between $+1$ and -1.

b) A positive value of r_s indicate a tendency for y to increase as x increases. A negative value of r_s indicates a tendency for y to decrease as x increases.

c) The closer r_s is to 1 or -1, the stronger the correlation.

For $n \geq 10$ tests of

H_0: No correlation exists between X and Y

H_1: A correlation exists between X and Y

are based on the t distribution with $\nu = n - 2$ d.f. where

$$t = \frac{r_s(n - 2)}{\sqrt{1 - r_s^2}} \, .$$

Example: Using $\alpha = .05$, test for a correlation between X and Y using the following data

x	1	2	2	3	4	5	5	6	7	8	9
y	12	10	14	16	20	18	22	25	25	27	30

First we rank the x's and y's separately

x	1	2	2	3	4	5	5	6	7	8	9
rank	1	2.5	2.5	4	5	6.5	6.5	8	9	10	11
y	10	12	14	16	18	20	22	25	25	27	30
rank	1	2	3	4	5	6	7	8.5	8.5	10	11

Now match up the ranks with the pairs of observations and compute the differences.

x	y	Rank of x	Rank of y	Difference of Ranks d	Square of Differences d^2
1	12	1	2	1	1
2	10	2.5	1	-1.5	2.25
2	14	2.5	3	-.5	.25
3	16	4	4	0	0
4	20	5	6	1	1
5	18	6.5	5	-1.5	2.25
5	22	6.5	7	.5	.25
6	25	8	8.5	.5	.25
7	25	9	8.5	-.5	.25
8	27	10	10	0	0
9	30	11	11	0	0

$$\Sigma\, d^2 = 7.50$$

$$r_s = 1 - \frac{6\,\Sigma\, d^2}{n(n^2 - 1)} = 1 - \frac{6(7.50)}{11(11^2 - 1)} = .97$$

$$t = \frac{r_s(n - 2)}{\sqrt{1 - r_s^2}} = \frac{(.97)(11 - 2)}{\sqrt{1 - (.97)^2}} = 35.91$$

For $\alpha = .05$ and $\nu = 11 - 2 = 9$, the two tailed critical values are $\pm t_{.025} = 2.262$.

Since $t = 35.91 > t_{.025}$ we conclude that a correlation exists.

Note that in computing the ranks, ties are assigned average ranks exactly as in the Mann Whitney computations.

9) The Kruskal-Wallis One Way Analysis of Variance Test

The Kruskal-Wallis test is a test based on ranks for testing the equality of several population means medians, or distributions.

The basic idea is to extend the Mann Whitney test to several populations by pooling all the samples, ranking the observations, and then finding the rank sum u_i for each sample.

The value h of the test statistic H is then computed by

$$h = \frac{12}{n(n+1)} \sum_{i=1}^{m} \left(\frac{u_i^2}{n_i} \right) - 3(n+1)$$

where   m   is the number of samples,

n_i is the size of the ith sample

u_i is the rank sum of the ith sample.

For $n \geq 5$ the distribution of   H   may be approximated by the chi–square distribution with $\nu = m - 1$ d.f.   provided all populations have the same distribution. Thus if $h > \chi_\alpha^2$ we reject

H_0:   The populations have identical distributions

and conclude

H_1:   At least two of the populations have distributions
that differ as to location (mean, or median).

Example: Samples of starting salaries (1000's of dollars) of engineering graduates from three areas are given. Using $\alpha = .05$, test for a difference in the distribution of starting salaries in these 3 fields.

Electrical (x)	Mechanical (y)	Chemical (z)
26	25	26
29	27	27
30	28	28
31	29	28
32	30	29
34	30	30
	33	31
		31

First combine the 3 samples and rank as shown

Observation	25	26	26	27	27	28	28	28	29	29	29	30	30	30	30	31	31	31
Rank	1	2.5	2.5	4.5	4.5	7	7	7	10	10	10	13.5	13.5	13.5	13.5	17	17	17
Sample	y	x	z	y	z	y	z	z	x	y	z	x	y	y	z	x	z	z

(continued on next page)

Observation	32 33 34
Rank	19 20 21
Sample	x y x

The samples and their ranks are shown below:

Electrical		Mechanical		Chemical	
x	rank	x	rank	x	rank
26	2.5	25	1	26	2.5
29	10	27	4.5	27	4.5
30	13.5	28	7	28	7
31	17	29	10	28	7
32	19	30	13.5	29	10
34	21	30	13.5	30	13.5
		33	20	31	17
$u_1 = 83$				31	17
$n_1 = 6$		$u_2 = 69.5$			
		$n_2 = 7$		$u_3 = 78.5$	
				$n_3 = 8$	

The value of the test statistic is

$$h = \frac{12}{21(21 + 1)} \left[\frac{(83)^2}{6} + \frac{(69.5)^2}{7} + \frac{(78.5)^2}{8} \right] - 3(21 + 1)$$

$$= 1.75.$$

For $\alpha = .05$ and $\nu = 3 - 1 = 2$, $\chi_{.05} = 5.99$.
Thus the null hypothesis of identical starting salary distributions
for these 3 fields is not rejected.

Notes:

a) In assigning ranks, averages are assigned to ties exactly as
   with the Mann Whitney computations.

b) For m = 2 samples, the Kurskal–Wallis test is equivalent to
   the Mann–Whitney test.

c) In the case of a great many ties, a more powerful test is obtained
   if

$$h' = \frac{h}{c}$$

   is used in place of h. Here

$$c = 1 - \frac{\Sigma(t_i^3 - t_i)}{n^3 - 1} \quad \text{where}$$

t_i = number of ties in the ith sample.

10) The Friedman Two Way Analysis of Variance Test

The Friedman test is appropriate with a randomized block design which is ordinarily treated by a two way analysis of variance. It makes use, however, only of the ranks of the observations and does not have any normality requirements. Assuming there are r blocks or rows, the observations in each row are first ranked. Next find the sum of the ranks in each of the c treatment columns. Finally the value χ_f^2 of the Friedman statistic is computed by

$$\chi_f^2 = \left[\frac{12}{rc(c + 1)} \sum_{i=1}^{c} u_i^2 \right] - 3r(c + 1).$$

Here $u_1, u_2, \ldots, u_c$ are the rank sums for the c columns. For $c \geq 5$, the Friedman statistic has approximately the chi-square distribution with $\nu = c - 1$ d.f.

Critical values for $c = 3$ and $c = 4$ are given in the following table:

c\r	2	3	4	5	6	7	8	9
3	4(.17)	6(.03)	6.5(.04)	6.4(.04)	6.3(.05)	6.0(.05)	6.3(.05)	6.2(.05)
4	6.0(.04)	7.0(.05)	7.5					

Critical values of χ_f^2 for $c = 3$ and $c = 4$
For $c = 3$ and $r \geq 10$ or $c = 4$ and $r \geq 5$ use
the chi-square distribution with $\nu = c - 1$ d.f.

Example: The entry $\chi^2 = 6$ for $r = 3$ and $c = 3$ is the critical value belonging to $\alpha = .03$.

Example: The three sections of a statistics course are taught by different instructors. Students enrolling in this course are matched by mathematical backgrounds in blocks of 3. Students who could not be blocked are not considered in the analysis. Each student in a block is then assigned to a different section. Using the students grade on the common final below, test for an effect due to the instructor.

	Section 1		Section 2		Section 3	
	Grade	Rank	Grade	Rank	Grade	Rank
Block 1	62	1	80	3	73	2
Block 2	75	1	94	3	86	2
Block 3	88	3	86	2	84	1
Block 4	72	2	83	3	64	1
Block 5	75	1.5	80	3	75	1.5
Block 6	90	3	83	2	71	1
Block 7	50	1	84	3	70	2

$$u_1 = 12.5 \qquad u_2 = 19 \qquad u_3 = 10.5$$

Using $r = 7$ and $c = 3$ and the above rank sums we find

$$x_f^2 = \frac{12}{7 \cdot 3(3 + 1)} [(12.5)^2 + (19)^2 + (10.5)^2] - 3 \cdot 7(3 + 1)$$

$$= 5.64$$

For $c = 3$ and $r = 7$, the above brief table yields 6.0 as the critical value for $\alpha = .05$. Thus we cannot reject at $\alpha = .05$ the null hypothesis that the 3 treatment distributions are equal, i.e., we do not reject the contention that the teachers are equally effective.

11) <u>The Run Test for Randomness</u>

Suppose we have a set of observations exactly in the order in which they were recorded. If an observation falls below the median, replace it by an F. If it falls above the median replace it with an S. Ignore any observation falling at the median.

Now count the number of runs r of F's and S's and compute

$$z = \frac{r - \left(\dfrac{2n_1 n_2}{n_1 + n_2} + 1 \right)}{\sqrt{\dfrac{2n_1 n_2 (2n_1 n_2 - n_1 - n_2)}{(n_1 + n_2)^2 (n_1 + n_2 - 1)}}}$$

where n_1 is the number of S's and n_2 is the number of F's in the sequence of S's and F's derived from the observations.

When n_1 and n_2 are at least 10, the standard normal distribution may be used. Specifically, if $z > z_{\alpha/2}$ or $z < -z_{\alpha/2}$, reject

H_0: The sequence of observations is random

and accept

H_1: The sequence is not random.

Example: Using $\alpha = .05$, test the following sequence for randomness:

12,15,7,3,27,14,9,20,18,17,8,6,2,9,10,15,23,18,26,4,5,8.

The median is 11. Replacing those observations of the above list by F if they fall below $\tilde{x} = 11$ and by S if they are above it, we obtain

S,S,F,F,S,S,F,S,S,S,F,F,F,F,F,S,S,S,S,F,F,F.

There are $r = 8$ runs (underlined) and $n_1 = 11$ S's and $n_2 = 11$ F's.
Thus

$$z = \frac{8 - \left(\frac{2 \cdot 11 \cdot 11}{11 + 11} + 1\right)}{\sqrt{\frac{2 \cdot 11 \cdot 11(2 \cdot 11 \cdot 11 - 11 - 11)}{(11 + 11)^2(11 + 11 - 1)}}} = -1.74.$$

For $\alpha = .05$, the critical values are $\pm z_{.025} = \pm 1.96$.
Thus the null hypothesis that the sequence is random is not rejected.

Note: The expected number of runs in a random sequence with n_1 S's and n_2 F's is

$$\mu_R = \frac{2 \cdot n_1 n_2}{n_1 + n_2} + 1.$$

Example: In a random sequence of $n_1 = n_2 = 11$ S's and F's (as in the last example), the expected number of runs is

$$\mu_R = \frac{2 \cdot 11 \cdot 11}{11 + 11} + 1 = 12. \quad r = 8 \text{ is not statistically different}$$

from this value.

Exercise Set XIII.1

1) a) 75

b) 40 (row 1 total)

c) 26 (column 2 total)

d) 8 (row 2, column 3 cell entry)

e)

	Poor	Satisfactory	Excellent	Total
Slow	8 / 12.27	14 / 13.87	18 / 13.87	40
Fast	15 / 10.73	12 / 12.13	8 / 12.13	35
Total	23	26	26	75

$$e_{11} = \frac{40 \cdot 23}{75} = 12.27 \qquad e_{12} = \frac{40 \cdot 26}{75} = 13.87 \qquad e_{13} = \frac{40 \cdot 26}{75} = 13.87$$

$$e_{21} = \frac{35 \cdot 23}{75} = 10.73 \qquad e_{22} = \frac{35 \cdot 26}{75} = 12.13 \qquad e_{23} = \frac{35 \cdot 26}{75} = 12.13$$

f) $\nu = (2 - 1)(3 - 1) = 2$

$\chi^2_{.05} = 5.99$

$$\chi^2 = \frac{(8 - 12.27)^2}{12.27} + \frac{(14 - 13.87)^2}{13.87} + \frac{(18 - 13.87)^2}{13.87}$$

$$+ \frac{(15 - 10.73)^2}{10.73} + \frac{(12 - 12.13)^2}{12.13} + \frac{(8 - 12.13)^2}{12.13} = 5.82$$

$\chi^2 = 5.82 < \chi^2_{.05}$. Do not reject the null hypothesis of independent classifications.

3) $\nu = (2 - 1)(3 - 1) = 2$

Choosing $\alpha = .05$, the critical value is $\chi^2_{.05} = 5.99$. The completed contingency table follows.

	I	II	III	Total
Male	9 / 10.22	12 / 15.56	19 / 14.22	40
Female	14 / 12.78	23 / 19.44	13 / 17.78	50
Total	23	35	32	90

$$e_{11} = \frac{40 \cdot 23}{90} = 10.22 \qquad e_{12} = \frac{40 \cdot 35}{90} = 15.56 \qquad e_{13} = \frac{40 \cdot 32}{90} = 14.22$$

$$e_{21} = \frac{50 \cdot 23}{90} = 12.78 \qquad e_{22} = \frac{50 \cdot 35}{90} = 19.44 \qquad e_{23} = \frac{50 \cdot 32}{90} = 17.78$$

$$\chi^2 = \frac{(9 - 10.22)^2}{10.22} + \frac{(12 - 15.56)^2}{15.56} + \frac{(19 - 14.22)^2}{14.22}$$

$$+ \frac{(14 - 12.78)^2}{12.78} + \frac{(23 - 19.44)^2}{19.44} + \frac{(13 - 17.78)^2}{17.78} = 4.62$$

Since $\chi^2 = 4.62 < 5.99$ we do not reject the null hypothesis of no relationship (independence) i.e. we cannot conclude (at $\alpha = .05$) that there is a relationship between the brand preferred and the sex of the student making the choice.

5) The low cell frequencies in the 3rd and 4th rows suggest we combine the last two row categories. The resulting contingency table follows:

	17–25	26–40	Over 40	Total
0	23 / 35.33	38 / 35.33	45 / 35.33	106
1	18 / 29.67	31 / 29.67	40 / 29.67	89
2	45 / 26.33	25 / 26.33	9 / 26.33	79
Over 2	14 / 8.67	6 / 8.67	6 / 8.67	26
Total	100	100	100	300

For this modified table, $\nu = (4 - 1)(3 - 1) = 6$.

Using $\alpha = .05$, the critical value is $\chi^2_{.05} = 12.59$.

Note that the contribution of the (3,1) cell to the χ^2 value is

$$\frac{(45 - 26.33)^2}{26.33} = 13.24.$$

Thus $\chi^2 > \chi^2_{.05} = 12.59$ and we conclude (at $\alpha = .05$) that a relationship exists between the number of tickets and the age of the driver. In fact $\chi^2 = 45.03 > \chi^2_{.005}$. Thus p value $< .005$.

7) $\nu = (2 - 1)(3 - 1) = 2$

For $\alpha = .05$, the critical value is $\chi^2_{.05} = 5.99$.

	I	II	III	Total
Planned to Return	215 / 193.47	137 / 128.98	251 / 280.54	603
Do Not Plan to Return	85 / 106.52	63 / 71.02	184 / 154.46	332
Total	300	200	435	935

$\chi^2 = 16.91$

Reject the null hypothesis of equal proportions for these three nationalities.

9) $\nu = (2 - 1)(4 - 1) = 3$

$\chi^2_{.05} = 7.81$

	1	2	3	4	Total
Recommended	36 / 39.50	42 / 39.50	49 / 39.50	31 / 39.50	158
Not Recommended	164 / 160.50	158 / 160.50	151 / 160.50	169 / 160.50	642
Total	200	200	200	200	800

$$\chi^2 = 5.71 < \chi^2_{.05}$$

We cannot reject the null hypothesis of equal recommended rates (equal proportions) for the four clinics.

11) a) $\nu = (2 - 1)(3 - 1) = 2$
         $\chi^2_{.05} = 5.99$

	Defend Homes	Leave to Police	No Opinion	Total
Black Sample	304 / 271.43	112 / 155.23	52 / 41.33	468
White Sample	287 / 319.57	226 / 182.77	38 / 48.67	551
Total	591	338	90	1019

Note that some "guesswork" is needed in arriving at the observed frequencies to do the severe rounding that was used in reporting the percents.

$$\chi^2 = 34.58$$

Reject the null hypothesis of homogeneity of whites and blacks with respect to attitudes on home defense.

b) i) Black Sample

• $\nu = 2$, $\chi^2_{.05} = 5.99$

	Defend Homes	Leave to Police	No Opinion	Total
Male	123 / 107.95	30 / 40.75	15 / 19.30	168
Female	179 / 194.05	84 / 73.25	39 / 34.70	302
Total	302	114	54	470

Note: Count is off due to rounding in original data.

$$\chi^2 = 9.17$$

Reject the null hypothesis of no relationship between sex and home defense orientation of blacks.

ii) White Sample
$\nu = 2,\ \chi^2_{.05} = 5.99$

	Defend Homes	Leave to Police	No Opinion	Total
Male	128 / 130.25	108 / 103.11	15 / 17.64	251
Female	160 / 157.75	120 / 124.89	24 / 21.36	304
Total	288	228	39	555

$\chi^2 = 1.19 < \chi^2_{.05}$

Do not reject the null hypothesis of no relationship between sex and home defense orientation of whites.

iii) Pooled Sample

	Defend Homes	Leave to Police	No Opinion	Total
Male	251 / 241.18	138 / 139.80	30 / 38.02	419
Female	339 / 348.82	204 / 202.20	63 / 54.98	606
Total	590	342	93	1025

$\chi^2 = 3.56 < \chi^2_{.05}$

Do not reject the null hypothesis of no relationship between sex and home defense orientation of individuals.

c) Black Group

	Defend Homes	Leave to Police	No Opinion	Total
N.E.	45 / 45.67	17 / 17.18	9 / 8.14	71
N.C.	115 / 99.71	31 / 37.52	9 / 17.77	155
S	129 / 142.17	59 / 53.49	33 / 25.34	221
W	14 / 15.44	7 / 5.80	3 / 2.70	24
Total	303	114	54	471

$\nu = (4 - 1)(3 - 1) = 6$

$\chi^2_{.05} = 12.59$

$\chi^2 = 12.41$. Do not reject the null hypothesis of no relationship between region and orientation.

White Group

	Defend Homes	Leave to Police	No Opinion	Total
N.E.	57 / 76.73	83 / 60.58	7 / 9.69	147
N.C.	77 / 75.58	56 / 59.34	11 / 9.49	144
S	106 / 74.64	30 / 58.93	7 / 9.43	143
W	45 / 58.46	56 / 46.15	11 / 7.38	112
Total	285	225	36	546

$\nu = 6$, $\chi^2_{.05} = 12.59$

$\chi^2 = 49.31$. Reject H_0 and conclude a relationship exists between region and orientation for whites.

d) The increasing sequence 27, 41, 49, 61 (%) indicates that support for the police role increases with the size of the city of residence.

Exercise Set XIII.2

1) a)

Observation	2,	3,	4,	4,	6,	7,	8,	10,	12
Rank	1	2	3.5	3.5	5	6	7	8	9
Variable	y	x	x	y	x	y	x	y	y

b)

x	rank
3	2
4	3.5
6	5
8	7
w_1 = 17.5	

y	rank
2	1
4	3.5
7	6
10	8
12	9
w_2 = 27.5	

c) minimum rank sum = $1 + 2 + 3 + 4 = \dfrac{4 \cdot 5}{2} = 10$

maximum rank sum = $5 + 6 + 7 + 8 + 9$

$$= 4 \cdot 5 + \dfrac{5(6)}{2} = 35$$

3)

Observation	38,	40,	44,	46,	47,	54,	56,	58,	62,	65,	66,	73
Rank	1	2	3	4	5	6	7	8	9	10	11	12
Sample	x	y	x	y	x	y	x	y	y	y	y	x

x	rank
38	1
44	3
47	5
56	7
73	12
w_1 = 28	
n_1 = 5	

y	rank
40	2
46	4
54	6
58	8
62	9
65	10
66	11
w_2 = 50	
n_2 = 7	

$$u_1 = 5 \cdot 7 + \dfrac{5(5 + 1)}{2} - 28 = 22$$

$$u_2 = 5 \cdot 7 + \dfrac{7(7 + 1)}{2} - 50 = 13$$

$u = 13$

For $\alpha = .05$, the critical value is $u_{.025} = 5$.

Since $u = 13 > u_{.025}$, do not reject the null hypothesis of equal mean room costs for the two areas.

5)

Left Handed			Right Handed	
x	rank		y	rank
64	1		66	2
78	6		68	3
82	9		70	4
84	10.5		75	5
88	14		81	7.5
90	15		81	7.5
91	17		84	10.5
91	17		86	12
			87	13
$w_1 = 89.5$			91	17
$n_1 = 8$			$w_2 = 81.5$	
			$n_2 = 10$	

$$u_1 = 8 \cdot 10 + \frac{8(8 + 1)}{2} - 89.5 = 26.5$$

$$u_2 = 8 \cdot 10 + \frac{10(10 + 1)}{2} - 81.5 = 53.5$$

$u = 26.5$

For $\alpha = .05$, the critical value is $u_{.025} = 17$.

Since $u = 26.5 > u_{.025}$, do not reject the null hypothesis of no difference in the distribution of mastery scores of left and right handed students, i.e. we cannot conclude that a difference in ability exists.

7)   590(x)   790(x)   830(x)   860(x)   870(x)   910(y)   920(y)
        1   2   3   4   5   6   7

      980(y)   1000(y)   1050(x)   1090(x)   1150(x)   1230(y)   1280(y)
        8   9   10   11   12   13   14

     1290(x)   1300(x)   1320(y)   1340(y)   1350(y)   1410(y)   1430(y)
       15   16   17   18   19   20   21.5

     1430(x)   1440(y)   1460(y)   1500(y)   1500(y)   1530(y)   1560(y)
      21.5   23   24   25.5   25.5   27   28

     1570(y)   1580(x)   1610(y)   1650(x)   1690(y)   1720(x)   1730(x)
       29   30   31   32   33   34   35

     1840(x)   1860(y)   1900(x)   2000(x)   2030(y)   2130(x)   2150(x)
       36   37   38   39   40   41   42

     2560(x)   2650(x)   2780(x)
       43   44   45

$w_1 = 559.5$, $w_2 = 475.5$

$n_1 = 23$, $n_2 = 22$

$u_1 = 23 \cdot 22 + \dfrac{23(23 + 1)}{2} - 559.5 = 222.5$

$u_2 = 23 \cdot 22 + \dfrac{22(22 + 1)}{2} - 475.5 = 283.5$

$u = 222.5$

For $\alpha = .05$ the critical values are $\pm z_{.025} = \pm 1.96$

$$z = \frac{222.5 - \dfrac{23 \cdot 22}{2}}{\sqrt{\dfrac{23 \cdot 22(23 + 22 + 1)}{2}}} = -.28$$

Do not reject H_0 (the hypothesis of equal mean expenditures for auto
accessories by men and women.

9) For $\alpha = .01$, the critical values are $\pm z_{.005} = \pm 2.58$

$u_1 = 25 \cdot 20 + \dfrac{25(25 + 1)}{2} - 335 = 490$

$u_2 = 25 \cdot 20 + \dfrac{20(20 + 1)}{2} - 700 = 10$

$u = 10$

$$z = \frac{10 - \dfrac{25 \cdot 20}{2}}{\sqrt{\dfrac{25 \cdot 20(25 + 20 + 1)}{2}}} = -2.24$$

Since $z > -2.58$, do not reject the null hypothesis of equal distribution of attitude scores for workers at the two factories, i.e., we cannot conclude a difference in attitude of workers at these two factories.

11) a)

7	9	9	10	11	13	15	30	108	810	810	810
1	2.5	2.5	4	5	6	7	8	9	11	11	11
x	x	y	y	y	y	x	y	y	x	y	y

(Mothers)			(Controls)	
x	rank		x	rank
7	1		9	2.5
9	2.5		10	4
15	7		11	5
810	11		13	6
			30	8
$w_1 = 21.5$			108	9
$n_1 = 4$			810	11
			810	11
			$w_2 = 56.5$	
			$n_2 = 8$	

$$u_1 = 4 \cdot 8 + \frac{4 \cdot (4 + 1)}{2} - 21.5 = 20.5$$

$$u_2 = 4 \cdot 8 + \frac{8(8 + 1)}{2} - 56.5 = 11.5$$

$u = 11.5$

b) Yes. $p = .003$ so results are significant for $\alpha < .003$ in a one tailed test.

c) Yes. $p = .006$

d) Yes

Exercise Set XII.3

1) a) For $\alpha = .05$ and $n = 15$, the two tailed critical value is $T_{.025} = 25$.
   Since $T = 40 > T_{.025}$, do not reject H_0.

   b) For $\alpha = .01$ and $n = 12$, the two tailed critical value is $T_{.005} = 7$.
   Since $T = 26 > T_{.005}$, do not reject H_0.

3) a)

x	y	d = x - y	|d|	rank	signed rank	
---	---	-----------	-------	------	-------------	
47	58	-11	11	7.5	-7.5	
63	61	2	2	2.5	2.5	
76	72	4	4	4.5	4.5	$T_+ = 10.5$
28	45	-17	17	10.0	-10	$T_- = 44.5$
42	53	-11	11	7.5	-7.5	$T = 10.5$
55	59	-4	4	4.5	-4.5	
38	49	-11	11	7.5	-7.5	
83	83	0	0	*	*	
37	35	2	2	2.5	2.5	
45	56	-11	11	7.5	-7.5	
81	80	1	1	1	1	

For $\alpha = .05$ and $n = 10$ (originally 11 reduced by 1 due to tie), the critical value is $T_{.025} = 8$. Since $T = 10.5 > T_{.025}$, do not reject H_0.

   b) Consider only the initial scores that were under 60.

x	y	d = x - y	|d|	rank	signed rank	
---	---	-----------	-------	------	-------------	
47	58	-11	11	4.5	-4.5	
28	45	-17	17	7	-7	
42	53	-11	11	4.5	-4.5	$T_+ = 1$
55	59	-4	4	2	-2	$T_- = 27$
38	49	-11	11	4.5	-4.5	$T = 1$
37	35	2	2	1	1	
45	56	-11	11	4.5	-4.5	

New alternate hypothesis: The new procedure improves the dexterity of those who are severly impaired. This will require a one tailed test.

For $\alpha = .05$ and $n = 7$, $T_{.05} = 4$.

Since $T = 1 < T_{.05}$ reject the null hypothesis of no improvement.

5)

x	78	84	65	98	56	28	70	66	55	87	90	61	70	83
y	74	81	73	98	60	13	58	74	59	88	93	66	88	90
d	4	3	-8	0	-4	15	12	-10	-4	-1	-3	-5	-18	-7
$\|d\|$	4	3	8	0	4	15	12	10	4	1	3	5	18	7
rank	5	2.5	9	*	5	12	11	10	5	1	2.5	7	13	8
signed rank	5	2.5	-9	*	-5	12	11	-10	-5	-1	-2.5	-7	-18	-8

$T_+ = 30.5$, $T_- = 60.5$, $T = 30.5$

For $\alpha = .01$ and $n = 13$ (one tie), the two tailed critical value is

$T_{.005} = 10$.

Since $T = 30.5 > T_{.005}$, do not reject H_0.

7) The critical values are $\pm z_{.025} = \pm 1.96$.

$$z = \frac{T - \dfrac{n(n+1)}{4}}{\sqrt{\dfrac{n(n+1)(2n+1)}{24}}} = \frac{160 - \dfrac{35(36)}{4}}{\sqrt{\dfrac{(35)(36)(71)}{24}}} = -2.54$$

Since $z = -2.54 < -z_{.025}$, reject H_0.

p value $= P(z < -2.54) = .0055$

9) a) $T = 40$, $n = 15$

$$z = \frac{40 - \dfrac{(15)(16)}{4}}{\sqrt{\dfrac{(15)(16)(31)}{24}}} = -1.14$$

p value $= p(z < -1.14) = .5000 - .3729 = .1271$

b) $T = 26$, $n = 12$

$$z = \frac{26 - \dfrac{(12)(13)}{4}}{\sqrt{\dfrac{(12)(13)(25)}{24}}} = -1.04$$

p value $= P(z < -1.04) = .5000 - .3508 = .1492$

11) $T_+ + T_- = 1 + 2 + 3 + \cdots + n = \dfrac{n(n + 1)}{2}$

We should expect each of these to cluster around

$$\frac{1}{2} \left(\frac{n(n + 1)}{2} \right) = \frac{n(n + 1)}{4}$$

when the 2 distributions are identical.

Exercise Set XIII.4

1) $r_s = .37$ and $n = 25$

   For $\nu = 25 - 2 = 23$ d.f.,   the ctitical values are $\pm t_{.025} = 2.069$

$$t = .37 \sqrt{\frac{25 - 2}{1 - (.37)^2}} = 1.91$$

   Since $t = 1.91 < t_{.025}$,   do not reject the null hypothesis of no correlation.

3)

x	y	d	d^2
3	1	2	4
8	5	3	9
6	7	-1	1
5	8	-3	9
4	3	1	1
1	4	-3	9
7	8	-1	1
10	2	8	64
9	6	3	9
2	7	-5	25

$$\Sigma d^2 = 132$$

   For $\alpha = .05$ and $\nu = 10 - 2 = 8$ d.f.,   the critical values are $\pm t_{.025} = \pm 2.306$

$$r_s = 1 = \frac{(6)(132)}{10(10^2 - 1)} = .200$$

$$t = .200 \sqrt{\frac{10 - 2}{1 - (.018)^2}} = .577.$$

   Do not reject the null hypothesis of no correlation.

5) $\Sigma d^2 = 2320$

$$r_s = 1 = \frac{6(2320)}{20(20^2 - 1)} = -.74$$

$$t = -.74 \sqrt{\frac{20 - 2}{1 - (.74)^2}} = -4.67.$$

Since $t = -4.70 < -t_{.025}$, reject the hypothesis of no correlation between sales and the size of their assigned territories. The negative correlation might be due to the salesmen being unable to adequately service their accounts if there are too many or if they are spread out too much geographically.

7)

x	y	d	d^2
2	1	1	1
5	8	-3	9
11	10	1	1
7	9	-2	4
6	5	1	1
9	7	2	4
10	11	-1	1
4	2	2	4
3	4	-1	1
8	6	2	4
1	3	-2	4

$$\Sigma\, d^2 = 34$$

$$r_s = 1 - \frac{6(34)}{11(11^2 - 1)} = .85$$

$$t = .85\sqrt{\frac{11 - 2}{1 - (.85)^2}} = 4.77$$

$t > t_{.025}$. Reject H_0 and conclude that a correlation exists between the two rankings.

9) a) First Assessment

Yes (at $\alpha = .001$) since the p value $< .001$ for all five dimensions.

Second Assessment

Yes at say $\alpha = .01$ since the p value $< .01$ for all 5 dimensions.

b) Yes. p value $< .001$ for both assessments

c) Hostility and Evaluation

d) The TPS scores by each rates were highly correlated for the two assessments. Both sets of raters scores on the two assessments proved to be correlated ($\alpha = .05$) for evaluation, pressuring and hostility dimensions.

Exercise Set XIII.5

1)

	I		II		III	
	x	rank	x	rank	x	rank
	38	1	63	7	50	2
	56	3	72	11	57	4
	71	10	74	12.5	61	5.5
	76	14	81	16.5	61	5.5
	77	15	81	16.5	65	8
	87	19	86	18	70	9
	88	20	90	21	74	12.5

$$n_1 = 7 \qquad n_2 = 7 \qquad n_3 = 7$$
$$u_1 = 82 \qquad u_2 = 102.5 \qquad u_3 = 46.5$$

$$h = \frac{12}{(21)(22)} \left[\frac{(82)^2}{7} + \frac{(102.5)^2}{7} + \frac{(46.5)^2}{7} \right] - 3(22) = 5.96$$

For $\nu = 3 - 1 = 2$ d.f. and $\alpha = .05$, the critical value is $\chi^2_{.05} = 5.99$.
Since $h = 5.96 < \chi^2_{.05}$, do not reject the null hypothesis.

3)

	East		Central		Western	
	x	rank	x	rank	x	rank
	9.20	4	6.40	1	8.80	3
	9.90	8	8.60	2	9.30	5
	10.30	10.5	9.40	6	10.00	9
	11.30	16	9.70	7	10.40	12
	12.20	17	10.30	10.5	11.10	14.5
	13.40	18	10.60	13	11.10	14.5

$$n_1 = 6 \qquad n_2 = 6 \qquad n_3 = 6$$
$$u_1 = 73.5 \qquad u_2 = 39.5 \qquad u_3 = 58$$

$$h = \frac{12}{(18)(19)} \left[\frac{(73.5)^2}{6} + \frac{(39.5)^2}{6} + \frac{(58)^2}{6} \right] - 3(19) = 3.39$$

For $\nu = 3 - 1 = 2$ d.f., $\chi^2_{.05} = 5.99$.
Since $h = 3.39 < \chi^2_{.05}$, do not reject the null hypothesis of identical distributions.

5)

	1		2		3		4
x	rank	x	rank	x	rank	x	rank
14.7	1	19.4	10	17.9	6	15.3	2
17.4	5	20.6	15	18.2	7	16.2	3
20.0	12	21.3	16	19.9	11	17.2	4
22.4	19	21.4	17	20.3	13	18.3	8
23.2	20	22.1	18	20.5	14	19.0	9

$$n_1 = 5 \qquad n_2 = 5 \qquad n_3 = 5 \qquad n_4 = 5$$
$$u_1 = 57 \qquad u_2 = 76 \qquad u_3 = 51 \qquad u_4 = 26$$

For $\alpha = .05$ and $\nu = 4 - 1 = 3$ d.f., $\chi^2_{.05} = 7.81$

$$h = \frac{12}{(20)(21)} \left[\frac{(57)^2}{5} + \frac{(76)^2}{5} + \frac{(51)^2}{5} + \frac{(26)^2}{5} \right] - 3(21)$$

$$= 7.29$$

Since $h = 7.29 < \chi^2_{.05}$, do not reject H_0. We cannot conclude a difference exists in the performance of the 4 dehumidifiers.

7) H_0: All four brands have the same mean sugar content
   H_1: At least one pair have different means

ranks(1)	ranks(2)	ranks(3)	ranks(4)
2	13	1	3.5
10.5	18.5	15.5	3.5
10.5	20	26.5	5
14	21.5	26.5	6
17	21.5	29	7
18.5	23	30	8.5
	24	31	8.5
	28	32	12
		33	15.5
			25

$$n_1 = 6 \qquad n_2 = 8 \qquad n_3 = 9 \qquad n_4 = 10$$
$$u_1 = 72.5 \qquad u_2 = 169.5 \qquad u_3 = 224.5 \qquad u_4 = 94.5$$

For $\nu = 4 - 1 = 3$ d.f., $\chi^2_{.05} = 7.81$

$$h = \frac{12}{(33)(34)} \left[\frac{(72.5)^2}{6} + \frac{(169.5)^2}{8} + \frac{(224.5)^2}{9} + \frac{(94.5)^2}{10} \right] - 3(34)$$

$$= 15.22$$

Since $h = 15.22 > \chi^2_{.05}$, reject H_0.

9) $m = 6$, $n = 30$, $\nu = 6 - 1 = 6$ d.f., $\chi^2_{.05} = 11.07$

$$h = \frac{12}{(30)(31)} \left[\frac{(40)^2}{5} + \frac{(130)^2}{5} + \frac{(110)^2}{5} + \frac{(89)^2}{5} + \frac{(26)^2}{5} + \frac{(70)^2}{5} \right] - 3(31)$$

$$= 20.80$$

Since $h = 20.80 > \chi^2_{.05}$, reject H_0.

Exercise Set XIII.6

1)

High		Medium		Low	
days	rank	days	rank	days	rank
197	3	164	1	172	2
239	3	186	2	145	1
412	3	154	1	286	2
275	3	243	2	201	1
314	3	205	1	214	2
536	3	181	2	152	1
102	1	176	3	150	2

$$u_1 = 19 \qquad u_2 = 12 \qquad u_3 = 11$$

Using r = 7 (blocks), c = 3 (columns), and the above ranks, we find

$$\chi_f^2 = \frac{12}{7 \cdot 3 \cdot 4} (19^2 + 12^2 + 11^2) - 3(7)(4)$$

$$= 5.43$$

For $\nu = 3 - 1 = 2$ d.f., $\chi_{.05}^2 = 5.99$.
Reject H_0 and conclude that a treatment effect exists.

3)

I		II		III	
time	rank	time	rank	time	rank
66.9	3	60.2	2	54.1	1
84.3	3	78.4	2	69.3	1
99.0	3	85.4	1	86.1	2
102.4	2	112.3	3	101.5	1

$$u_1 = 11 \qquad u_2 = 8 \qquad u_3 = 5$$

For r = 4 and c = 3, the critical value of 6.5 is found corresponding
to $\alpha = .04$ (Table 13.12).

$$\chi_f^2 = \frac{12}{(4)(3)(4)} [11^2 + 8^2 + 5^2] - 3(4)(4)$$

$$= 4.5$$

Since $\chi_f^2 = 6.5$, do not reject H_0 at $\alpha = .04$.

5)

	1		2		3
grade	rank	grade	rank	grade	rank
123	2	135	3	118	1
145	2	141	1	155	3
125	1.5	130	3	125	1.5
150	1	160	2	170	3
120	2	115	1	130	3
143	1	159	3	148	2

$$u_1 = 9.5 \qquad u_2 = 13 \qquad u_3 = 13.5$$

For $r = 6$ and $c = 3$, the critical value of 6.3 is found in Table 13.12 corresponding to $\alpha = .05$.

$$\chi_f^2 = \frac{12}{(6)(3)(4)} [(9.5)^2 + (13)^2 + (13.5)^2] - 3(6)(4)$$
$$= 1.58.$$

Since $\chi_f^2 < 6.3$, do not reject H_0. We cannot conclude an effect on grades due to the class assignment.

7)

	1		2		3
time	rank	time	rank	time	rank
19.6	3	11.2	2	7.8	1
5.4	3	3.9	2	3.3	1
25.9	3	17.2	2	10.3	1
65.8	1	74.7	2	103.5	3
55.4	3	50.3	2	37.9	1
45.8	3	33.9	2	30.0	1
65.0	3	25.8	2	15.8	1
37.2	3	32.0	1	33.5	2

$$u_1 = 22 \qquad u_2 = 15 \qquad u_3 = 11$$

For $c = 3$ and $r = 8$, Table 13.12 yields the critical value 6.3 corresponding to $\alpha = .05$.

$$\chi_f^2 = \frac{12}{(8)(3)(4)} [22^2 + 15^2 + 11^2] - 3(8)(4)$$
$$= 7.75$$

Since $\chi_f^2 > 6.3$, reject H_0 and conclude that there is a difference in the (mean) operating speeds of these 3 computers.

1) $\underset{1}{\underline{TTTT}}\ \underset{2}{\underline{FFFFF}}\ \underset{3}{\underline{TTT}}\ \underset{4}{\underline{FF}}\ \underset{5}{\underline{TTT}}\ \underset{6}{\underline{F}}\ \underset{7}{\underline{T}}\ \underset{8}{\underline{F}}\ \underset{9}{\underline{T}}\ \underset{10}{\underline{FFF}}\ \underset{11}{\underline{TTT}}\ \underset{12}{\underline{FF}}\ \underset{13}{\underline{TT}}\ \underset{14}{\underline{FFFF}}\ \underset{15}{\underline{TT}}\ \underset{16}{\underline{FF}}\ \underset{17}{\underline{T}}\ \underset{18}{\underline{FF}}$

For $\alpha = .05$, the critical values are $\pm z_{.025} = \pm 1.96$.
Using $r = 18$, $n_1 = 20$ and $n_2 = 22$ we find

$$z = \frac{18 - \left(\dfrac{2(20)(22)}{20 + 22} + 1 \right)}{\sqrt{\dfrac{2(20)(22)[2(20)(22) - 20 - 22]}{(20 + 22)^2(20 + 22 - 1)}}}$$

$$= -1.23$$

Do not reject the null hypothesis of a random sequence.

3) $\underset{1}{\underline{A}}\ \underset{2}{\underline{B}}\ \underset{3}{\underline{A}}\ \underset{4}{\underline{BBB}}\ \underset{5}{\underline{AA}}\ \underset{6}{\underline{BBBBB}}\ \underset{7}{\underline{AA}}\ \underset{8}{\underline{BBB}}\ \underset{9}{\underline{AA}}\ \underset{10}{\underline{B}}\ \underset{11}{\underline{A}}\ \underset{12}{\underline{B}}\ \underset{13}{\underline{A}}\ \underset{14}{\underline{BBBBB}}\ \underset{15}{\underline{AAAAAAAA}}\ \underset{16}{\underline{B}}\ \underset{17}{\underline{A}}\ \underset{18}{\underline{BB}}\ \underset{19}{\underline{A}}\ \underset{20}{\underline{BBB}}$

$\underset{21}{\underline{AAAA}}\ \underset{22}{\underline{BB}}$

For $\alpha = .05$, the critical values are $\pm z_{.025} = \pm 1.96$.
Using $r = 22$, $n_1 = 24$ and $n_2 = 27$, we find

$$z = \frac{22 - \left(\dfrac{2(24)(27)}{24 + 27} + 1 \right)}{\sqrt{\dfrac{2(24)(27)[2(24)(27) - 24 - 27]}{(24 + 27)^2(24 + 27 - 1)}}}$$

$$= -1.25$$

Since $z < -1.96$, do not reject H_0.

5) The median is $\tilde{x} = 71.5$.

88,	74,	96,	15,	62,	77,	91,	100,	45,	72,	75,	69,	71,	65,	89,	96,	23,	45,	41,
S	S	S	F	F	S	S	S	F	S	S	F	F	F	S	S	F	F	F

56,	71,	90,	54,	82,	79,	86,	31,	20
F	F	S	F	S	S	S	F	F

The runs above and below the median are underlined.

$$\underset{1}{\underline{\text{SSS}}} \ \underset{2}{\underline{\text{FF}}} \ \underset{3}{\underline{\text{SSS}}} \ \underset{4}{\underline{\text{F}}} \ \underset{5}{\underline{\text{SS}}} \ \underset{6}{\underline{\text{FFF}}} \ \underset{7}{\underline{\text{SS}}} \ \underset{8}{\underline{\text{FFFFF}}} \ \underset{9}{\underline{\text{S}}} \ \underset{10}{\underline{\text{F}}} \ \underset{11}{\underline{\text{SSS}}} \ \underset{12}{\underline{\text{FF}}}$$

$r = 12$, $n_1 = 14$, $n_2 = 14$, $\pm z_{.025} = \pm 1.96$

$$z = \frac{12 - \left(\dfrac{2(14)(14)}{14 + 14} + 1 \right)}{\sqrt{\dfrac{2(14)(14)[2(14)(14) - 14 - 14]}{(14 + 14)^2 (14 + 14 - 1)}}} = -1.16.$$

Do not reject the null hypothesis of randomness.

7) From the formula for z, we infer

$$\mu = \frac{2n_1 n_2}{n_1 + n_2} + 1$$

$$= \frac{2(50)(50)}{50 + 50} + 1 = 51$$

9) a) Two

   b) Five

   c) $\mu = \dfrac{2n_1 n_2}{n_1 + n_2} + 1 = \dfrac{2 \cdot 3 \cdot 2}{3 + 2} + 1 = 3.4$

Chapter Test

1)

	I	II	III	Total
Definite Reduction	23 / 25.94	34 / 32.68	12 / 10.38	69
Otherwise	27 / 24.06	29 / 30.32	8 / 9.62	64
Total	50	63	20	133

$\chi^2 = 1.32 < \chi^2_{.05}$ (with $\nu = 2$ d.f.)

Do not reject the hypothesis that the same percentage of reductions in tumor size occur with the 3 therapies.

Note: An alternate approach is to pool all 3 samples to obtain an estimate of the proportion who experience a reduction. Then carry out a goodness of fit test. $\hat{p} = .5188$, $e_1 = 25.94$, $e_2 = 32.68$, $e_3 = 10.38$ and $\chi^2 < 1$ is found.

2)

	For	Against	No Opinion	Total
Freshman	40 / 66.00	50 / 37.20	30 / 16.80	120
Sophomore	50 / 55.00	35 / 31.00	15 / 14.00	100
Junior	70 / 66.00	40 / 37.20	10 / 16.80	120
Senior	75 / 55.00	18 / 31.00	7 / 14.00	100
Graduate	40 / 33.00	12 / 18.60	8 / 8.40	60
Total	275	155	70	500

$\chi^2 = 46.98$

For $\nu = (5 - 1)(3 - 1) = 8$ d.f., the critical value is $\chi^2_{.05} = 15.51$. Since $\chi^2 > \chi^2_{.05}$, reject the null hypothesis and conclude that attitude and class standing are related.

3) $u_1 = 15$, $u_2 = 12$, $u_3 = 13$, and $u_4 = 10$

The critical value for $\nu = 4 - 1 = 3$ d.f. is $\chi^2_{.05} = 7.81$

$$\chi^2_f = \frac{12}{(5)(4)(5)} [15^2 + 12^2 + 13^2 + 10^2] - 3(5)(5)$$

$$= 1.56$$

Since $\chi^2_f < \chi^2_{.05}$, do not reject H_0.

4) The Freidman ANOVA test is appropriate.

1		2		3		4	
fade	rank	fade	rank	fade	rank	fade	rank
8.2	4	7.0	3	4.3	1	5.2	2
6.1	3	8.4	4	3.1	2	2.9	1
7.4	3	7.5	4	5.6	2	4.9	1
9.0	4	6.4	2	6.5	3	5.3	1
8.3	4	7.5	3	6.0	2	5.8	1
$u_1 = 18$		$u_2 = 16$		$u_3 = 10$		$u_4 = 6$	

For $\nu = 4 - 1 = 3$ d.f., the critical value is $\chi^2_{.05} = 7.81$

$$\chi^2_f = \frac{12}{(5)(4)(5)} [18^2 + 16^2 + 10^2 + 6^2] - 3(5)(5)$$

$$= 10.92.$$

Reject the null hypothesis of identical populations and conclude a difference exists in the resistance to fading by these 4 colors.

5)

Observations	20,	21,	24,	24,	27,	28,	28,	30,	30,	31,	32,	33,	34,	35,	36,	38
Variable	y	x	x	y	y	x	y	x	y	y	x	y	x	x	x	y
Rank	1	2	3.5	3.5	5	6.5	6.5	8	9	10	11	12	13	14	15	16

x	rank	y	rank
21	2	20	1
24	3.5	24	3.5
28	6.5	27	5
30	8	28	6.5
32	11	30	9
34	13	31	10
35	14	33	12
36	15	38	16
$w_1 = 73$		$w_2 = 63$	

$$u_1 = (8)(8) + \frac{8 \cdot 9}{2} - 73 = 27$$

$$u_2 = (8)(8) + \frac{8 \cdot 9}{2} - 63 = 37$$

$u = 27$

For $\alpha = .05$, the (two tailed) critical value is $u_{.025} = 13$.

Do not reject the null hypothesis.

A one tailed test is more appropriate. For it $u_{.05} = 15$. Still H_0 is not rejected.

6)

rank(x)	rank(y)	d	d^2
10	8	2	4
4	5	-1	1
7	6	1	1
3	3	0	0
6	9	-3	9
9	7	2	4
2	1	1	1
5	4	1	1
1	2	-1	1
8	10	2	4
		$\Sigma d^2 = 26$	

$$r_s = 1 - \frac{6(26)}{10(99)} = .84$$

For $\nu = 10 - 2 = 8$, and $\alpha = .05$, the <u>one</u> tailed critical value is $t_{.05} = 1.860$

$$t = .84 \sqrt{\frac{10 - 2}{1 - (.84)^2}} = 4.38$$

Reject the null hypothesis of no correlation and conclude a positive correlation exists. p value < .005.

7) The Wilcoxon matched pair signed rank test is appropriate.

Initial Score x	Final Score y	$d = x - y$	$\lvert d \rvert$	rank	Signed rank	
18	19	-1	1	2	-2	
12	9	3	3	6.5	6.5	
15	15	0	0	*	*	$T_+ = 23.5$
20	21	-1	1	2	-2	$T_- = 12.5$
4	2	2	2	4.5	4.5	$T = 12.5$
17	20	-3	3	6.5	-6.5	
11	12	-1	1	2	-2	
14	10	4	4	8	8	
23	23	0	0	*	*	
13	11	2	2	4.5	4.5	

For $\alpha = .05$ and $n = 8$, the critical value for a two tailed test is $T_{.025} = 4$.
Since $T = 12.5 \geq T_{.025}$, do not reject H_0. We cannot conclude an effect on attitude due to the film.

8)

	I		II		III	
	Observation	rank	Observation	rank	Observation	rank
	20	1	30	4.5	25	2.5
	25	2.5	45	8	30	4.5
	35	6	50	9	40	7
	65	12	60	11	55	10
	70	14	70	14	70	14
	$u_1 = 35.5$		$u_2 = 46.5$		$u_3 = 38$	

$n_1 = 5$, $n_2 = 5$, $n_3 = 5$, $n = 15$, $m = 3$, and $\nu = 3 - 1 = 2$.

For $\alpha = .05$, the critical value is $\chi^2_{.05} = 5.99$

$$h = \frac{12}{(15)(16)} \left[\frac{(35.5)^2}{5} + \frac{(46.5)^2}{5} + \frac{(38)^2}{5} \right] - 3(16)$$

$$= .665$$

Since $h < \chi^2_{.05}$, do not reject the null hypothesis of identical locations or distributions for the 3 populations.

9) 0, 2, 3, 4, 4, 5, 5, 6, 6, 7, 7, 7, 7, 7, 8, 8, 9, 9, 9, 9, 10, 10, 10, 10,

11, 12, 12, 14, 15

The median is $\tilde{x} = 7.5$.

2, 0, 4, 3, 5, 5, 7, 6, 7, 10, 4, 3, 8, 9, 10, 6, 9, 12, 8, 7, 9, 11, 15,
F F F F F F F F F S   F F S S S   F S S   S F S S   S

10, 7, 10, 12, 7, 9, 14
S   F S   S   F S S

$$\frac{\text{FFFFFFFF}}{1} \ \frac{\text{S}}{2} \ \frac{\text{FF}}{3} \ \frac{\text{SSS}}{4} \ \frac{\text{F}}{5} \ \frac{\text{SSS}}{6} \ \frac{\text{F}}{7} \ \frac{\text{SSSS}}{8} \ \frac{\text{F}}{9} \ \frac{\text{SS}}{10} \ \frac{\text{F}}{11} \ \frac{\text{SS}}{12}$$

$r = 12$, $n_1 = 15$, $n_2 = 15$

The critical values are $\pm z_{.025} = \pm 1.96$

$$z = \frac{12 - \left(\frac{2(15)(15)}{15 + 15} + 1 \right)}{\sqrt{\frac{2(15)(15)[2(15)(15) - 15 - 15]}{(15 + 15)^2(15 + 15 - 1)}}} = -1.49.$$

Do not reject H_0.

10) We choose, somewhat arbitrarily, to combine the $4 - 5$ and > 5 admission categories.

	0-1	2-3	> 3	Total
< 40	8 / 3.12	6 / 5.25	7 / 12.62	21
40-60	25 / 6.25	8 / 10.50	9 / 25.24	42
60-70	9 / 10.13	29 / 17.00	30 / 40.87	68
> 70	17 / 39.48	56 / 66.25	192 / 159.27	265
Total	59	99	238	396

$\nu = (4 - 1)(3 - 1) = 6$

$\chi^2_{.05} = 12.6$

$\chi^2 = 107.4 > \chi^2_{.05}$

Reject the null hypothesis that the 2 age and admission classifications are independent.

Appendix
The Use of Statistical Packages on Hand Held Calculators

Using a Calculator Statistical Package

Statistical packages found on hand held calculators have a great deal of similarity. In the following we limit our discussion to the Sharp's model EL-512 and the Texas Instrument's model 55. First we consider the Sharp's model EL-512.

Using the Sharp's EL-512

The statistical mode is initiated by the $\boxed{2nd}$ $\boxed{STAT}$ keystroke sequence. Data for a single sample is input with the $\boxed{DATA}$ or $\boxed{M+}$ key, no $\boxed{2ndF}$ key being needed. The mean and standard deviation of a sample are obtained by using the $\boxed{\bar{x}}$ and $\boxed{S_x}$ keys in combination with the $\boxed{2ndF}$ key. The standard deviation S_x is computed using a divisor of $n - 1$. If a divisor of n is wanted, use the $\boxed{\sigma_x}$ key.

Calculator Usage Illustration

Find the mean and standard deviation of the sample 3, 4, 5, 8.

Method: Enter the data and use the $\boxed{\bar{x}}$ and $\boxed{S_x}$ keys

$\boxed{2nd}$ $\boxed{STAT}$ 3 $\boxed{DATA}$ 4 $\boxed{DATA}$ 5 $\boxed{DATA}$ 8 $\boxed{DATA}$

$\boxed{2nd}$ $\boxed{\bar{x}}$ 5 $\boxed{2nd}$ $\boxed{S_x}$ 2.160247

Frequencies are entered using the $\boxed{\times}$ key when in the statistical mode on the EL-512.

Calculator Usage Illustration

Find the mean and standard deviation of the following distribution:

x	2	5	7
f	4	3	1

Method: Enter the data and associated frequencies, then use the $\boxed{\overline{x}}$ and $\boxed{S_x}$ keys.

$\boxed{2ndF}$ $\boxed{STAT}$ 2 $\boxed{\times}$ 4 $\boxed{DATA}$ 5 $\boxed{\times}$ 3 $\boxed{DATA}$ 7 $\boxed{DATA}$

$\boxed{2ndF}$ $\boxed{\overline{x}}$ <u>3.75</u> $\boxed{2ndF}$ $\boxed{S_x}$ <u>1.982062</u>

Note: The frequency f = 1 need not be entered.

Data pairs are entered on the model EL-512 using the $\boxed{(x,y)}$ and $\boxed{DATA}$ keys following x and y respectively. The correlation coefficient r is obtained using the $\boxed{2ndF}$ and $\boxed{r}$ keys. In a similar fashion, the $\boxed{2ndF}$ $\boxed{a}$ and $\boxed{2ndF}$ $\boxed{b}$ keystroke sequences produce the intercept b_0 and slope b_1 of the least squares line. Finally predicted y' values are obtained by following the entry of the number with the $\boxed{2ndF}$ $\boxed{y'}$ sequence of keystrokes.

Calculator Usage Illustration

Find the correlation coefficient r and the slope and intercept of the least squares line for the data (1,2), (3,4), and (5,4).

Method: Enter data, then use $\boxed{r}$, $\boxed{b}$, and $\boxed{a}$ keys.

$\boxed{2ndF}$ $\boxed{STAT}$ 1 $\boxed{(x,y)}$ 2 $\boxed{DATA}$ 3 $\boxed{(x,y)}$ 4 $\boxed{DATA}$

5 $\boxed{(x,y)}$ 4 $\boxed{DATA}$ $\boxed{2ndF}$ $\boxed{r}$ <u>.866025</u> $\boxed{2ndF}$

$\boxed{a}$ <u>1.833333</u> $\boxed{2ndF}$ $\boxed{b}$ <u>.5</u>

continued on next page

Note: r = .866025 and the equation of the least squares line is
y = 1.83 + .5x when the coefficients are rounded.

Calculator Usage Illustration

Find the y' values predicted for x = 1, x = 3, and x = 5 by the
least squares line of the last example.

Method: Input the x value, then key $\boxed{\text{2nd}}$ and $\boxed{y'}$.

(Continuing the above sequence of keystrokes)

$$1 \;\boxed{\text{2ndF}}\;\boxed{y'}\; \underline{2.333333} \quad 3 \;\boxed{\text{2ndF}}\;\boxed{y'}$$

$$\underline{3.333333} \quad 5 \;\boxed{\text{2ndF}}\;\boxed{y'}\; \underline{4.333333}$$

Note: $y'_1 = 2.33$, $y'_2 = 3.33$, and $y'_3 = 4.33$ (rounded).

Sums of products and squares of x's and y's are produced on the EL-512
using the $\boxed{\Sigma xy}$, $\boxed{\Sigma x^2}$ and $\boxed{\Sigma y^2}$ keys in combination with the $\boxed{\text{2ndF}}$ keys.
The ability to accumulate such sums proves to be extremely useful in ANOVA
and regression analysis.

Calculator Usage Illustration

Find Σx^2, Σy^2, and Σxy for the data (2,5), (3,7), and (5,1).

Method: Enter the data pairs, then use the $\boxed{\Sigma xy}$, $\boxed{\Sigma x^2}$, $\boxed{\Sigma y^2}$, and $\boxed{\text{2ndF}}$
         keys.

$$2 \;\boxed{(x,y)}\; 5 \;\boxed{\text{DATA}}\; 3 \;\boxed{(x,y)}\; 7 \;\boxed{\text{DATA}}\; 5 \;\boxed{(x,y)}\; 1 \;\boxed{\text{DATA}}$$

$$\boxed{\text{2ndF}}\;\boxed{\Sigma xy}\; \underline{36.} \;\boxed{\text{2ndF}}\;\boxed{\Sigma x^2}\; \underline{38.} \;\boxed{\text{2ndF}}\;\boxed{\Sigma y^2}\; \underline{75.}$$

Note: The sums of squares for a single variable may be obtained by
entering the data with the $\boxed{\text{DATA}}$ key as with the mean and standard
deviation

Using the Texas Instruments' Model 55

The statistics mode on the model 55 is initiated automatically by the $\boxed{\Sigma+}$ key on data entry or by a combination of the $\boxed{\text{2nd}}$ and $\boxed{\text{FRQ}}$ keys. Data from a single sample is input with the $\boxed{\Sigma+}$ key. The mean and standard deviation of a sample are obtained using the $\boxed{\text{MEAN}}$ and $\boxed{\sigma_{n-1}}$ keys in combination with the $\boxed{\text{2nd}}$ key. The $n-1$ subscript on σ indicates a divisor of $n-1$ is being used in computing the standard deviation. The $\boxed{\sigma_n^2}$ key uses a divisor of n.

Calculator Usage Illustration

Find the mean and standard deviation of the sample 3, 4, 5, 8.

Method: Enter the data and use the $\boxed{\text{MEAN}}$ and $\boxed{\sigma_{n-1}}$ keys.

$3\ \boxed{\Sigma+}\ 4\ \boxed{\Sigma+}\ 5\ \boxed{\Sigma+}\ 8\ \boxed{\Sigma+}\ \boxed{\text{2nd}}\ \boxed{\text{MEAN}}$

$\underline{\underline{5.}}\ \boxed{\text{2nd}}\ \boxed{\sigma_{n-1}}\ \underline{\underline{2.160247}}$

Frequencies are entered using the $\boxed{\text{FRQ}}$ and $\boxed{\text{2nd}}$ keys.

Calculator Usage Illustration

Find the mean and standard deviation of the distribution below:

x	2	5	7
y	4	3	1

Method: Input the data and associated frequencies, then use the $\boxed{\text{MEAN}}$ and $\boxed{\sigma_{n-1}}$ keys.

$2\ \boxed{\text{2nd}}\ \boxed{\text{FRQ}}\ 4\ \boxed{\Sigma+}\ 5\ \boxed{\text{2nd}}\ \boxed{\text{FRQ}}\ 3\ \boxed{\Sigma+}\ 7\ \boxed{\Sigma+}$

$\boxed{\text{2nd}}\ \boxed{\text{MEAN}}\ \underline{\underline{3.75}}\ \boxed{\text{2nd}}\ \boxed{\sigma_{n-1}}\ \underline{\underline{1.982062}}$

Note: The frequency $f = 1$ need not be entered.

Data pairs are entered on the T.I. model 55 using the $\boxed{(x \leftrightarrow y)}$ and $\boxed{\Sigma +}$ keys following x and y respectively. The correlation coefficient r is obtained using the $\boxed{2nd}$ and $\boxed{CORR}$ keys. The slope of the least squares line is obtained using the sequence $\boxed{2nd}\,\boxed{b/a}\,\boxed{x \leftrightarrow y}$. Its intercept is found using $\boxed{2nd}\,\boxed{b/a}$. Finally predicted y' values are found by following the entry of x with the $\boxed{2nd}\,\boxed{y'}$ sequence of keystrokes.

Calculator Usage Illustration

Find the correlation coefficient r as well as the intercept and slope of the least squares line for the data (1,2), (3,4), and (5,4).

Method: Enter the data, then use the $\boxed{2nd}\,\boxed{CORR}$, $\boxed{2nd}\,\boxed{b/a}$ and $\boxed{x \leftrightarrow y}$ sequences of keystrokes.

1 $\boxed{x \leftrightarrow y}$ 2 $\boxed{\Sigma +}$ 3 $\boxed{x \leftrightarrow y}$ 4 $\boxed{\Sigma +}$ 5 $\boxed{x \leftrightarrow y}$

4 $\boxed{\Sigma +}$ $\boxed{2nd}$ $\boxed{CORR}$ .866025 $\boxed{2nd}$ $\boxed{b/a}$ 1.833333

$x \leftrightarrow y$ .5

Note: r = .866025 and the least squares line is y = 1.83 + .5x (coefficients rounded).

Calculator Usage Illustration

Find the y' values predicted for x = 1, x = 3, and x = 5 by the least squares line of the previous example.

Method: Input x, then key $\boxed{2nd}$ and $\boxed{y'}$.

(Continuing the sequence of keystrokes of the last example.)

1 $\boxed{2nd}$ $\boxed{y'}$ 2.333333 3 $\boxed{2nd}$ $\boxed{y'}$ 3.333333 5 $\boxed{2nd}$ $\boxed{y'}$ 4.333333

Note: $y'_1 = 2.33$, $y'_2 = 3.33$, and $y'_3 = 4.33$ (rounded).